FEDERAL EXECUTIVE TEAM

Acting Director, Climate Change Science Program: ....William J. Brennan

Director, Climate Change Science Program Office:...............................Peter A. Schultz

Lead Agency Principal Representative to CCSP;
Deputy Under Secretary of Commerce for Oceans and Atmosphere,
National Oceanic and Atmospheric Administration:..............................Mary M. Glackin

Product Lead, Director, National Climatic Data Center,
National Oceanic and Atmospheric Administration:..............................Thomas R. Karl

Synthesis and Assessment Product Advisory
Group Chair; Associate Director, EPA National
Center for Environmental Assessment:..Michael W. Slimak

Synthesis and Assessment Product Coordinator,
Climate Change Science Program Office: ....Fabien J.G. Laurier

Special Advisor, National Oceanic
and Atmospheric Administration...Chad A. McNutt

EDITORIAL AND PRODUCTION TEAM

Co-Chairs... Thomas R. Karl, NOAA
                                                                                                              Gerald A. Meehl, NCAR
Federal Advisory Committee Designated Federal Official....................... Christopher D. Miller, NOAA
Senior Editor.. Susan J. Hassol, STG, Inc.
Associate Editors... Christopher D. Miller, NOAA
                                                                                                              William L. Murray, STG, Inc.
                                                                                                              Anne M. Waple, STG, Inc.
Technical Advisor... David J. Dokken, USGCRP
Graphic Design Lead.. Sara W. Veasey, NOAA
Graphic Design Co-Lead... Deborah B. Riddle, NOAA
Designer... Brandon Farrar, STG, Inc.
Designer... Glenn M. Hyatt, NOAA
Designer... Deborah Misch, STG, Inc.
Copy Editor.. Anne Markel, STG, Inc.
Copy Editor.. Lesley Morgan, STG, Inc.
Copy Editor.. Mara Sprain, STG, Inc.
Technical Support... Jesse Enloe, STG, Inc.
                                                                                                              Adam Smith, NOAA

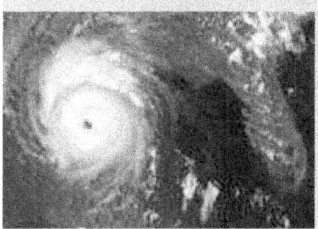

Weather and Climate Extremes in a Changing Climate

Regions of Focus: North America, Hawaii, Caribbean, and U.S. Pacific Islands

Synthesis and Assessment Product 3.3
Report by the U.S. Climate Change Science Program
and the Subcommittee on Global Change Research

EDITED BY:
Thomas R. Karl, Gerald A. Meehl, Christopher D. Miller, Susan J. Hassol,
Anne M. Waple, and William L. Murray

June, 2008

Members of Congress:

On behalf of the National Science and Technology Council, the U.S. Climate Change Science Program (CCSP) is pleased to transmit to the President and the Congress this Synthesis and Assessment Product (SAP), *Weather and Climate Extremes in a Changing Climate, Regions of Focus: North America, Hawaii, Caribbean, and U.S. Pacific Islands.* This is part of a series of 21 SAPs produced by the CCSP aimed at providing current assessments of climate change science to inform public debate, policy, and operational decisions. These reports are also intended to help the CCSP develop future program research priorities.

The CCSP's guiding vision is to provide the Nation and the global community with the science-based knowledge needed to manage the risks and capture the opportunities associated with climate and related environmental changes. The SAPs are important steps toward achieving that vision and help to translate the CCSP's extensive observational and research database into informational tools that directly address key questions being asked of the research community.

This SAP assesses the state of our knowledge concerning changes in weather and climate extremes in North America and U.S. territories. It was developed with broad scientific input and in accordance with the Guidelines for Producing CCSP SAPs, the Federal Advisory Committee Act, the Information Quality Act, Section 515 of the Treasury and General Government Appropriations Act for fiscal year 2001 (Public Law 106-554), and the guidelines issued by the Department of Commerce and the National Oceanic and Atmospheric Administration pursuant to Section 515.

We commend the report's authors for both the thorough nature of their work and their adherence to an inclusive review process.

Sincerely,

Carlos M. Gutierrez
Secretary of Commerce
Chair, Committee on Climate Change
Science and Technology Integration

Samuel W. Bodman
Secretary of Energy
Vice Chair, Committee on Climate
Change Science and Technology
Integration

John H. Marburger
Director, Office of Science and
Technology Policy
Executive Director, Committee
on Climate Change Science and
Technology Integration

TABLE OF CONTENTS

Synopsis...V

Preface..IX

Executive Summary..I

CHAPTER

1 11

Why Weather and Climate Extremes Matter
1.1 Weather And Climate Extremes Impact People, Plants, And Animals.......................12
1.2 Extremes Are Changing..16
1.3 Nature And Society Are Sensitive To Changes In Extremes................................19
1.4 Future Impacts Of Changing Extremes Also Depend On Vulnerability...................21
1.5 Systems Are Adapted To The Historical Range Of Extremes So Changes In
        Extremes Pose Challenges...28
1.6 Actions Can Increase Or Decrease The Impact Of Extremes..............................29
1.7 Assessing Impacts Of Changes In Extremes Is Difficult....................................31
1.8 Summary And Conclusions..33

2 35

Observed Changes in Weather and Climate Extremes
2.1 Background...37
2.2 Observed Changes And Variations In Weather And Climate Extremes...................37
    2.2.1 Temperature Extremes..37
    2.2.2 Precipitation Extremes..42
        2.2.2.1 Drought..42
        2.2.2.2 Short Duration Heavy Precipitation..46
        2.2.2.3 Monthly to Seasonal Heavy Precipitation...50
        2.2.2.4 North American Monsoon...50
        2.2.2.5 Tropical Storm Rainfall in Western Mexico.......................................52
        2.2.2.6 Tropical Storm Rainfall in the Southeastern United States...................53
        2.2.2.7 Streamflow...53
    2.2.3 Storm Extremes...53
        2.2.3.1 Tropical Cyclones...53
        2.2.3.2 Strong Extratropical Cyclones Overview...62
        2.2.3.3 Coastal Waves: Trends of Increasing Heights and Their Extremes...........68
        2.2.3.4 Winter Storms...73
        2.2.3.5 Convective Storms..75
2.3 Key Uncertainties Related To Measuring Specific Variations And Change................78
    2.3.1 Methods Based on Counting Exceedances Over a High Threshold...................78
    2.3.2 The GEV Approach..79

TABLE OF CONTENTS

3 81

Causes of Observed Changes in Extremes and Projections of Future Changes

3.1 Introduction..82

3.2 What Are The Physical Mechanisms Of Observed Changes In Extremes?..............82

   3.2.1 Detection and Attribution: Evaluating Human Influences on Climate Extremes Over North America..82

      3.2.1.1 Detection and Attribution: Human-Induced Changes in Average Climate That Affect Climate Extremes..83

      3.2.1.2 Changes in Modes of Climate-system Behavior Affecting Climate Extremes..85

   3.2.2 Changes in Temperature Extremes..87

   3.2.3 Changes in Precipitation Extremes..89

      3.2.3.1 Heavy Precipitation..89

      3.2.3.2 Runoff and Drought..90

   3.2.4 Tropical Cyclones..92

      3.2.4.1 Criteria and Mechanisms For tropical cyclone development..92

      3.2.4.2 Attribution Preamble..94

      3.2.4.3 Attribution of North Atlantic Changes..95

   3.2.5 Extratropical Storms..97

   3.2.6 Convective Storms..98

3.3 Projected Future Changes in Extremes, Their Causes, Mechanisms, and Uncertainties..99

   3.3.1 Temperature..99

   3.3.2 Frost..101

   3.3.3 Growing Season Length..101

   3.3.4 Snow Cover and Sea Ice..102

   3.3.5 Precipitation..102

   3.3.6 Flooding and Dry Days..103

   3.3.7 Drought..104

   3.3.8 Snowfall..105

   3.3.9 Tropical Cyclones (Tropical Storms and Hurricanes)..105

      3.3.9.1 Introduction..105

      3.3.9.2 Tropical Cyclone Intensity..107

      3.3.9.3 Tropical Cyclone Frequency and Area of Genesis..110

      3.3.9.4 Tropical Cyclone Precipitation..113

      3.3.9.5 Tropical Cyclone Size, Duration, Track, Storm Surge, and Regions of Occurrence..114

      3.3.9.6 Reconciliation of Future Projections and Past Variations..114

      3.3.10 Extratropical Storm..115

      3.3.11 Convective Storms..116

4 117

Measures To Improve Our Understanding of Weather and Climate Extremes

Appendix A...127

Example 1: Cold Index Data (Section 2.2.1)...128

Example 2: Heat Wave Index Data (Section 2.2.1 and Fig. 2.3(a))........................129

Example 3: 1-day Heavy Precipitation Frequencies (Section 2.1.2.2)....................130

Example 4: 90-day Heavy Precipitation Frequencies (Section 2.1.2.3 and Fig. 2.9).......131

Example 5: Tropical cyclones in the North Atlantic (Section 2.1.3.1)....................131

Example 6: U.S. Landfalling Hurricanes (Section 2.1.3.1)......................................132

Glossary and Acronyms..133

References...137

TABLE OF CONTENTS

Preface

Authors: Thomas R. Karl, NOAA; Gerald A. Meehl, NCAR; Christopher D. Miller, NOAA; William L. Murray, STG, Inc.

Executive Summary

Convening Lead Authors: Thomas R. Karl, NOAA; Gerald A. Meehl, NCAR
Lead Authors: Thomas C. Peterson, NOAA; Kenneth E. Kunkel, Univ. Ill. Urbana-Champaign, Ill. State Water Survey; William J. Gutowski, Jr., Iowa State Univ.; David R. Easterling, NOAA
Editors: Susan J. Hassol, STG, Inc.; Christopher D. Miller, NOAA; William L. Murray, STG, Inc.; Anne M. Waple, STG, Inc.

Chapter 1

Convening Lead Author: Thomas C. Peterson, NOAA
Lead Authors: David M. Anderson, NOAA; Stewart J. Cohen, Environment Canada and Univ. of British Columbia; Miguel Cortez-Vázquez, National Meteorological Service of Mexico; Richard J. Murnane, Bermuda Inst. of Ocean Sciences; Camille Parmesan, Univ. of Tex. at Austin; David Phillips, Environment Canada; Roger S. Pulwarty, NOAA; John M.R. Stone, Carleton Univ.
Contributing Authors: Tamara G. Houston, NOAA; Susan L. Cutter, Univ. of S.C.; Melanie Gall, Univ. of S.C.

Chapter 2

Convening Lead Author: Kenneth E. Kunkel, Univ. Ill. Urbana-Champaign, Ill. State Water Survey
Lead Authors: Peter D. Bromirski, Scripps Inst. Oceanography, UCSD; Harold E. Brooks, NOAA; Tereza Cavazos, Centro de Investigación Científica y de Educación Superior de Ensenada, Mexico; Arthur V. Douglas, Creighton Univ.; David R. Easterling, NOAA; Kerry A. Emanuel, Mass. Inst. Tech.; Pavel Ya. Groisman, Univ. Corp. Atmos. Res.; Greg J. Holland, NCAR; Thomas R. Knutson, NOAA; James P. Kossin, Univ. Wis., Madison, CIMSS; Paul D. Komar, Oreg. State Univ.; David H. Levinson, NOAA; Richard L. Smith, Univ. N.C., Chapel Hill
Contributing Authors: Jonathan C. Allan, Oreg. Dept. Geology and Mineral Industries; Raymond A. Assel, NOAA; Stanley A. Changnon, Univ. Ill. Urbana-Champaign, Ill. State Water Survey; Jay H. Lawrimore, NOAA; Kam-biu Liu, La. State Univ., Baton Rouge; Thomas C. Peterson, NOAA

Chapter 3 **Convening Lead Author:** William J. Gutowski, Jr., Iowa State Univ.
Lead Authors: Gabriele C. Hegerl, Univ. Edinburgh; Greg J. Holland, NCAR; Thomas R. Knutson, NOAA; Linda O. Mearns, NCAR; Ronald J. Stouffer, NOAA; Peter J. Webster, Ga. Inst. Tech.; Michael F. Wehner, Lawrence Berkeley National Laboratory; Francis W. Zwiers, Environment Canada
Contributing Authors: Harold E. Brooks, NOAA; Kerry A. Emanuel, Mass. Inst. Tech.; Paul D. Komar, Oreg. State Univ.; James P. Kossin, Univ. Wisc., Madison; Kenneth E. Kunkel, Univ. Ill. Urbana-Champaign, Ill. State Water Survey; Ruth McDonald, Met Office, United Kingdom; Gerald A. Meehl, NCAR; Robert J. Trapp, Purdue Univ.

Chapter 4 **Convening Lead Author:** David R. Easterling, NOAA
Lead Authors: David M. Anderson, NOAA; Stewart J. Cohen, Environment Canada and Univ. of British Columbia; William J. Gutowski, Jr., Iowa State Univ.; Greg J. Holland, NCAR; Kenneth E. Kunkel, Univ. Ill. Urbana-Champaign, Ill. State Water Survey; Thomas C. Peterson, NOAA; Roger S. Pulwarty, NOAA; Ronald J. Stouffer, NOAA; Michael F. Wehner, Lawrence Berkeley National Laboratory

Appendix A **Author:** Richard L. Smith, Univ. N.C., Chapel Hill

ACKNOWLEDGEMENT

CCSP Synthesis and Assessment Product 3.3 (SAP 3.3) was developed with the benefit of a scientifically rigorous, first draft peer review conducted by a committee appointed by the National Research Council (NRC). Prior to their delivery to the SAP 3.3 Author Team, the NRC review comments, in turn, were reviewed in draft form by a second group of highly qualified experts to ensure that the review met NRC standards. The resultant NRC Review Report was instrumental in shaping the final version of SAP 3.3, and in improving its completeness, sharpening its focus, communicating its conclusions and recommendations, and improving its general readability.

We wish to thank the members of the NRC Review Committee: John Gyakum (Co-Chair), McGill University, Montreal, Quebec; Hugh Willoughby (Co-Chair), Florida International University, Miami; Cortis Cooper, Chevron, San Ramon, California; Michael J. Hayes, University of Nebraska, Lincoln; Gregory Jenkins, Howard University, Washington, DC; David Karoly, University of Oklahoma, Norman; Richard Rotunno, National Center for Atmospheric Research, Boulder, Colorado; and Claudia Tebaldi, National Center for Atmospheric Research, Boulder Colorado, and Visiting Scientist, Stanford University, Stanford, California; and also the NRC Staff members who coordinated the process: Chris Elfring, Director, Board on Atmospheric Sciences and Climate; Curtis H. Marshall, Study Director; and Katherine Weller, Senior Program Assistant.

We also thank the individuals who reviewed the NRC Report in its draft form: Walter F. Dabberdt, Vaisala Inc., Boulder, Colorado; Jennifer Phillips, Bard College, Annandale-on-Hudson, New York; Robert Maddox, University of Arizona, Tucson; Roland Madden, Scripps Institution of Oceanography, La Jolla, California; John Molinari, The State University of New York, Albany; and also George L. Frederick, Falcon Consultants LLC, Georgetown, Texas, the overseer of the NRC review.

We would also like to thank the NOAA Research Council for coordinating a review conducted in preparation for the final clearance of this report. This review provided valuable comments from the following internal NOAA reviewers:

Henry Diaz (Earth System Research Laboratory)
Randy Dole (Earth System Research Laboratory)
Michelle Hawkins (Office of Program Planning and Integration)
Isaac Held (Geophysical Fluid Dynamics Laboratory)
Wayne Higgins (Climate Prediction Center)
Chris Landsea (National Hurricane Center)

The review process for SAP 3.3 also included a public review of the Second Draft, and we thank the individuals who participated in this cycle. The Author Team carefully considered all comments submitted, and a substantial number resulted in further improvements and clarity of SAP 3.3.

Finally, it should be noted that the respective review bodies were not asked to endorse the final version of SAP 3.3, as this was the responsibility of the National Science and Technology Council.

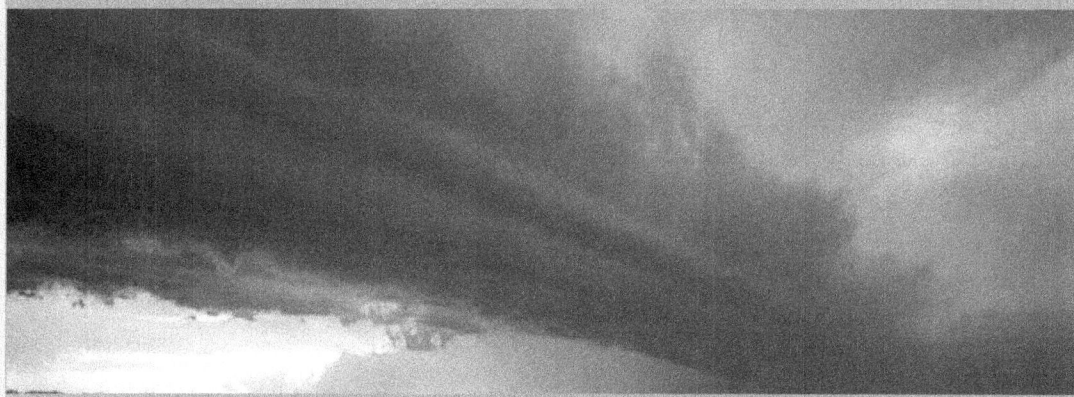

Changes in extreme weather and climate events have significant impacts and are among the most serious challenges to society in coping with a changing climate.

Many extremes and their associated impacts are now changing. For example, in recent decades most of North America has been experiencing more unusually hot days and nights, fewer unusually cold days and nights, and fewer frost days. Heavy downpours have become more frequent and intense. Droughts are becoming more severe in some regions, though there are no clear trends for North America as a whole. The power and frequency of Atlantic hurricanes have increased substantially in recent decades, though North American mainland land-falling hurricanes do not appear to have increased over the past century. Outside the tropics, storm tracks are shifting northward and the strongest storms are becoming even stronger.

It is well established through formal attribution studies that the global warming of the past 50 years is due primarily to human-induced increases in heat-trapping gases. Such studies have only recently been used to determine the causes of some changes in extremes at the scale of a continent. Certain aspects of observed increases in temperature extremes have been linked to human influences. The increase in heavy precipitation events is associated with an increase in water vapor, and the latter has been attributed to human-induced warming. No formal attribution studies for changes in drought severity in North America have been attempted. There is evidence suggesting a human contribution to recent changes in hurricane activity as well as in storms outside the tropics, though a confident assessment will require further study.

In the future, with continued global warming, heat waves and heavy downpours are very likely to further increase in frequency and intensity. Substantial areas of North America are likely to have more frequent droughts of greater severity. Hurricane wind speeds, rainfall intensity, and storm surge levels are likely to increase. The strongest cold season storms are likely to become more frequent, with stronger winds and more extreme wave heights.

Current and future impacts resulting from these changes depend not only on the changes in extremes, but also on responses by human and natural systems.

RECOMMENDED CITATIONS

For the Report as a whole:

CCSP, 2008: *Weather and Climate Extremes in a Changing Climate. Regions of Focus: North America, Hawaii, Caribbean, and U.S. Pacific Islands.* A Report by the U.S. Climate Change Science Program and the Subcommittee on Global Change Research. [Thomas R. Karl, Gerald A. Meehl, Christopher D. Miller, Susan J. Hassol, Anne M. Waple, and William L. Murray (eds.)]. Department of Commerce, NOAA's National Climatic Data Center, Washington, D.C., USA, 164 pp.

For the Preface:

Karl, T.R., G.A. Meehl, C.D. Miller, W.L. Murray, 2008: Preface in *Weather and Climate Extremes in a Changing Climate. Regions of Focus: North America, Hawaii, Caribbean, and U.S. Pacific Islands.* T.R. Karl, G.A. Meehl, C.D. Miller, S.J. Hassol, A.M. Waple, and W.L. Murray (eds.). A Report by the U.S. Climate Change Science Program and the Subcommittee on Global Change Research, Washington, DC.

For the Executive Summary:

Karl, T.R., G.A. Meehl, T.C. Peterson, K.E. Kunkel, W.J. Gutowski, Jr., D.R. Easterling, 2008: Executive Summary in *Weather and Climate Extremes in a Changing Climate. Regions of Focus: North America, Hawaii, Caribbean, and U.S. Pacific Islands.* T.R. Karl, G.A. Meehl, C.D. Miller, S.J. Hassol, A.M. Waple, and W.L. Murray (eds.). A Report by the U.S. Climate Change Science Program and the Subcommittee on Global Change Research, Washington, DC.

For Chapter 1:

Peterson, T.C., D.M. Anderson, S.J. Cohen, M. Cortez-Vázquez, R.J. Murnane, C. Parmesan, D. Phillips, R.S. Pulwarty, J.M.R. Stone, 2008: Why Weather and Climate Extremes Matter in *Weather and Climate Extremes in a Changing Climate. Regions of Focus: North America, Hawaii, Caribbean, and U.S. Pacific Islands.* T.R. Karl, G.A. Meehl, C.D. Miller, S.J. Hassol, A.M. Waple, and W.L. Murray (eds.). A Report by the U.S. Climate Change Science Program and the Subcommittee on Global Change Research, Washington, DC.

For Chapter 2:

Kunkel, K.E., P.D. Bromirski, H.E. Brooks, T. Cavazos, A.V. Douglas, D.R. Easterling, K.A. Emanuel, P.Ya. Groisman, G.J. Holland, T.R. Knutson, J.P. Kossin, P.D. Komar, D.H. Levinson, R.L. Smith, 2008: Observed Changes in Weather and Climate Extremes in *Weather and Climate Extremes in a Changing Climate. Regions of Focus: North America, Hawaii, Caribbean, and U.S. Pacific Islands.* T.R. Karl, G.A. Meehl, C.D. Miller, S.J. Hassol, A.M. Waple, and W.L. Murray (eds.). A Report by the U.S. Climate Change Science Program and the Subcommittee on Global Change Research, Washington, DC.

For Chapter 3:

Gutowski, W.J., G.C. Hegerl, G.J. Holland, T.R. Knutson, L.O. Mearns, R.J. Stouffer, P.J. Webster, M.F. Wehner, F.W. Zwiers, 2008: Causes of Observed Changes in Extremes and Projections of Future Changes in *Weather and Climate Extremes in a Changing Climate. Regions of Focus: North America, Hawaii, Caribbean, and U.S. Pacific Islands.* T.R. Karl, G.A. Meehl, C.D. Miller, S.J. Hassol, A.M. Waple, and W.L. Murray (eds.). A Report by the U.S. Climate Change Science Program and the Subcommittee on Global Change Research, Washington, DC.

For Chapter 4:

Easterling, D.R., D.M. Anderson, S.J. Cohen, W.J. Gutowski, G.J. Holland, K.E. Kunkel, T.C. Peterson, R.S. Pulwarty, R.J. Stouffer, M.F. Wehner, 2008: Measures to Improve Our Understanding of Weather and Climate Extremes in *Weather and Climate Extremes in a Changing Climate. Regions of Focus: North America, Hawaii, Caribbean, and U.S. Pacific Islands.* T.R. Karl, G.A. Meehl, C.D. Miller, S.J. Hassol, A.M. Waple, and W.L. Murray (eds.). A Report by the U.S. Climate Change Science Program and the Subcommittee on Global Change Research, Washington, DC.

For Appendix A:

Smith, R.L., 2008: Statistical Trend Analysis in *Weather and Climate Extremes in a Changing Climate. Regions of Focus: North America, Hawaii, Caribbean, and U.S. Pacific Islands.* T.R. Karl, G.A. Meehl, C.D. Miller, S.J. Hassol, A.M. Waple, and W.L. Murray (eds.). A Report by the U.S. Climate Change Science Program and the Subcommittee on Global Change Research, Washington, DC.

Report Motivation and Guidance for Using this Synthesis/Assessment Report

Authors:
Thomas R. Karl, NOAA; Gerald A. Meehl, NCAR; Christopher D. Miller, NOAA;
William L. Murray, STG, Inc.

According to the National Research Council, "an essential component of any research program is the periodic synthesis of cumulative knowledge and the evaluation of the implications of that knowledge for scientific research and policy formulation." The U.S. Climate Change Science Program (CCSP) is helping to meet that fundamental need through a series of 21 "synthesis and assessment products" (SAPs). A key component of the CCSP Strategic Plan (released July 2003), the SAPs integrate research results focused on important science issues and questions frequently raised by decision makers.

The SAPs support informed discussion and decisions by policymakers, resource managers, stakeholders, the media, and the general public. They are also used to help define and set the future direction and priorities of the program. The products help meet the requirements of the Global Change Research Act of 1990. The law directs agencies to "produce information readily usable by policymakers attempting to formulate effective strategies for preventing, mitigating, and adapting to the effects of global change" and to undertake periodic scientific assessments. This SAP (3.3) provides an in-depth assessment of the state of our knowledge about changes in weather and climate extremes in North America (and U.S. territories), where we live, work, and grow much of our food.

The impact of weather and climate extremes can be severe and wide-ranging although, in some cases, the impact can also be beneficial. Weather and climate extremes affect all sectors of the economy and the environment, including human health and well-being. During the period 1980-2006, the U.S. experienced 70 weather-related disasters in which overall damages exceeded $1 billion at the time of the event. Clearly, the direct impact of extreme weather and climate events on the U.S. economy is substantial.

There is scientific evidence that a warming world will be accompanied by changes in the intensity, duration, frequency, and spatial extent of weather and climate extremes. The Intergovernmental Panel on Climate Change (IPCC) Fourth Assessment Report has evaluated extreme weather and climate events on a global basis in the context of observed and projected changes in climate. However, prior to SAP 3.3 there has not been a specific assessment of observed and projected changes in weather and climate extremes across North America (including the U.S. territories in the Caribbean Sea and the Pacific Ocean), where observing systems are among the best in the world, and the extremes of weather and climate are some of the most notable occurring across the globe.

The term "weather extremes," as used in SAP 3.3, signifies individual weather events that are unusual in their occurrence (minimally, the event must lie in the upper or lower ten percentile of the distribution) or have destructive potential, such as hurricanes and tornadoes. The term "climate extremes" is used to represent the same type of event, but viewed over seasons (e.g., droughts), or longer periods. In this assessment we are particularly interested in whether climate extremes are changing in terms of a variety of characteristics, including intensity, duration, frequency, or spatial extent, and how they are likely to evolve in the future, although, due to data limitations and the scarcity of published analyses, there is little that can be said about extreme events in Hawaii, the Caribbean, or the Pacific Islands outside of discussion of tropical cyclone intensity and frequency. It is often very difficult to attribute a particular climate or weather extreme, such as a single drought episode or a single severe hurricane, to a specific cause. It is more feasible to attribute the changing "risk" of extreme events to specific causes. For this reason, this assessment focuses on the possible changes of past and future statistics of weather and climate extremes.

In doing any assessment, it is helpful to precisely convey the degree of certainty of important findings. For this reason, a lexicon expressing the likelihood of each key finding is presented below and used throughout this report. There is often considerable confusion as to what likelihood statements really represent. Are they statistical in nature? Do they consider the full spectrum of uncertainty or certainty? How reliable are they? Do they actually represent the true probability of occurrence, that is, when the probability states a 90% chance, does the event actually occur nine out of ten times?

There have been numerous approaches to address the problem of uncertainty. We considered a number of previously used methods, including the lexicon used in the IPCC Fourth Assessment (AR4), the US National Assessment of 2000, and previous Synthesis and Assessment Products, in particular SAP 1.1. SAP 1.1 was the first assessment to point out the importance of including both the statistical uncertainty related to finite samples and the "structural" uncertainty" related to the assumptions and limitations of physical and statistical models. This SAP adopted an approach very similar to that used in SAP 1.1 and the US National Assessment of 2000, with some small modifications (Preface Figure 1).

The likelihood scale in Figure 1 has fuzzy boundaries and is less discrete than the scale used in AR4. This is because the science of studying changes in climate extremes is not as well-developed as the study of changes in climate means over large space scales. The latter is an important topic addressed in IPCC. In addition, the AR4 adopted a confidence terminology which ranged from low confidence to medium confidence (5 chances in 10) to high confidence. As discussed in AR4, in practice, the confidence and likelihood statements are often linked. This is due in part to the limited opportunities we have in climate science to assess the confidence in our likelihood statements, in contrast to daily weather forecasts, where the reliability of forecasts

based on expert judgment has been shown to be quite good. For example, the analysis of past forecasts have shown it does actually rain nine of ten times when a 90% chance of rain is predicted

It is important to consider both the uncertainty related to limited samples and the uncertainty of alternatives to fundamental assumptions. Because of these factors, and taking into account the proven reliability of weather forecast likelihood statements based on expert judgment, this SAP relies on the expert judgment of the authors for its likelihood statements.

Statements made without likelihood qualifiers, such as "will occur", are intended to indicate a high degree of certainty, *i.e.*, approaching 100%.

DEDICATION

This Climate Change and Synthesis Product is dedicated to the memory of our colleague, friend, and co-author Dr. Miguel Cortez-Vázquez whose untimely passing during the writing of the report was a loss to us all, both professionally and personally.

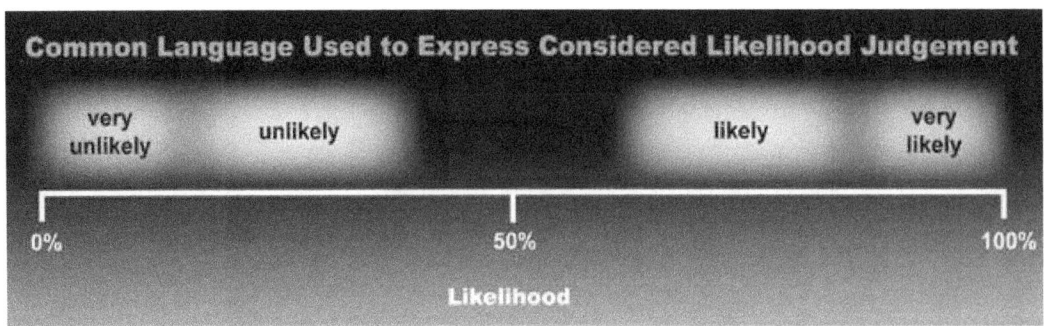

Figure P.1 Language in this Synthesis and Assessment Product used to express the team's expert judgment of likelihood.

Convening Lead Authors: Thomas R. Karl, NOAA; Gerald A. Meehl, NCAR
Lead Authors: Thomas C. Peterson, NOAA; Kenneth E. Kunkel, Univ. Ill. Urbana-Champaign, Ill. State Water Survey; William J. Gutowski, Jr., Iowa State Univ.; David R. Easterling, NOAA
Editors: Susan J. Hassol, STG, Inc.; Christopher D. Miller, NOAA; William L. Murray, STG, Inc.; Anne M. Waple, STG, Inc.

Synopsis

Changes in extreme weather and climate events have significant impacts and are among the most serious challenges to society in coping with a changing climate.

Many extremes and their associated impacts are now changing. For example, in recent decades most of North America has been experiencing more unusually hot days and nights, fewer unusually cold days and nights, and fewer frost days. Heavy downpours have become more frequent and intense. Droughts are becoming more severe in some regions, though there are no clear trends for North America as a whole. The power and frequency of Atlantic hurricanes have increased substantially in recent decades, though North American mainland land-falling hurricanes do not appear to have increased over the past century. Outside the tropics, storm tracks are shifting northward and the strongest storms are becoming even stronger.

It is well established through formal attribution studies that the global warming of the past 50 years is due primarily to human-induced increases in heat-trapping gases. Such studies have only recently been used to determine the causes of some changes in extremes at the scale of a continent. Certain aspects of observed increases in temperature extremes have been linked to human influences. The increase in heavy precipitation events is associated with an increase in water vapor, and the latter has been attributed to human-induced warming. No formal attribution studies for changes in drought severity in North America have been attempted. There is evidence suggesting a human contribution to recent changes in hurricane activity as well as in storms outside the tropics, though a confident assessment will require further study.

In the future, with continued global warming, heat waves and heavy downpours are very likely to further increase in frequency and intensity. Substantial areas of North America are likely to have more frequent droughts of greater severity. Hurricane wind speeds, rainfall intensity, and storm surge levels are likely to increase. The strongest cold season storms are likely to become more frequent, with stronger winds and more extreme wave heights.

Current and future impacts resulting from these changes depend not only on the changes in extremes, but also on responses by human and natural systems.

I. WHAT ARE EXTREMES AND WHY DO THEY MATTER?

Weather and climate extremes (Figure ES1) have always posed serious challenges to society. Changes in extremes are already having impacts on socioeconomic and natural systems, and future changes associated with continued warming will present additional challenges. Increased frequency of heat waves and drought, for example, could seriously affect human health, agricultural production, water availability and quality, and other environmental conditions (and the services they provide) (Chapter 1, section 1.1).

Extremes are a natural part of even a stable climate system and have associated costs (Figure ES.2) and benefits. For example, extremes are essential in some systems to keep insect pests under control. While hurricanes cause significant disruption, including death, injury, and damage, they also provide needed rainfall

Recent and projected changes in climate and weather extremes have primarily negative impacts.

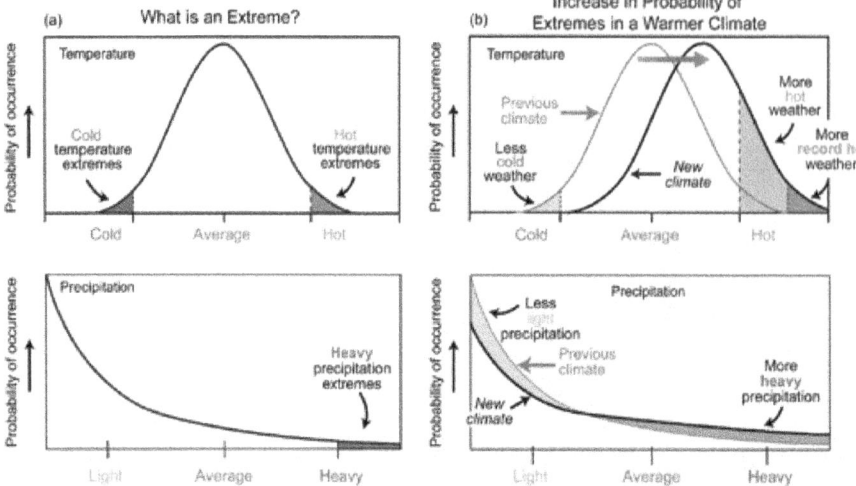

Many currently rare extreme events will become more commonplace.

Figure ES.1 Most measurements of temperature (top) will tend to fall within a range close to average, so their probability of occurrence is high. A very few measurements will be considered extreme and these occur very infrequently. Similarly, for rainfall (bottom), there tends to be more days with relatively light precipitation and only very infrequently are there extremely heavy precipitation events, meaning their probability of occurrence is low. The exact threshold for what is classified as an extreme varies from one analysis to another, but would normally be as rare as, or rarer than, the top or bottom 10% of all occurrences. A relatively small shift in the mean produces a larger change in the number of extremes for both temperature and precipitation (top right, bottom right). Changes in the shape of the distribution (not shown), such as might occur from the effects of a change in atmospheric circulation, could also affect changes in extremes. For the purposes of this report, all tornadoes and hurricanes are considered extreme.

to certain areas, and some tropical plant communities depend on hurricane winds toppling tall trees, allowing more sunlight to rejuvenate low-growing trees. But on balance, because systems have adapted to their historical range of extremes, the majority of events outside this range have primarily negative impacts (Chapter 1, section 1.4 and 1.5).

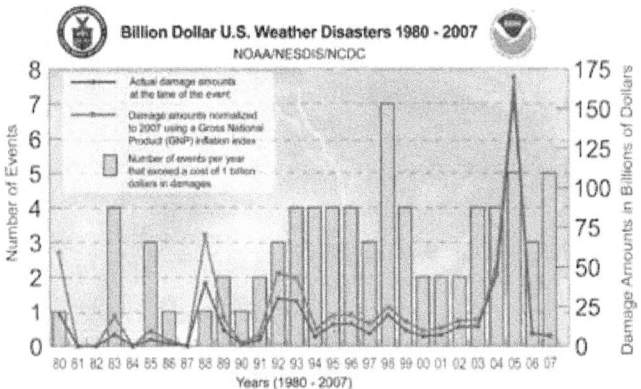

Figure ES.2 The blue bars show the number of events per year that exceed a cost of 1 billion dollars (these are scaled to the left side of the graph). The blue line (actual costs at the time of the event) and the red line (costs adjusted for wealth/inflation) are scaled to the right side of the graph, and depict the annual damage amounts in billions of dollars. This graphic does not include losses that are non-monetary, such as loss of life.

The impacts of changes in extremes depend on both changes in climate and ecosystem and societal vulnerability. The degree of impacts are due, in large part, to the capacity of society to respond. Vulnerability is shaped by factors such as population dynamics and economic status as well as adaptation measures such as appropriate building codes, disaster preparedness, and water use efficiency. Some short-term actions taken to lessen the risk from extreme events can lead to increases in vulnerability to even larger extremes. For example, moderate flood control measures on a river can stimulate development in a now "safe" floodplain, only to see those new structures damaged when a very large flood occurs (Chapter 1, section 1.6).

Human-induced warming is known to affect climate variables such as temperature and precipitation. Small changes in the averages of many variables result in larger changes in their extremes. Thus, within a changing climate system, some of what are now considered to be extreme events will occur more frequently, while others will occur less frequently (e.g., more heat waves and fewer cold snaps [Figures

ES.1, ES.3, ES.4]). Rates of change matter since these can affect, and in some cases overwhelm, existing societal and environmental capacity. More frequent extreme events occurring over a shorter period reduce the time available for recovery and adaptation. In addition, extreme events often occur in clusters. The cumulative effect of compound or back-to-back extremes can have far larger impacts than the same events spread out over a longer period of time. For example, heat waves, droughts, air stagnation, and resulting wildfires often occur concurrently and have more severe impacts than any of these alone (Chapter 1, section 1.2).

2. TEMPERATURE-RELATED EXTREMES

Observed Changes

Since the record hot year of 1998, six of the last ten years (1998-2007) have had annual average temperatures that fall in the hottest 10% of all years on record for the U.S. Accompanying a general rise in the average temperature, most of North America is experiencing more unusually hot days and nights. The number of heat waves (extended periods of extremely hot weather) also has been increasing over the past fifty years (see Table ES.1). However, the heat waves of the 1930s remain the most severe in the U.S. historical record (Chapter 2, section 2.2.1).

There have been fewer unusually cold days during the last few decades. The last 10 years have seen fewer severe cold snaps than for any other 10-year period in the historical record, which dates back to 1895. There has been a decrease in frost days and a lengthening of the frost-free season over the past century (Chapter 2, section 2.2.1).

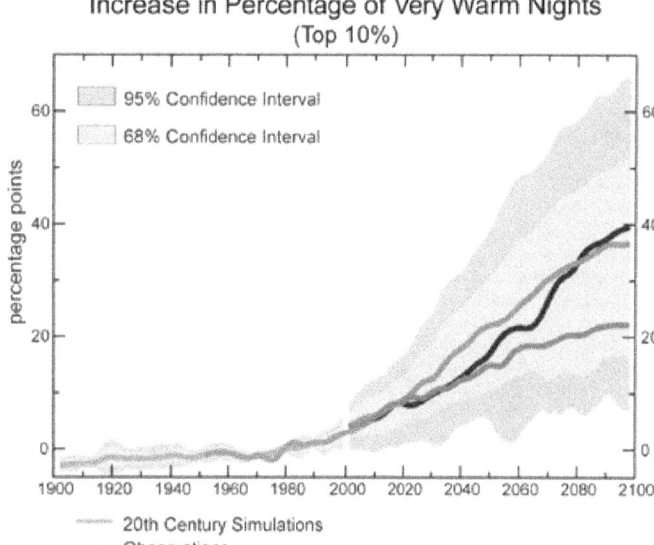

Increase in Percentage of Very Warm Nights
(Top 10%)

- 95% Confidence Interval
- 68% Confidence Interval

percentage points

---- 20th Century Simulations
---- Observations
—— Emission Scenario A2*: High at 2100
---- Emission Scenario A1B*: High at 2050, mid-range at 2100
---- Emission Scenario B1*: Low at 2100

Figure ES.3 Increase in the percent of days in a year over North America in which the daily low temperature is unusually warm (falling in the top 10% of annual daily lows, using 1961 to 1990 as a baseline). Under the lower emissions scenario[a], the percentage of very warm nights increases about 20% by 2100 whereas under the higher emissions scenarios, it increases by about 40%. Data for this index at the continental scale are available since 1950.

In summary, there is a shift towards a warmer climate with an increase in extreme high temperatures and a reduction in extreme low temperatures. These changes have been especially apparent in the western half of North America (Chapter 2, section 2.2.1).

Attribution of Changes

Human-induced warming has likely caused much of the average temperature increase in North America over the past fifty years and, consequently, changes in temperature extremes. For example, the increase in human-induced

Abnormally hot days and nights and heat waves are very likely to become more frequent.

The footnote below refers to Figures 3, 4, and 7.

* Three future emission scenarios from the IPCC Special Report on Emissions Scenarios:

B1 blue line: emissions increase very slowly for a few more decades, then level off and decline

A2 black line: emissions continue to increase rapidly and steadily throughout this century

A1B red line: emissions increase very rapidly until 2030, continue to increase until 2050, and then decline.

More details on the above emissions scenarios can be found in the IPCC Summary for Policymakers (IPCC, 2007)

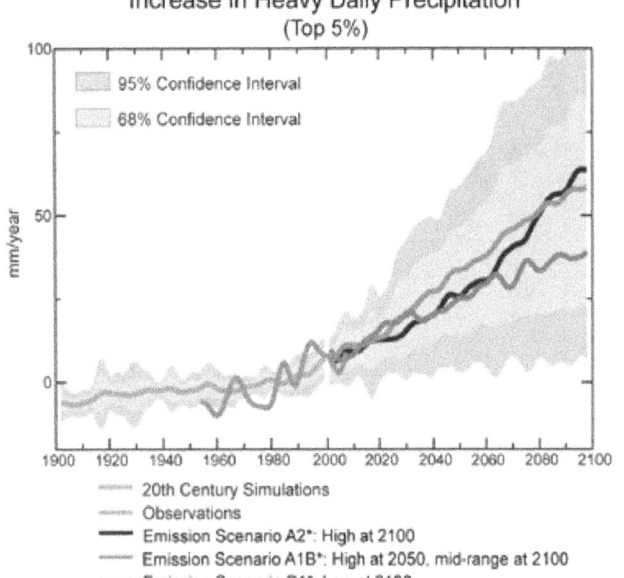

Figure ES.4 Increase in the amount of daily precipitation over North America that falls in heavy events (the top 5% of all precipitation events in a year) compared to the 1961-1990 average. Various emission scenarios are used for future projections*. Data for this index at the continental scale are available only since 1950.

In the U.S., the heaviest 1% of daily precipitation events increased by 20% over the past century.

In the future, precipitation is likely to be less frequent but more intense.

emissions of greenhouse gases is estimated to have substantially increased the risk of a very hot year in the U.S., such as that experienced in 2006 (Chapter 3, section 3.2.1 and 3.2.2). Additionally, other aspects of observed increases in temperature extremes, such as changes in warm nights and frost days, have been linked to human influences (Chapter 3, section 3.2.2).

Projected Changes
Abnormally hot days and nights (Figure ES.3) and heat waves are very likely to become more frequent. Cold days and cold nights are very likely to become much less frequent (see Table ES.1). The number of days with frost is very likely to decrease (Chapter 3, section 3.3.1 and 3.3.2).

Climate models indicate that currently rare extreme events will become more commonplace. For example, for a mid-range scenario of future greenhouse gas emissions, a day so hot that it is currently experienced only once every 20 years would occur every three years by the middle of the century over much of the continental U.S. and every five years over most of Canada. By the end of the century, it would occur every other year or more (Chapter 3, section 3.3.1).

Episodes of what are now considered to be unusually high sea surface temperature are very likely to become more frequent and widespread. Sustained (*e.g.*, months) unusually high temperatures could lead, for example, to more coral bleaching and death of corals (Chapter 3, section 3.3.1).

Sea ice extent is expected to continue to decrease and may even disappear entirely in the Arctic Ocean in summer in the coming decades. This reduction of sea ice increases extreme coastal erosion in Arctic Alaska and Canada due to the increased exposure of the coastline to strong wave action (Chapter 3, section 3.3.4 and 3.3.10).

3. PRECIPITATION EXTREMES

Observed Changes
Extreme precipitation episodes (heavy downpours) have become more frequent and more intense in recent decades over most of North America and now account for a larger percentage of total precipitation. For example, intense precipitation (the heaviest 1% of daily precipitation totals) in the continental U.S. increased by 20% over the past century while total precipitation increased by 7% (Chapter 2, section 2.2.2.2).

The monsoon season is beginning about 10 days later than usual in Mexico. In general, for the summer monsoon in southwestern North America, there are fewer rain events, but the events are more intense (Chapter 2, section 2.2.2.3).

Attribution of Changes
Heavy precipitation events averaged over North America have increased over the past 50 years, consistent with the observed increases in atmospheric water vapor, which have been associated

with human-induced increases in greenhouse gases (Chapter 3, section 3.2.3).

On average, precipitation is likely to be less frequent but more intense (Figure ES.4), and precipitation extremes are very likely to increase (see Table ES.1; Figure ES.5). For example, for a mid-range emission scenario, daily precipitation so heavy that it now occurs only once every 20 years is projected to occur approximately every eight years by the end of this century over much of Eastern North America (Chapter 3, section 3.3.5).

4. DROUGHT

Observed Changes

Drought is one of the most costly types of extreme events and can affect large areas for long periods of time. Drought can be defined in many ways. The assessment in this report focuses primarily on drought as measured by the Palmer Drought Severity Index, which represents multi-seasonal aspects of drought and has been extensively studied (Box 2.1).

Averaged over the continental U.S. and southern Canada the most severe droughts occurred in the 1930s and there is no indication of an overall trend in the observational record, which dates back to 1895. However, it is more meaningful to consider drought at a regional scale, because as one area of the continent is dry, often another is wet. In Mexico and the U.S. Southwest, the 1950s were the driest period, though droughts in the past 10 years now rival the 1950s drought. There are also recent regional tendencies toward more severe droughts in parts of Canada and Alaska (Chapter 2, section 2.2.2.1).

Attribution of Changes

No formal attribution studies for greenhouse warming and changes in drought severity in North America have been attempted. Other attribution studies have been completed that link the location and severity of droughts to the geographic pattern of sea surface temperature variations, which appears to have been a factor in the severe droughts of the 1930s and 1950s (Chapter 3, section 3.2.3).

Weather and Climate Extremes in a Changing Climate
Regions of Focus: North America, Hawaii, Caribbean, and U.S. Pacific Islands

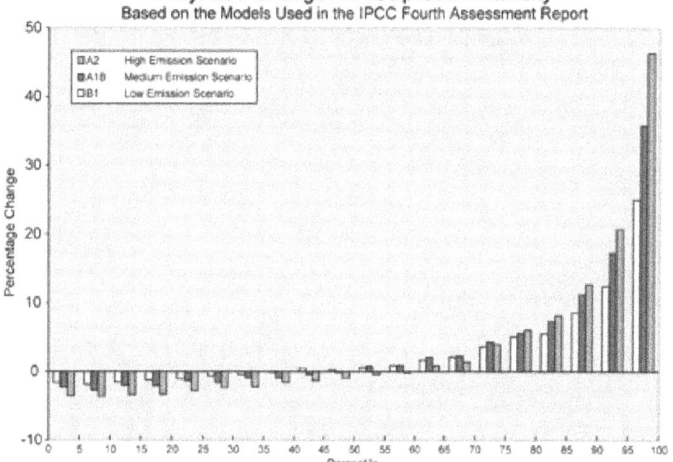

Figure ES.5 Projected changes in the intensity of precipitation, displayed in 5% increments, based on a suite of models and three emission scenarios. As shown here, the lightest precipitation is projected to decrease, while the heaviest will increase, continuing the observed trend. The higher emission scenarios yield larger changes. Figure courtesy of Michael Wehner.

Projected Changes

A contributing factor to droughts becoming more frequent and severe is higher air temperatures increasing evaporation when water is available. It is likely that droughts will become more severe in the southwestern U.S. and parts of Mexico in part because precipitation in the winter rainy season is projected to decrease (see Table ES.1). In other places where the increase in precipitation cannot keep pace with increased evaporation, droughts are also likely to become more severe (Chapter 3, section 3.3.7).

It is likely that droughts will continue to be exacerbated by earlier and possibly lower spring snowmelt run-off in the mountainous West, which results in less water available in late summer (Chapter 3, section 3.3.4 and 3.3.7).

5. STORMS

Hurricanes and Tropical Storms
Observed Changes

Atlantic tropical storm and hurricane destructive potential as measured by the Power Dissipation Index (which combines storm intensity, duration, and frequency) has increased (see Table ES.1). This increase is substantial since about 1970, and is likely substantial since the 1950s and 60s, in association with warming Atlantic sea surface temperatures (Figure ES.6) (Chapter 2, section 2.2.3.1).

A contributing factor to droughts becoming more frequent and severe is higher air temperatures increasing evaporation when water is available.

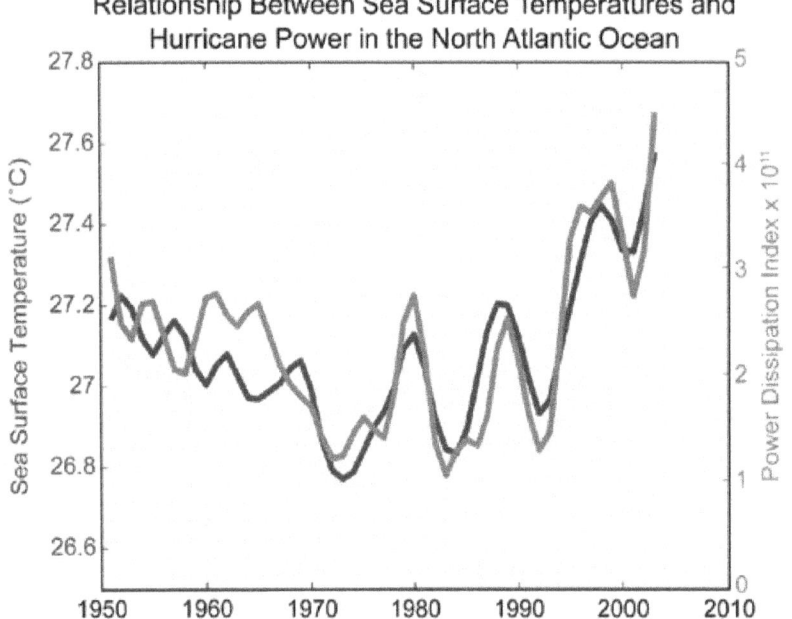

Figure ES.6 Sea surface temperatures (blue) and the Power Dissipation Index for North Atlantic hurricanes (Emanuel, 2007).

There have been fluctuations in the number of tropical storms and hurricanes from decade to decade and data uncertainty is larger in the early part of the record compared to the satellite era beginning in 1965. Even taking these factors into account, it is likely that the annual numbers of tropical storms, hurricanes and major hurricanes in the North Atlantic have increased over the past 100 years, a time in which Atlantic sea surface temperatures also increased. The evidence is not compelling for significant trends beginning in the late 1800s. Uncertainty in the data increases as one proceeds back in time. There is no observational evidence for an increase in North American mainland land-falling hurricanes since the late 1800s (Chapter 2, section 2.2.3.1). There is evidence for an increase in extreme wave height characteristics over the past couple of decades, associated with more frequent and more intense hurricanes (Chapter 2 section 2.2.3.3.2).

Hurricane intensity shows some increasing tendency in the western north Pacific since 1980. It has decreased since 1980 in the eastern Pacific, affecting the Mexican west coast and shipping lanes. However, coastal station observations show that rainfall from hurricanes has nearly doubled since 1950, in part due to slower moving storms (Chapter 2, section 2.2.3.1).

It is likely that hurricane rainfall and wind speeds will increase in response to human-caused warming.

ATTRIBUTION OF CHANGES

It is very likely that the human-induced increase in greenhouse gases has contributed to the increase in sea surface temperatures in the hurricane formation regions. Over the past 50 years there has been a strong statistical connection between tropical Atlantic sea surface temperatures and Atlantic hurricane activity as measured by the Power Dissipation Index (which combines storm intensity, duration, and frequency). This evidence suggests a human contribution to recent hurricane activity. However, a confident assessment of human influence on hurricanes will require further studies using models and observations, with emphasis on distinguishing natural from human-induced changes in hurricane activity through their influence on factors such as historical sea surface temperatures, wind shear, and atmospheric vertical stability (Chapter 3, section 3.2.4.3).

PROJECTED CHANGES

For North Atlantic and North Pacific hurricanes, it is likely that hurricane rainfall and wind speeds will increase in response to human-caused warming. Analyses of model simulations suggest that for each 1°C (1.8°F) increase in tropical sea surface temperatures, core rainfall rates will increase by 6-18% and the surface wind speeds of the strongest hurricanes will increase by about 1-8% (Chapter 3, section 3.3.9.2 and 3.3.9.4). Storm surge levels are likely to increase due to projected sea level rise, though the degree of projected increase has not been adequately studied. It is presently unknown how late 21st century tropical cyclone frequency in the Atlantic and North Pacific basins will change compared to the historical period (~1950-2006) (Chapter 3, section 3.3.9.3).

Other Storms
OBSERVED CHANGES

There has been a northward shift in the tracks of strong low-pressure systems (storms) in both the North Atlantic and North Pacific over the past fifty years. In the North Pacific, the strongest

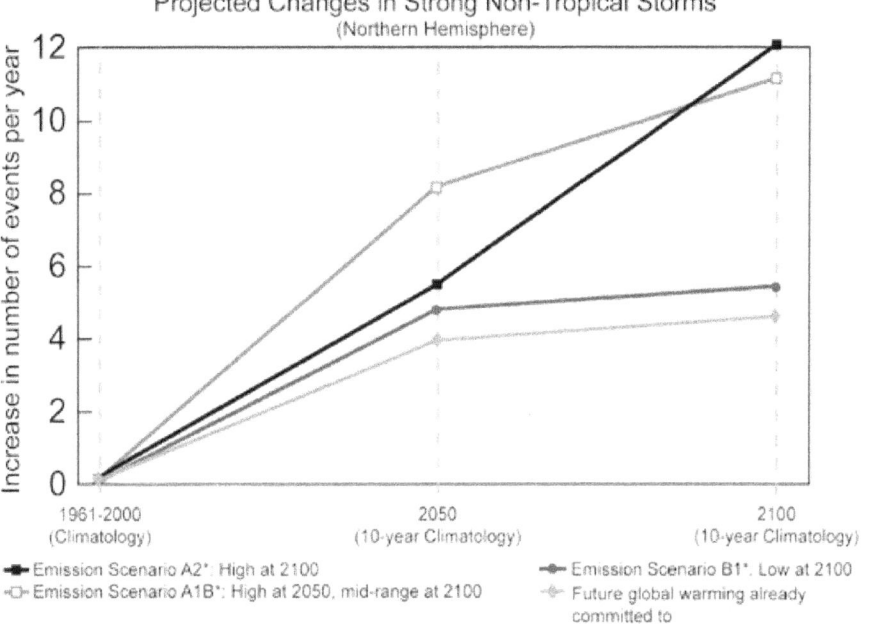

There are likely to be more frequent strong storms outside the Tropics, with stronger winds and more extreme wave heights.

Figure ES.7 The projected change in intense low pressure systems (strong storms) during the cold season for the Northern Hemisphere for various emission scenarios* (adapted from Lambert and Fyfe; 2006).

storms are becoming even stronger. Evidence in the Atlantic is insufficient to draw a conclusion about changes in storm strength (Chapter 2, section 2.2.3.2)

Increases in extreme wave heights have been observed along the Pacific Northwest coast of North America based on three decades of buoy data, and are likely a reflection of changes in cold season storm tracks (Chapter 2, section 2.2.3.3).

Over the 20th century, there has been considerable decade-to-decade variability in the frequency of snow storms (six inches or more). Regional analyses suggest that there has been a decrease in snow storms in the South and Lower Midwest of the U.S., and an increase in snow storms in the Upper Midwest and Northeast. This represents a northward shift in snow storm occurrence, and this shift, combined with higher temperature, is consistent with a decrease in snow cover extent over the U.S. In northern Canada, there has also been an observed increase in heavy snow events (top 10% of storms) over the same time period. Changes in heavy snow events in southern Canada are dominated by decade-to-decade variability (Chapter 2, section 2.2.3.4).

The pattern of changes in ice storms varies by region. The data used to examine changes in the frequency and severity of tornadoes and severe thunderstorms are inadequate to make definitive statements about actual changes (Chapter 2, section 2.2.3.5).

ATTRIBUTION OF CHANGES
Human influences on changes in atmospheric pressure patterns at the surface have been detected over the Northern Hemisphere and this reflects the location and intensity of storms (Chapter 3, section 3.2.5).

PROJECTED CHANGES
There are likely to be more frequent deep low-pressure systems (strong storms) outside the Tropics, with stronger winds and more extreme wave heights (Figure ES.7) (Chapter 3, section 3.3.10).

Observed changes in North American extreme events, assessment of human influence for the observed changes, and likelihood that the changes will continue through the 21st century[1].

Phenomenon and direction of change	Where and when these changes occurred in past 50 years	Linkage of human activity to observed changes	Likelihood of continued future changes in this century
Warmer and fewer cold days and nights	Over most land areas, the last 10 years had lower numbers of severe cold snaps than any other 10-year period	Likely warmer extreme cold days and nights, and fewer frosts[2]	Very likely[4]
Hotter and more frequent hot days and nights	Over most of North America	Likely for warmer nights[2]	Very likely[4]
More frequent heat waves and warm spells	Over most land areas, most pronounced over northwestern two thirds of North America	Likely for certain aspects, e.g., night-time temperatures; & linkage to record high annual temperature[2]	Very likely[4]
More frequent and intense heavy downpours and higher proportion of total rainfall in heavy precipitation events	Over many areas	Linked indirectly through increased water vapor, a critical factor for heavy precipitation events[3]	Very likely[4]
Increases in area affected by drought	No overall average change for North America, but regional changes are evident	Likely, Southwest USA.[3] Evidence that 1930's & 1950's droughts were linked to natural patterns of sea surface temperature variability	Likely in Southwest U.S.A., parts of Mexico and Carribean[4]
More intense hurricanes	Substantial increase in Atlantic since 1970; Likely increase in Atlantic since 1950s; increasing tendency in W. Pacific and decreasing tendency in E. Pacific (Mexico West Coast) since 1980[5]	Linked indirectly through increasing sea surface temperature, a critical factor for intense hurricanes[5]; a confident assessment requires further study[3]	Likely[4]

[1]Based on frequently used family of IPCC emission scenarios
[2]Based on formal attribution studies and expert judgment
[3]Based on expert judgment
[4]Based on model projections and expert judgment
[5]As measured by the Power Dissipation Index (which combines storm intensity, duration and frequency)

6. WHAT MEASURES CAN BE TAKEN TO IMPROVE THE UNDERSTANDING OF WEATHER AND CLIMATE EXTREMES?

Drawing on the material presented in this report, opportunities for advancement are described in detail in Chapter 4. Briefly summarized here, they emphasize the highest priority areas for rapid and substantial progress in improving understanding of weather and climate extremes.

1. The continued development and maintenance of high quality climate observing systems will improve our ability to monitor and detect future changes in climate extremes.

2. Efforts to digitize, homogenize and analyze long-term observations in the instrumental record with multiple independent experts and analyses improve our confidence in detecting past changes in climate extremes.

3. Weather observing systems adhering to standards of observation consistent with the needs of both the climate and the weather research communities improve our ability to detect observed changes in climate extremes.

4. Extended reconstructions of past climate using weather models initialized with homogenous surface observations would help improve our understanding of strong extratropical cyclones and other aspects of climate variabilty.

5. The creation of annually-resolved, regional-scale reconstructions of the climate for the past 2,000 years would help improve our understanding of very long-term regional climate variability.

6. Improvements in our understanding of the mechanisms that govern hurricane intensity would lead to better short and long-term predictive capabilities.

7. Establishing a globally consistent wind definition for determining hurricane intensity would allow for more consistent comparisons across the globe.

8. Improvements in the ability of climate models to recreate the recent past as well as make projections under a variety of forcing scenarios are dependent on access to both computational and human resources.

9. More extensive access to high temporal resolution data (daily, hourly) from climate model simulations both of the past and for the future would allow for improved understanding of potential changes in weather and climate extremes.

10. Research should focus on the development of a better understanding of the physical processes that produce extremes and how these processes change with climate.

11. Enhanced communication between the climate science community and those who make climate-sensitive decisions would strengthen our understanding of climate extremes and their impacts.

12. A reliable database that links weather and climate extremes with their impacts, including damages and costs under changing socioeconomic conditions, would help our understanding of these events.

CHAPTER 1

Why Weather and Climate Extremes Matter

Convening Lead Author: Thomas C. Peterson, NOAA

Lead Authors: David M. Anderson, NOAA; Stewart J. Cohen, Environment Canada and Univ. of British Columbia; Miguel Cortez-Vázquez, National Meteorological Service of Mexico; Richard J. Murnane, Bermuda Inst. of Ocean Sciences; Camille Parmesan, Univ. of Tex. at Austin; David Phillips, Environment Canada; Roger S. Pulwarty, NOAA; John M.R. Stone, Carleton Univ.

Contributing Author: Tamara G. Houston, NOAA; Susan L. Cutter, Univ. of S.C.; Melanie Gall, Univ. of S.C.

KEY FINDINGS

- Climate extremes expose existing human and natural system vulnerabilities.

- Changes in extreme events are one of the most significant ways socioeconomic and natural systems are likely to experience climate change.
    - Systems have adapted to their historical range of extreme events.
    - The impacts of extremes in the future, some of which are expected to be outside the historical range of experience, will depend on both climate change and future vulnerability. Vulnerability is a function of the character, magnitude, and rate of climate variation to which a system is exposed, the sensitivity of the system, and its adaptive capacity. The adaptive capacity of socioeconomic systems is determined largely by such factors as poverty and resource availability.

- Changes in extreme events are already observed to be having impacts on socioeconomic and natural systems.
    - Two or more extreme events that occur over a short period reduce the time available for recovery.
    - The cumulative effect of back-to-back extremes has been found to be greater than if the same events are spread over a longer period.

- Extremes can have positive or negative effects. However, on balance, because systems have adapted to their historical range of extremes, the majority of the impacts of events outside this range are expected to be negative.

- Actions that lessen the risk from small or moderate events in the short-term, such as construction of levees, can lead to increases in vulnerability to larger extremes in the long-term, because perceived safety induces increased development.

1.1 WEATHER AND CLIMATE EXTREMES IMPACT PEOPLE, PLANTS, AND ANIMALS

Extreme events cause property damage, injury, loss of life, and threaten the existence of some species. Observed and projected warming of North America has direct implications for the occurrence of extreme weather and climate events. It is very unlikely that the average climate could change without extremes changing as well. Extreme events drive changes in natural and human systems much more than average climate (Parmesan *et al.*, 2000; Parmesan and Martens, 2008).

Society recognizes the need to plan for the protection of communities and infrastructure from extreme events of various kinds, and engages in risk management. More broadly, responding to the threat of climate change is quintessentially a risk management problem. Structural measures (such as engineering works), governance measures (such as zoning and building codes), financial instruments (such as insurance and contingency funds), and emergency practices are all risk management measures that have been used to lessen the impacts of extremes. To the extent that changes in extremes can be anticipated, society can engage in additional risk management practices that would encourage proactive adaptation to limit future impacts.

Global and regional climate patterns have changed throughout the history of our planet. Prior to the Industrial Revolution, these changes occurred due to natural causes, including variations in the Earth's orbit around the Sun, volcanic eruptions, and fluctuations in the Sun's energy. Since the late 1800s, the changes have been due more to increases in the atmospheric concentrations of carbon dioxide and other trace greenhouse gases (GHG) as a result of human activities, such as fossil-fuel combustion and land-use change. On average, the world has warmed by 0.74°C (1.33°F) over the last century with most of that occurring in the last three decades, as documented by instrument-based observations of air temperature over land and ocean surface temperature (IPCC, 2007a; Arguez, 2007; Lanzante *et al.*, 2006). These observations are corroborated by, among many examples, the shrinking of mountain glaciers

Extreme events drive changes in natural and human systems much more than average climate.

(Barry, 2006), later lake and river freeze dates and earlier thaw dates (Magnuson *et al.*, 2000), earlier blooming of flowering plants (Cayan *et al.*, 2001), earlier spring bird migrations (Sokolov, 2006), thawing permafrost and associated shifts in ecosystem functioning, shrinking sea ice (Arctic Climate Impact Assessment, 2004), and shifts of plant and animal ranges both poleward and up mountainsides, both within the U.S. (Parmesan and Galbraith, 2004) and globally (Walther *et al.*, 2002; Parmesan and Yohe, 2003; Root *et al.*, 2003; Parmesan, 2006). Most of the recent warming observed around the world very likely has been due to observed changes in GHG concentrations (IPCC, 2007a). The continuing increase in GHG concentration is projected to result in additional warming of the global climate by 1.1 to 6.4°C (2.0 to 11.5°F) by the end of this century (IPCC, 2007a).

Extremes are already having significant impacts on North America. Examination of Figure 1.1 reveals that it is an unusual year when the United States does not have any billion dollar weather- and climate-related disasters. Furthermore, the costs of weather-related disasters in the U.S. have been increasing since 1960, as shown in Figure 1.2. For the world as a whole, "weather-related [insured] losses in recent years have been trending upward much faster than population, inflation, or insurance penetration, and faster than non-weather-related events" (Mills, 2005a). Numerous studies indicate that both the climate and the socioeconomic vulnerability to weather and climate extremes are changing (Brooks and Doswell, 2001; Pielke *et al.*, 2008; Downton *et al.*, 2005), although these factors' relative contributions to observed increases in disaster costs are subject to debate. For example, it is not easy to quantify the extent to which increases in coastal building damage is due to increasing wealth and population growth[1] in vulnerable locations versus an increase in storm intensity. Some authors (*e.g.*, Pielke *et al.*, 2008) divide damage costs by a wealth factor in order to "normalize" the damage costs. However, other factors such as changes in building codes, emergency response, warning systems, *etc.* also need to be taken into account. At this time, there is no universally

[1] Since 1980, the U.S. coastal population growth has generally reflected the same rate of growth as the entire nation (Crossett *et al.*, 2004).

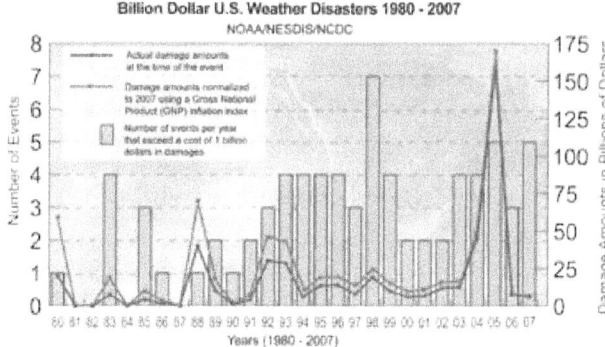

Figure 1.1 U.S. Billion Dollar Weather Disasters. The blue bars show number of events per year that exceed a cost of one billion dollars (these are scaled to the left side of the graph). The red line (costs adjusted for wealth/inflation) is scaled to the right side of the graph, and depicts the annual damage amounts in billions of dollars. This graphic does not include losses that are non-monetary, such as loss of life (Lott and Ross, 2006).

accepted approach to normalizing damage costs (Guha-Sapir *et al.*, 2004). Though the causes of the current damage increases are difficult to quantitatively assess, it is clear that any change in extremes will have a significant impact.

The relative costs of the different weather phenomena are presented in Figure 1.3 with tropical cyclones (hurricanes) being the most costly (Box 1.1). About 50% of the total tropical cyclone damages since 1960 occurred in 2005. Partitioning losses into the different categories is often not clear-cut. For example, tropical storms also contribute to damages that were categorized as flooding and coastal erosion. Based on data from 1940 to 1995, the annual mean loss of life from weather extremes in the U.S. exceeded 1,500 per year (Kunkel *et al.*, 1999), not including such factors as fog-related traffic fatalities. Approximately half of these deaths were related to hypothermia due to extreme cold, with extreme heat responsible for another one-fourth of the fatalities. For the period 1999 through 2003, the Centers for Disease Control reported an annual average of 688 deaths in the U.S. due to exposure to extreme heat (Luber *et al.*, 2006). From 1979 to 1997, there appears to be no trend in the number of deaths from extreme weather (Goklany and Straja, 2000). However, these statistics were compiled before the 1,400 hurricane-related fatalities in 2004-2005 (Chowdhury and Leatherman, 2007).

Natural systems display complex vulner-abilities to climate change that sometimes are

not evident until after the event. According to van Vliet and Leemans (2006), "the unexpected rapid appearance of ecological responses throughout the world" can be explained largely by the observed changes in extremes over the last few decades. Insects in particular have the ability to respond quickly to climate warming by increasing in abundances and/or increasing numbers of generations per year, which has resulted in widespread mortality of previously healthy trees (Logan *et al.*, 2003) (Box 1.2). The observed warming-related biological changes may have direct adverse effects on biodiversity, which in turn have been shown to impact ecosystem stability, resilience, and ability to provide societal goods and services (Parmesan and Galbraith, 2004; Arctic Climate Impact Assessment, 2004). The greater the change in global mean temperature, the greater will be the change in extremes and their consequent impacts on species and systems.

The costs of weather-related disasters in the U.S. have been increasing since 1960.

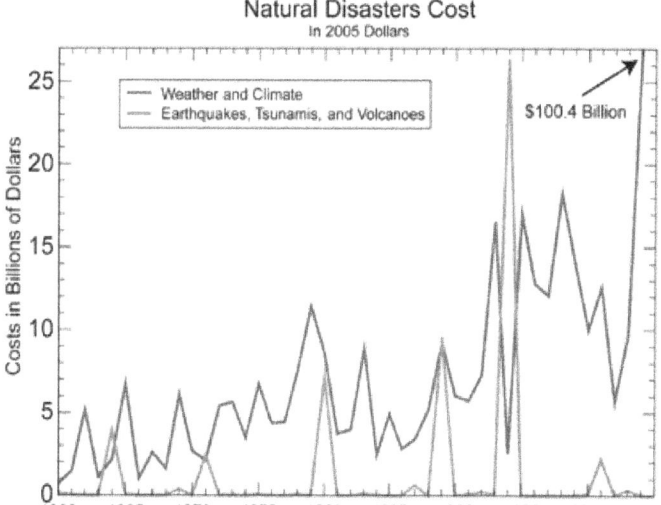

Figure 1.2 Costs from the SHELDUS database (Hazards and Vulnerability Research Institute, 2007) for weather and climate disasters and non-weather-related natural disasters in the U.S. The value for weather and climate damages in 2005 is off the graph at $100.4 billion. Weather and climate related damages have been increasing since 1960.

BOX 1.1: Damage Due to Hurricanes

There are substantial vulnerabilities to hurricanes along the Atlantic and Gulf Coasts of the United States. Four major urban areas represent concentrations of economic vulnerability (with capital stock greater than $100 billion)—the Miami coastal area, New Orleans, Houston, and Tampa. Three of these four areas have been hit by major storms in the last fifteen years (Nordhaus, 2006). A simple extrapolation of the current trend of doubling losses every ten years suggests that a storm like the 1926 Great Miami Hurricane could result in perhaps $500 billion in damages as early as the 2020s (Pielke *et al.*, 2008; Collins and Lowe, 2001).

Property damages are well-correlated with hurricane intensity (ISRTC, 2007). Kinetic energy increases with the square of its speed. So, in the case of hurricanes, faster winds have much more energy, dramatically increasing damages, as shown in Figure Box 1.1. Only 21% of the hurricanes making landfall in the United States are in Saffir-Simpson categories 3, 4, or 5, yet they cause 83% of the damage (Pielke and Landsea, 1998). Nordhaus (2006) argues that hurricane damage does not increase with the square of the wind speed, as kinetic energy does, but rather, damage appears to rise faster, with the eighth power of maximum wind speed. The 2005 total hurricane economic damage of $159 billion was primarily due to the cost of Katrina ($125 billion) (updated from Lott and Ross, 2006). As Nordhaus (2006) notes, 2005 was an economic outlier not because of extraordinarily strong storms but because the cost as a function of hurricane strength was high.

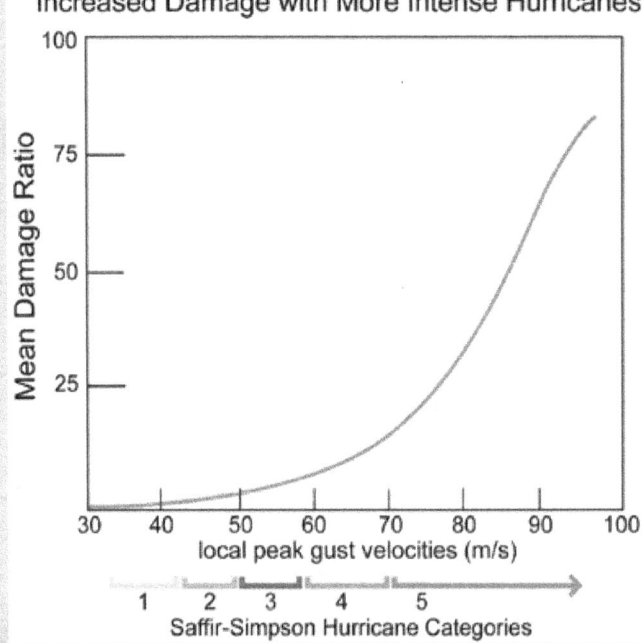

Figure Box 1.1 More intense hurricanes cause much greater losses. Mean damage ratio is the average expected loss as a percent of the total insured value. Adapted from Meyer *et al.* (1997).

A fundamental problem within many economic impact studies lies in the unlikely assumption that there are no other influences on the macro-economy during the period analyzed for each disaster (Pulwarty *et al.*, 2008). More is at work than aggregate indicators of population and wealth. It has long been known that different social groups, even within the same community, can experience the same climate event quite differently. In addition, economic analysis of capital stocks and densities does not capture the fact that many cities, such as New Orleans, represent unique corners of American culture and history

(Kates *et al.*, 2006). Importantly, the implementation of past adaptations (such as levees) affects the degree of present and future impacts (Pulwarty *et al.*, 2003). At least since 1979, the reduction of mortality over time has been noted, including mortality due to floods and hurricanes in the United States. On the other hand, the effectiveness of past adaptations in reducing property damage is less clear because aggregate property damages have risen along with increases in the population, material wealth, and development in hazardous areas.

BOX 1.2: Cold Temperature Extremes and Forest Beetles

Forest beetles in western North America have been responding to climate change in ways that are destroying large areas of forests (Figure Box 1.2). The area affected is 50 times larger than the area affected by forest fire with an economic impact nearly five times as great (Logan *et al.*, 2003). Two separate responses are contributing to the problem. The first is a response to warmer summers, which enable the mountain pine beetle (*Dendroctonus ponderosae*), in the contiguous United States, to produce two generations in a year, when previously it had only one (Logan *et al.*, 2003). In south-central Alaska, the spruce beetle (*Dendroctonus rufipennis*) is maturing in one year, where previously it took two years (Berg *et al.*, 2006).

The second response is to changes in winter temperatures, specifically the lack of extremely cold winter temperatures, which strongly regulate over-winter survival of the spruce beetle in the Yukon (Berg *et al.*, 2006) and the mountain pine beetle in British Columbia, Canada. The supercooling threshold (about -40°C/F), is the temperature at which the insect freezes and dies (Werner *et al.*, 2006). Recent warming has limited the frequency of sub -40°C (-40°F) occurrences, reducing winter mortality of mountain pine beetle larvae in British Columbia. This has led to an explosion of the beetle population, killing trees covering an area of 8.7 million hectares (21.5 million acres) in 2005, a doubling since 2003, and a 50-fold increase since 1999 (British Columbia Ministry of Forests and Range, 2006a). It is estimated that at the current rate of spread, 80% of British Columbia's mature lodgepole pine trees, the province's most abundant commercial tree species, will be dead by 2013 (Natural Resources Canada, 2007). Similarly in Alaska, approximately 847,000 hectares (2.1 million acres) of south-central Alaska spruce forests were infested by spruce beetles from 1920 to 1989 while from 1990 to 2000, an extensive outbreak of spruce beetles caused mortality of spruce across 1.19 million hectares (2.9 million acres), approximately 40% more forest area than had been infested in the state during the previous 70 years (Werner *et al.*, 2006). The economic loss goes well beyond the lumber value (millions of

Figure Box 1.2 Photograph of a pine forest showing pine trees dying (red) from beetle infestation in the Quesnel-Prince George British Columbia area. Fewer instances of extreme cold winter temperatures that winterkill beetle larvae have contributed a greater likelihood of beetle infestations. Copyright © Province of British Columbia. All rights reserved. Reprinted with permission of the Province of British Columbia. www.ipp.gov.bc.ca

board-feet) of the trees, as tourism revenue is highly dependent on having healthy, attractive forests. Hundreds of millions of dollars are being spent to mitigate the impacts of beetle infestation in British Columbia alone (British Columbia Ministry of Forests and Range, 2006b).

Adding further complexity to the climate-beetle-forest relationship in the contiguous United States, increased beetle populations have increased incidences of a fungus they transmit (pine blister rust, *Cronartium ribicola*) (Logan *et al.*, 2003). Further, in British Columbia and Alaska, long-term fire suppression activities have allowed the area of older forests to double. Older trees are more susceptible to beetle infestation. The increased forest litter from infected trees has, in turn, exacerbated the forest fire risk. Forest managers are struggling to keep up with changing conditions brought about by changing climate extremes.

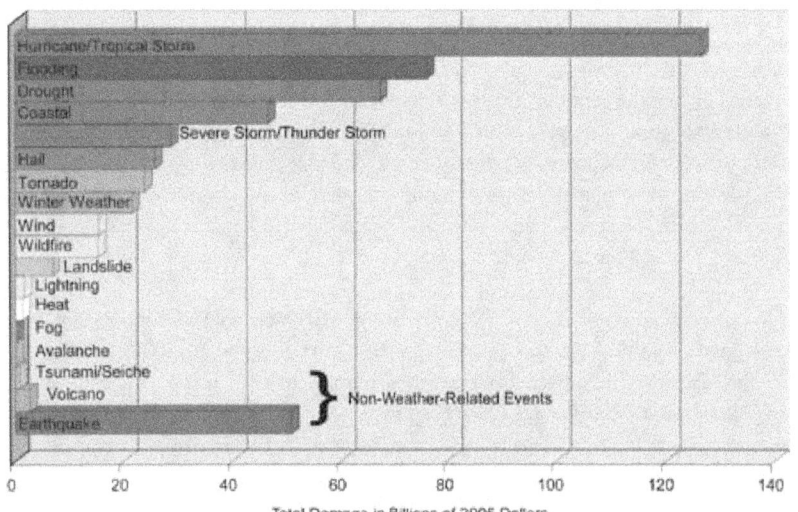

U.S. Natural Disaster Costs from 1960 to 2005

Figure 1.3 The magnitude of total U.S. damage costs from natural disasters over the period 1960 to 2005, in 2005 dollars. The data are from the SHELDUS data base (Hazards and Vulnerability Research Institute, 2007). SHELDUS is an event-based data set that does not capture drought well. Therefore, the drought bar was extended beyond the SHELDUS value to a more realistic estimate for drought costs. This estimate was calculated by multiplying the SHELDUS hurricane/tropical storm damage value by the fraction of hurricane/tropical storm damages (52%) relative to drought that occurs in the Billion Dollar Weather Disasters assessment (Lott and Ross, 2006). The damages are direct damage costs only. Note that weather- and climate-related disaster costs are 7.5 times those of non-weather natural disasters. Approximately 50% of the total hurricane losses were from the 2005 season. All damages are difficult to classify given that every classification is artificial and user- and database-specific. For example, SHELDUS' coastal classification includes damages from storm surge, coastal erosion, rip tide, tidal flooding, coastal floods, high seas, and tidal surges. Therefore, some of the coastal damages were caused by hurricanes just as some landslide damages are spawned by earthquakes.

This introductory chapter addresses various questions that are relevant to the complex relationships just described. Section 1.2 focuses on defining characteristics of extremes. Section 1.3 discusses the sensitivities of socioeconomic and natural systems to changes in extremes. Factors that influence the vulnerability of systems to changes in extremes are described in Section 1.4. As systems are already adapted to particular morphologies (forms) of extremes, Section 1.5 explains why changes in extremes usually pose challenges. Section 1.6 describes how actions taken in response to those challenges can either increase or decrease future impacts of extremes. Lastly, in Section 1.7, the difficulties in assessing extremes are discussed. The chapter also includes several boxes that highlight a number of topics related to particular extremes and their impacts, as well as analysis tools for assessing impacts.

The greater the change in global mean temperature, the greater will be the change in extremes and their consequent impacts on species and systems.

1.2 EXTREMES ARE CHANGING

When most people think of extreme weather or climate events, they focus on short-term intense episodes. However, this perspective ignores longer-term, more cumulative events, such as droughts. Thus, rather than defining extreme events solely in terms of how long they last, it is useful to look at them from a statistical point of view. If one plots all values of a particular variable, such as temperature, the values most likely will fall within a typical bell-curve with many values near average and fewer occurrences far away from the average. Extreme temperatures are in the tails of such distributions, as shown in the top panel of Figure 1.4.

According to the Glossary of the Intergovernmental Panel on Climate Change (IPCC) Fourth Assessment Report (IPCC, 2007a), "an extreme weather event is an event that is rare at a particular place and time of year." Here, as in the IPCC, we define rare as at least less common than the lowest or highest 10% of occurrences. For example, the heavy downpours that make up the top 5% of daily rainfall observations in a region would be classified as extreme precipitation events. By definition, the characteristics of extreme weather may vary from place to place in an absolute sense. When a pattern of extreme weather persists for some time, such as a season, it may be classed as an extreme climate event, especially if it yields an average or total that is itself extreme (*e.g.*, drought or heavy rainfall over a season). Extreme climate events, such as drought, can often be viewed as occurring in the tails of a distribution similar to the temperature distribution.

Daily precipitation, however, has a distribution that is very different from the temperature distribution. For most locations in North America, the majority of days have no precipitation at all. Of the days where some rain or snow does fall, many have very light precipitation, while only

What Is an Extreme?

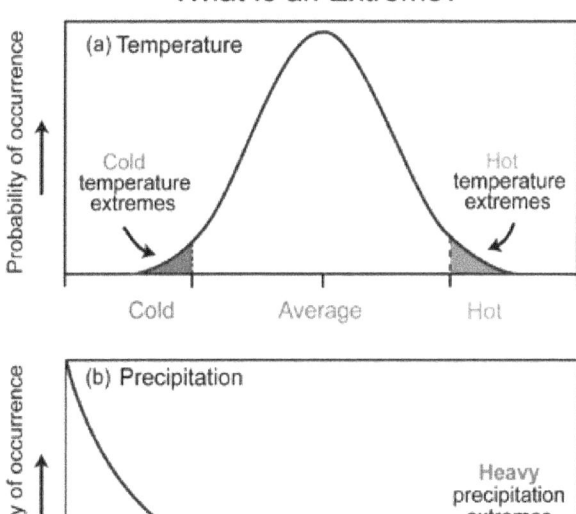

Figure 1.4 Probability distributions of daily temperature and precipitation. The higher the black line, the more often weather with those characteristics occurs.

This is because most extreme wind events are generated by special meteorological conditions that are well known. All tornadoes and hurricanes are considered extreme events. Extreme wind events associated with other phenomena, such as blizzards or nor'easters, tend to be defined by thresholds based on impacts, rather than statistics, or the wind is just one aspect of the measure of intensity of these storms.

An extreme weather event is an event that is rare at a particular place and time of year.

Most considerations of extreme weather and climate events are limited to discrete occurrences. However, in some cases, events that occur in rapid succession can have impacts greater than the simple sum of the individual events. For example, the ice storms that occurred in eastern Ontario and southern Quebec in 1998 were the most destructive and disruptive in Canada in recent memory. This was a series of three storms that deposited record amounts of freezing rain (more than 80 mm/3 in) over a record number of hours during January 5-10, 1998. Further, the storm brutalized an area extending nearly 1000 km² (380 mi²), which included one of the largest urban areas of Canada, leaving more than four million people freezing in the dark for hours and

a few have heavy precipitation, as illustrated by the bottom panel of Figure 1.4. Extreme value theory is a branch of statistics that fits a probability distribution to historical observations. The tail of the distribution can be used to estimate the probability of very rare events. This is the way the 100-year flood level can be estimated using 50 years of data. One problem with relying on historical data is that some extremes are far outside the observational record. For example, the heat wave that struck Europe in 2003 was so far outside historical variability (Figure 1.5) that public health services were unprepared for the excess mortality. Climate change is likely to increase the severity and frequency of extreme events for both statistical and physical reasons.

Wind is one parameter where statistics derived from all observations are not generally used to define what an extreme is.

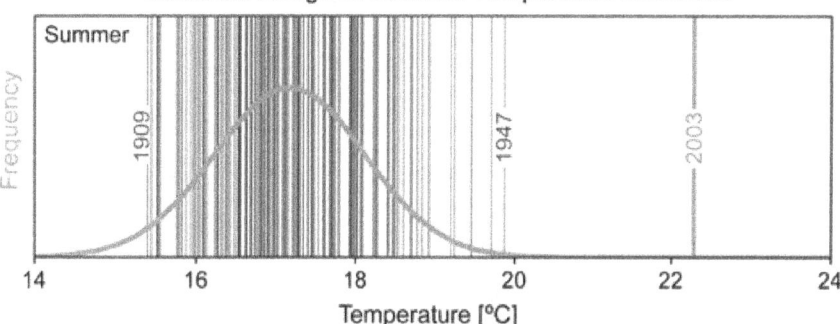

Figure 1.5 Like the European summer temperature of 2003, some extremes that are more likely to be experienced in the future will be far outside the range of historical observations. Each vertical line represents the mean summer temperature for a single year from the average of four stations in Switzerland over the period 1864 through 2003. Extreme values from the years 1909, 1947, and 2003 identified. [From Schär et al., 2004.]

Events that occur in rapid succession can have impacts greater than the simple sum of the individual events.

even days. The conditions were so severe that no clean-up action could be taken between storms. The ice built up, stranding even more people at airports, bringing down high-tension transmission towers, and straining food supplies. Damage was estimated to exceed $4 billion, including losses to electricity transmission infrastructure, agriculture, and various electricity customers (Lecomte *et al.*, 1998; Kerry *et al.*, 1999). Such cumulative events need special consideration.

Also, compound extremes are events that depend on two or more elements. For example, heat waves have greater impacts on human health when they are accompanied by high humidity. Additionally, serious impacts due to one extreme may only occur if it is preceded by a different extreme. For example, if a wind storm is preceded by drought, it would result in far more wind-blown dust than the storm would generate without the drought.

As the global climate continues to adjust to increasing concentrations of greenhouse gases in the atmosphere, many different aspects of extremes have the potential to change as well (Easterling *et al.*, 2000a,b). The most commonly considered aspect is frequency. Is the extreme occurring more frequently? Will currently rare events become commonplace in 50 years? Changes in intensity are as important as changes in frequency. For example, are hurricanes becoming more intense? This is important because, as explained in Box 1.1, hurricane damage increases exponentially with the speed of the wind, so an intense hurricane causes much more destruction than a weak hurricane.

Frequency and intensity are only two parts of the puzzle. There are also temporal considerations, such as time of occurrence and duration. For example, the timing of peak snowmelt in the western mountains has shifted to earlier in the spring (Johnson *et al.*, 1999; Cayan *et al.*, 2001). Earlier snowmelt in western mountains means a longer dry season with far-reaching impacts on the ecologies of plant and animal communities, fire threat, and human water resources. Indeed, in the American West, wildfires are strongly associated with increased spring and summer temperatures and correspondingly earlier spring snowmelt in the mountains (Westerling *et al.*, 2006). In Canada, anthropogenic (human-induced) warming of summer temperatures has increased the area burned by forest fires in recent decades (Gillett *et al.*, 2004). Changing the timing and/or number of wildfires might pose threats to certain species by overlapping with their active seasons (causing increased deaths) rather than occurring during a species' dormant phase (when they are less vulnerable). Further, early snowmelt reduces summer water resources, particularly in California where summer rains are rare. Also of critical importance to Southern California wildfires are the timing and intensity of Santa Ana winds, which may be sensitive to future global warming (Miller and Schlegel, 2006). The duration of extreme events (such as heat waves, flood-inducing rains, and droughts) is also potentially subject to change. Spatial characteristics need to be considered. Is the size of the impact area changing? In addition to the size of the individual events, the location is subject to change. For example, is the region susceptible to freezing rain moving farther north?

Therefore, the focus of this assessment is not only the meteorology of extreme events, but how climate change might alter the characteristics of extremes. Figure 1.6 illustrates how the tails of the distribution of temperature and precipitation are anticipated to change in a warming world.

For temperature, both the average (mean) and the tails of the distributions are expected to warm. While the change in the number of average days may be small, the percentage change in the number of very warm and very cold days can be quite large. For precipitation, model and observational evidence points to increases in the number of heavy rain events and decreases in the number of light precipitation events.

1.3 NATURE AND SOCIETY ARE SENSITIVE TO CHANGES IN EXTREMES

Sensitivity to climate is defined as the degree to which a system is affected by climate-related stimuli. The effect may be direct, such as crop yield changing due to variations in temperature or precipitation, or indirect, such as the decision to build a house in a location based on insurance rates, which can change due to flood risk caused by sea level rise (IPCC, 2007b). Indicators of sensitivity to climate can include changes in the timing of life events (such as the date a plant flowers) or distributions of individual species, or alteration of whole ecosystem functioning (Parmesan and Yohe, 2003; Parmesan and Galbraith, 2004).

Sensitivity to climate directly impacts the vulnerability of a system or place. As a result, managed systems, both rural and urban, are constantly adjusting to changing perceptions of risks and opportunities (OFCM, 2005). For example, hurricane destruction can lead to the adoption of new building codes (or enforcement of existing codes) and the implementation of new construction technology, which alter the future sensitivity of the community to climate. Further, artificial selection and genetic engineering of crop plants can adjust agricultural varieties to changing temperature and drought conditions. Warrick (1980) suggested that the impacts of extreme events would gradually decline because of improved planning and early warning systems. Ausubel (1991) went further, suggesting that irrigation, air conditioning, artificial snow making, and other technological improvements, were enabling society to become more climate-proof. While North American society is not as sensitive to extremes as it was 400 years ago – for example, a megadrought in Mexico in the mid-to-late 1500s created conditions that may have

altered rodent-human interactions and thereby contributed to tremendous population declines as illustrated by Figure 1.7 – socioeconomic systems are still far from being climate-proof.

Society is clearly altering relationships between climate and society and thereby sensitivities to climate. However, this is not a unidirectional change. Societies make decisions that alter regional-scale landscapes (urban expansion, pollution, land-use change, water withdrawals) which can increase or decrease both societal and ecosystem sensitivities (*e.g.*, Mileti, 1999; Glantz, 2003). Contrary to the possible gradual decline in impacts mentioned above, recent droughts have resulted in increased economic losses and conflicts (Riebsame *et al.*, 1991; Wilhite, 2005). The increased concern about El Niño's impacts reflect a heightened awareness of its effects on extreme events worldwide, and growing concerns about the gap between scientific information and adaptive responses by communities and governments (Glantz, 1996). In the U.S. Disaster Mitigation Act of 2000, Congress specifically wrote that a "greater emphasis needs to be placed on . . . implementing adequate mea-

> While North American society is not as sensitive to extremes as it was 400 years ago, socioeconomic systems are still far from being climate-proof.

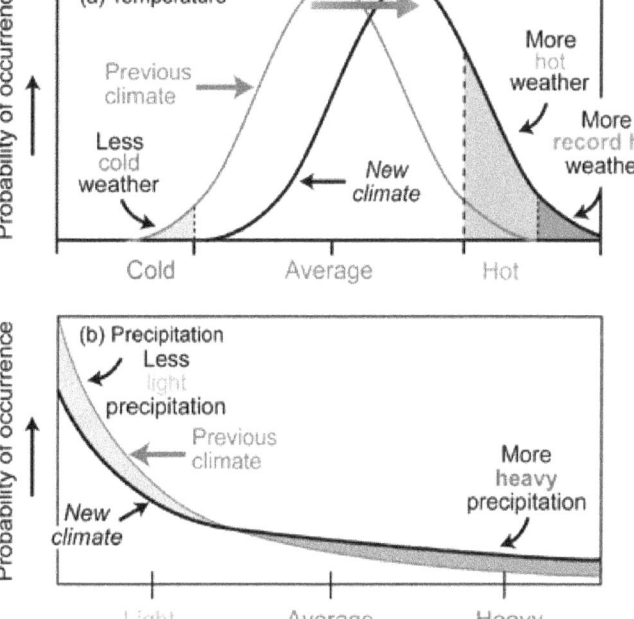

Increase in Probability of Extremes in a Warmer Climate

Figure 1.6 Simplified depiction of the changes in temperature and precipitation in a warming world.

Drought and Population Collapse in Mexico

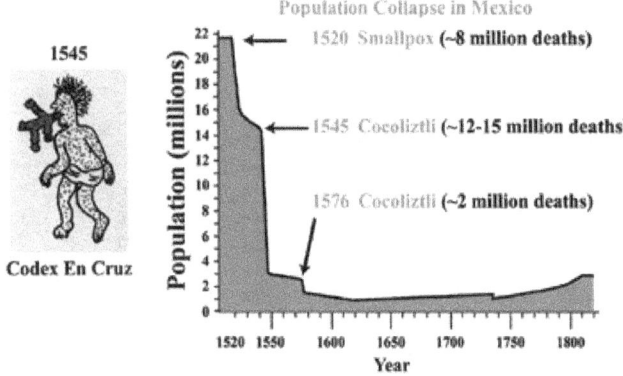

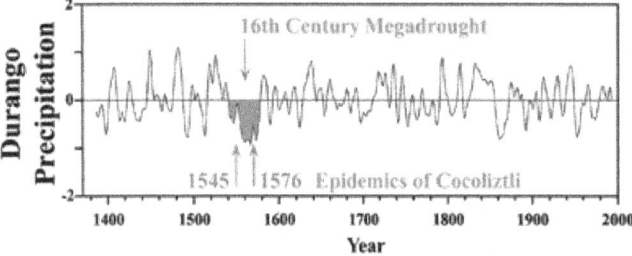

Figure 1.7 Megadrought and megadeath in 16th century Mexico. Four hundred years ago, the Mexican socioeconomic and natural systems were so sensitive to extremes that a megadrought in Mexico led to massive population declines (Acuna-Soto et al., 2002). The 1545 Codex En Cruz depicts the effects of the cocoliztli epidemic, which has symptoms similar to rodent-borne hantavirus hemorrhagic fever.

function are impacted by major disturbance events, such as tornadoes, floods, and hurricanes (Pickett and White, 1985; Walker, 1999). Warming winters, with a sparse snow cover at lower elevations, have led to false springs (an early warming followed by a return to normal colder winter temperatures) and subsequent population declines and extirpation (local extinction) in certain butterfly species (Parmesan, 1996, 2005).

By far, most of the documented impacts of global warming on natural systems have been ecological in nature. While ecological trends are summarized in terms of changes in mean biological and climatological traits, many detailed studies have implicated extreme weather events as the mechanistic drivers of these broad ecological responses to long-term climatic trends (Inouye, 2000; Parmesan et al., 2000). Observed ecological responses to local, regional, and continental warming include changes in species' distributions, changes in species' phenologies (the timing of the different phases of life events), and alterations of ecosystem functioning (Walther et al., 2002; Parmesan and Yohe, 2003; Root et al., 2003; Parmesan and Galbraith, 2004; Parmesan, 2006; IPCC, 2007b). Changes in species' distributions include a northward and upward shift in the mean location of populations of the Edith's checkerspot butterfly in western North America consistent with expectations from the observed 0.7°C (1.3°F) warming—about 100 kilometers (60 mi) northward and 100 meters (330 ft) upslope (Parmesan, 1996; Karl et al., 1996). Phenological (e.g., timing) changes include lilac blooming 1.5 days earlier per decade and honeysuckle blooming 3.5 days earlier per decade since the 1960s in the western U.S. (Cayan et al., 2001). In another example, tree swallows across the U.S. and southern Canada bred about 9 days earlier from 1959 to 1991, mirroring a gradual increase in mean May temperatures (Dunn and Winkler, 1999). One example of the impacts of warming on the functioning of a whole ecosystem comes from the Arctic tundra, where warming trends have been considerably stronger than in the contiguous U.S. Thawing of the permafrost layer has caused an increase in decomposition rates of dead organic matter during winter, which in some areas has already resulted in a shift from the tundra being a carbon sink to being a carbon source (Oechel et al., 1993; Oechel et al., 2000).

sures to reduce losses from natural disasters." Many biological processes undergo sudden shifts at particular thresholds of temperature or precipitation (Precht et al., 1973; Weiser, 1973; Hoffman and Parsons, 1997). The adult male/female sex ratios of certain reptile species such as turtles and snakes are determined by the extreme maximum temperature experienced by the growing embryo (Bull, 1980; Bull and Vogt, 1979; Janzen, 1994). A single drought year has been shown to affect population dynamics of many insects, causing drastic crashes in some species (Singer and Ehrlich, 1979; Ehrlich et al., 1980; Hawkins and Holyoak, 1998) and population booms in others (Mattson and Haack, 1987); see Box 1.3 on drought for more information. The nine-banded armadillo (*Dasypus novemcinctus*) cannot tolerate more than nine consecutive days below freezing (Taulman and Robbins, 1996). The high sea surface temperature event associated with El Niño in 1997-98 ultimately resulted in the death of 16% of the world's corals (Hoegh-Guldberg, 1999, 2005; Wilkinson, 2000); see Box 1.4 on coral bleaching for more information. Further, ecosystem structure and

Many biological processes undergo sudden shifts at particular thresholds of temperature or precipitation.

While many changes in timing have been observed (*e.g.*, change in when species breed or migrate), very few changes in other types of behaviors have been seen. One of these rare examples of behavioral changes is that some sooty shearwaters, a type of seabird, have shifted their migration pathway from the coastal California current to a more central Pacific pathway, apparently in response to a warming-induced shift in regions of high fish abundance during their summer flight (Spear and Ainley, 1999; Oedekoven *et al.*, 2001). Evolutionary studies of climate change impacts are also few (largely due to dearth of data), but it is clear that genetic responses have already occurred (Parmesan, 2006). Genetic changes in local populations have taken place resulting in much higher frequencies of individuals who are warm-adapted (*e.g.*, for fruit flies; Rodriguez-Trelles and Rodriguez, 1998; Levitan, 2003; Balanya *et al.*, 2006), or can disperse better (*e.g.*, for the bush cricket; Thomas *et al.*, 2001). For species-level evolution to occur, either appropriate novel mutations or novel genetic architecture (*i.e.*, new gene complexes) would have to emerge to allow a response to selection for increased tolerance to more extreme climate than the species is currently adapted to (Parmesan *et al.*, 2000; Parmesan *et al.*, 2005). However, so far there is no evidence for change in the absolute climate tolerances of a species, and, hence, no indication that evolution at the species level is occurring, nor that it might occur in the near future (Parmesan, 2006).

Ecological impacts of climate change on natural systems are beginning to have carry-over impacts on human health (Parmesan and Martens, 2008). The best example comes from bacteria which live in brackish rivers and sea water and use a diversity of marine life as reservoirs, including many shellfish, some fish, and even water hyacinth. Weather influences the transport and dissemination of these microbial agents via rainfall and runoff, and the survival and/or growth through factors such as temperature (Rose *et al,.* 2001). Two-hundred years of observational records reveal strong repeated patterns in which extreme high water temperatures cause algae blooms, which then promote rapid increases in zooplankton abundances and, hence, also in their associated bacteria (Colwell, 1996). Additionally, dengue is currently endemic in several cities in Texas and the mosquito vector (carrier) species is distributed across the Gulf Coast states (Brunkard *et al.*, 2007; Parmesan and Martens, 2008). Thus, climate related changes in ecosystems can also affect human health.

1.4 FUTURE IMPACTS OF CHANGING EXTREMES ALSO DEPEND ON VULNERABILITY

Climate change presents a significant risk management challenge, and dealing with weather and climate extremes is one of its more demanding aspects. In human terms, the importance of extreme events is demonstrated when they expose the vulnerabilities of communities and the infrastructure on which they rely. Extreme weather and climate events are not simply hydrometeorological occurrences. They impact socioeconomic systems and are often exacerbated by other stresses, such as social inequalities, disease, and conflict. Extreme events can threaten our very well-being. Understanding vulnerabilities from weather and climate extremes is a key first step in managing the risks of climate change.

According to IPCC (2007b), "vulnerability to climate change is the degree to which...systems are susceptible to, and unable to cope with, adverse impacts." Vulnerability is a function of the character, magnitude, and rate of climate change to which a system is exposed, its sensitivity, and its adaptive capacity. A system can be sensitive to change but not be vulnerable, such as some aspects of agriculture in North America, because of the rich adaptive capacity; or relatively insensitive but highly vulnerable. An example of the latter is incidence of diarrhea (caused by a variety of water-borne organisms) in less developed countries. Diarrhea is not correlated with temperatures in the U.S. because of highly-developed sanitation facilities. However, it does show a strong correlation with high temperatures in Lima, Peru (Checkley *et al.*, 2000; WHO, 2003, 2004). Thus, vulnerability is highly dependent on the robustness of societal infrastructures. For example, water-borne diseases have been shown to significantly increase following extreme precipitation events in the U.S. (Curriero *et al.*, 2001) and Canada (O'Connor, 2002) because water management systems failed (Box 1.5). Systems that normally survive are

Ecological impacts of climate change on natural systems are beginning to have carry-over impacts on human health.

BOX I.3: Drought

Drought should not be viewed as merely a physical phenomenon. Its impacts on society result from the interplay between a physical event (*e.g.*, less precipitation than expected) and the demands people place on water supply. Human beings often exacerbate the impact of drought. Recent droughts in both developing and developed countries and the resulting economic and environmental impacts and personal hardships have underscored the vulnerability of all societies to this natural hazard (National Drought Mitigation Center, 2006).

Over the past century, the area affected by severe and extreme drought in the United States each year averages around 14% with the affected area as high as 65% in 1934. In recent years, the drought-affected area ranged between 35 and 40% as shown in Figure Box 1.3. FEMA (1995) estimates average annual drought-related losses at $6-8 billion (based on relief payments alone). Losses were as high as $40 billion in 1988 (Riebsame *et al.*, 1991). Available economic estimates of the impacts of drought are difficult to reproduce. This problem has to do with the unique nature of drought relative to other extremes, such as hurricanes. The onset of drought is slow. Further, the secondary impacts may be larger than the immediately visible impacts and often occur past the lifetime of the event (Wilhite and Pulwarty, 2005).

In recent years, the western United States has experienced considerable drought impacts, with 30% of the region under severe drought since 1995. Widespread declines in springtime snow water equivalent in the U.S. West have occurred over the period 1925–2000, especially since mid-century. While non-climatic factors, such as the growth of forest canopy, might be partly responsible,

the primary cause is likely the changing climate because the patterns of climatic trends are spatially consistent and the trends are dependent on elevation (Mote *et al.*, 2005). Increased temperature appears to have led to increasing drought (Andreadis and Lettenmaier, 2006). In the Colorado River Basin, the 2000-2004 period had an average flow of 9.9 million acre feet[1] (maf) per year, lower than the driest period during the Dust Bowl years of 1931-1935 (with 11.4 maf), and the 1950s (with 10.2 maf) (Pulwarty *et al.*, 2005). For the winter of 2004-2005, average precipitation in the Basin was around 100% of normal. However, the combination of low antecedent soil moisture (absorption into soil), depleted high mountain aquifers, and the warmest January-July period on record (driving evaporation) resulted in a reduced flow of 75% of average.

At the same time, states in the U.S. Southwest experienced some of the most rapid economic and population growth in the country, with attendant demands on water resources and associated conflicts. It is estimated that as a result of the 1999-2004 drought and increased water resources extraction, Lake Mead and Lake Powell[2] will take 13 to 15 years of average flow conditions to refill. In the Colorado River Basin, high-elevation snow pack contributes approximately 70% of the annual runoff. Because the Colorado River Compact[3] prioritizes the delivery of water to the Lower Basin states of Arizona,

---

[1] One acre foot is equal to 325,853 U.S. gallons or 1,233.5 cubic meters. It is the amount of water needed to cover one acre with a foot of water.
[2] Lake Mead and Lake Powell are reservoirs on the Colorado River. Lake Mead is the largest man-made lake in the United States.
[3] The Colorado River Compact is a 1922 agreement among seven U.S. states in the basin of the Colorado River which governs the allocation of the river's water.

Area of the United States in Severe and Extreme Drought

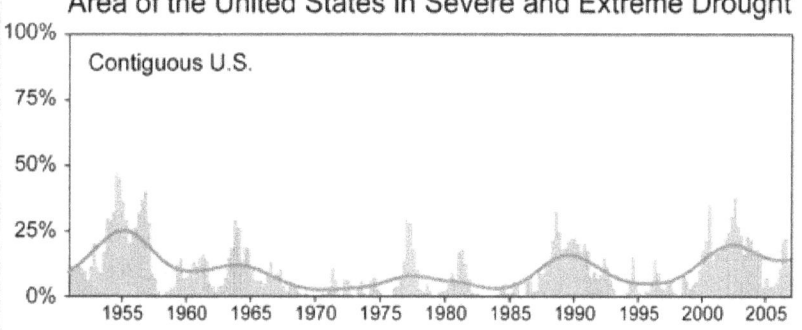

Contiguous U.S.

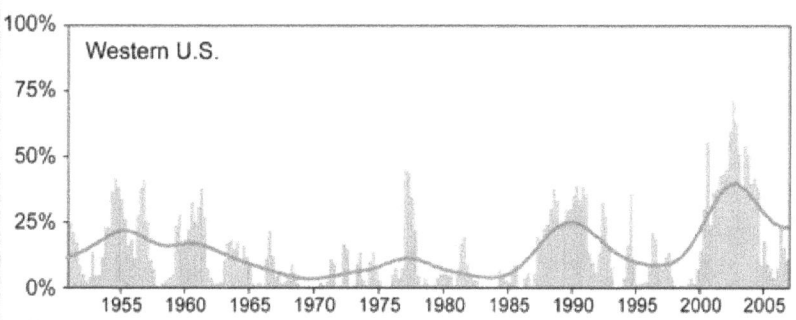

Western U.S.

Figure Box 1.3 Percent of area in the contiguous U.S. and western U.S. affected by severe and extreme drought as indicated by Palmer Drought Severity Index (PDSI) values of less than or equal to -3. Data from NOAA's National Climatic Data Center.

response was complex, "forest patches within the shift zone became much more fragmented, and soil erosion greatly accelerated," which may be the underlying reason why this boundary shift persisted over the next 40 years.

In the Sierra Nevada Mountains of California, increased frequency of fires has been shown to be an important element in local forest dynamics (Swetnam, 1993; Stephenson and Parsons, 1993; Westerling et al., 2006). Fire frequency is correlated with temperature, fuel loads (related to tree species composition and age structure), and fuel moisture. Periods of drought followed by weeks of extreme heat and low humidity provide ideal conditions for fire, which are, ironically, often sparked by lightning associated with thunderstorms at the drought's end.

California, and Nevada, the largest impacts may be felt in the Upper Basin states of Wyoming, Utah, Colorado, and New Mexico. With increased global warming, the compact requirements may only be met 59% to 75% of the time (Christensen et al., 2004).

Severe droughts in the western U.S. have had multiple impacts on wild plants and animals. The 1975-1977 severe drought over California caused the extinction of 5 out of 21 surveyed populations of Edith's checkerspot butterfly (Ehrlich et al., 1980; Singer and Ehrlich, 1979). A widespread drought in 1987-1988 caused simultaneous crashes of insect populations across the U.S., affecting diverse taxa from butterflies to sawflies to grasshoppers (Hawkins and Holyoak, 1998). Conversely, drought can be related to population booms in other insects (e.g., certain beetles, aphids, and moths) (Mattson and Haack, 1987). An extended drought in New Mexico in the 1950s caused mass mortality in semiarid ponderosa pine forests, causing an overall upslope shift in the boundary between pine forests and piñon/juniper woodland of as much as 2,000 meters (6,500 feet) (Allen and Breshears, 1998). The ecosystem

While there are multi-billion dollar estimates for annual agricultural losses (averaging about $4 billion a year over the last ten years), it is unclear whether these losses are directly related to crop production alone or other factors. Wildfire suppression costs to the United States Department of Agriculture (USDA) alone have surpassed $1 billion each of the last four years, though it is unclear how much of this is attributable to dry conditions. Little or no official loss estimates exist for the energy, recreation/tourism, timber, livestock, or environmental sectors, although the drought impacts within these sectors in recent years is known to be large. Better methods to quantify the cumulative direct and indirect impacts associated with drought need to be developed. The recurrence of a drought today of equal or similar magnitude to major droughts experienced in the past will likely result in far greater economic, social, and environmental losses and conflicts between water users.

BOX 1.4: High Temperature Extremes and Coral Bleaching

Corals are marine animals that obtain much of their nutrients from symbiotic[1] single-celled algae that live protected within the coral's calcium carbonate skeleton. Sea surface temperatures (SST), 1°C above long-term summer averages lead to the loss of symbiotic algae resulting in bleaching of tropical corals (Hoegh-Guldberg, 1999) (Figure Box 1.4). While global SST has risen an average of 0.13°C (0.23°F) per decade from 1979 to 2005 (IPCC, 2007a), a more acute problem for coral reefs is the increase in episodic warming events such as El Niño. High SSTs associated with the strong El Niño event in 1997-98 caused bleaching in every ocean basin (up to 95% of corals bleached in the Indian Ocean), ultimately resulting in 16% of corals dying globally (Hoegh-Guldberg, 1999, 2005; Wilkinson, 2000).

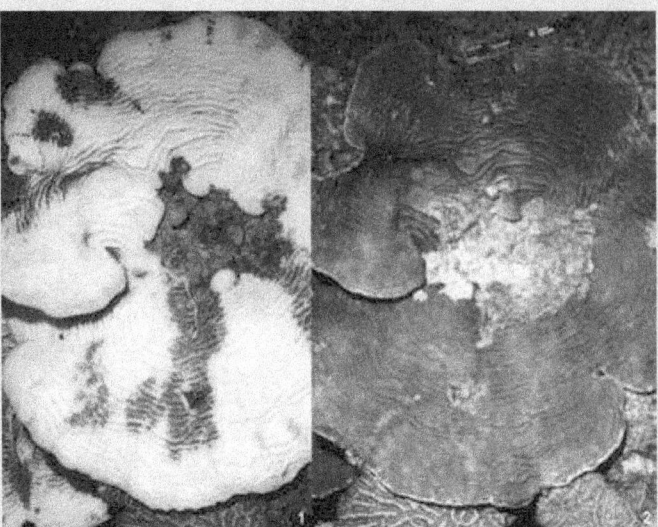

Recent evidence for genetic variation in temperature thresholds among the relevant symbiotic algae suggests that some evolutionary response to higher water temperatures may be possible (Baker, 2001; Rowan, 2004). Increased frequency of high temperature-tolerant symbiotic algae appear to have occurred within some coral populations between

Figure Box 1.4 An Agaricia coral colony shown: 1) bleached, and 2) almost fully recovered, from a bleaching event. Photos courtesy of Andy Bruckner, NOAA's National Marine Fisheries Service.

the mass bleaching events of 1997/1998 and 2000/2001 (Baker *et al.*, 2004). However, other studies indicate that many entire reefs are already at their thermal tolerance limits (Hoegh-Guldberg, 1999). Coupled with poor dispersal of symbiotic algae between reefs, this has led several researchers to conclude that local evolutionary responses are unlikely to mitigate the negative impacts of future temperature rises (Donner *et al.*, 2005; Hoegh-Guldberg *et al.*, 2002). Interestingly, though, hurricane-induced ocean cooling can temporarily alleviate thermal stress on coral reefs (Manzello *et al.*, 2007).

Examining coral bleaching in the Caribbean, Donner *et al.* (2007) concluded that "the observed warming trend in the region of the 2005 bleaching event is unlikely to be due to natural climate variability alone." Indeed, "simulation of background climate variability suggests that human-caused warming may have increased the probability of occurrence of significant thermal stress events for corals in this region by an order of magnitude. Under scenarios of future greenhouse gas emissions, mass coral bleaching in the eastern Caribbean may become a biannual event in 20–30 years." As coral reefs make significant contributions to attracting tourists to the Caribbean, coral bleaching has adverse socioeconomic impacts as well as ecological impacts.

[1] A symbiotic relationship between two living things is one that benefits both.

those well adapted to the more frequent forms of low-damage events. On the other hand, the less frequent high-damage events can overwhelm the ability of any system to recover quickly.

The adaptive capacity of socioeconomic systems is determined largely by characteristics such as poverty and resource availability, which often can be managed. Communities with little adaptive capacities are those with limited economic resources, low levels of technology, weak information systems, poor infrastructure, unstable or weak institutions, and uneven access to resources. Enhancement of social capacity, effectively addressing some of the exacerbating stresses, represents a practical means of coping with changes and uncertainties in climate. However, despite advances in knowledge and technologies, costs appear to be a major factor in limiting the adoption of adaptation measures (White *et al.*, 2001).

Communities can often achieve significant reductions in losses from natural disasters by adopting land-use plans that avoid the hazards, *e.g.*, by not allowing building in a floodplain. Building codes are also effective for reducing disaster losses, but they need to be enforced. For example, more than 25% of the damage from Hurricane Andrew could have been prevented if the existing building codes had been enforced (Board on Natural Disasters, 1999). One of the first major industry sectors to publicly show its concern about the threats posed by climate change was the insurance industry, in 1990 (Peara and Mills, 1999). Since then, the industry has recognized the steady increase in claims paralleling an increase in the number and severity of extreme weather and climate events—a trend that is expected to continue. The insurance industry, in fact, has an array of instruments/levers that can stimulate policyholders to take actions to adapt to future extremes. These possibilities are increasingly being recognized by governments. When such measures take effect, the same magnitude event can have less impact, as illustrated by the top panel of Figure 1.8.

Extreme events themselves can alter vulnerability and expose underlying stresses. There are various response times for recovery from the effects of any extreme weather or climate event—ranging from several decades in cases of significant loss of life, to years for the salinization of agricultural land following a tropical storm, to several months for stores to restock after a hurricane. A series of extreme events that occurs in a shorter period than the time needed for recovery can exacerbate the impacts, as illustrated in the bottom panel of Figure 1.8. For example, in 2004, a series of hurricanes made landfall in Florida; these occurred close enough in time and space that it often proved impossible to recover from one hurricane before the next arrived (Pielke *et al.*, 2008). Hardware stores and lumberyards were not able to restock quickly enough for residents to complete repairs to their homes which then led to further damage in the next storm. A

multitude or sequence of extreme events can also strain the abilities of insurance and re-insurance companies to compensate victims.

Extremes can also initiate adaptive responses. For example, droughts in the 1930s triggered waves of human migration that altered the population distribution of the United States. After the 1998 eastern Canadian ice storm, the design criteria for freezing rain on high-voltage power and transmission lines were changed to accommodate radial ice accretion of 25 mm (1 inch) in the Great Lakes region to 50 mm (2 inches) for Newfoundland and Labrador (Canadian Standards Association, 2001).

Factors such as societal exposure, adaptive capacity, and sensitivity to weather and climate can play a significant role in determining whether an event is considered extreme. In fact, an extreme weather or climate event, defined solely using statistical properties, may not be perceived to be an extreme if it affects something (*e.g.*, a building, city, *etc.*) that is designed to withstand that extreme. Conversely, a weather or climate event that is not extreme in a statistical sense might still be considered an extreme event because of the resultant impacts. Case in point, faced with an extended dry spell,

> Extreme events themselves can alter vulnerability and expose underlying stresses.

Extreme Events and Recovery of a System

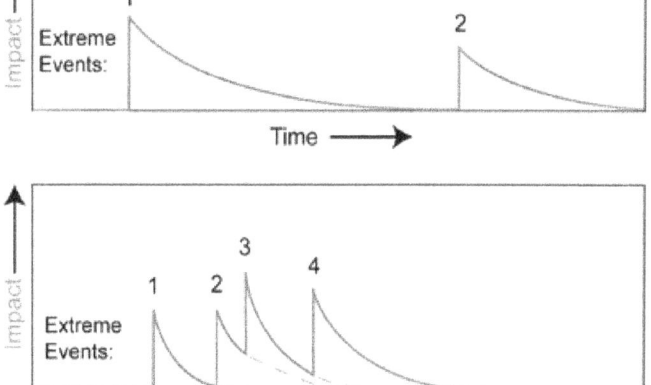

Figure 1.8 Extreme events such as hurricanes can have significant sudden impacts that take some time to recover from. Top: Two similar magnitude events take place but after the first one, new adaptation measures are undertaken, such as changes in building codes, so the second event doesn't have as great an impact. Bottom: An extreme that occurs before an area has completely recovered from the previous extreme can have a total impact in excess of what would have occurred in isolation.

consider the different effects and responses in a city with a well-developed water supply infrastructure and a village in an underdeveloped region with no access to reservoirs. These differences also highlight the role of adaptive capacity in a society's response to an extreme event. Wealthy societies will be able to devote the resources needed to construct a water supply system that can withstand an extended drought.

Given the relationship between extreme events and their resultant socioeconomic impacts, it would seem that the impacts alone would provide a good way to assess changes in extremes. Unfortunately, attempts to quantify trends in the impacts caused by extreme events are hindered by the difficulty in obtaining loss-damage records. As a result, there have been many calls for improvements in how socioeconomic data are collected (Changnon, 2003; Cutter and Emrich, 2005; National Research Council, 1999). However, there is no government-level coordinated mechanism for collecting data on all losses or damage caused by extreme events. A potentially

There is no government-level coordinated mechanism for collecting data on all losses or damages caused by extreme events.

valuable effort, led by the Hazards Research Lab at the University of South Carolina, is the assembly of the Spatial Hazard Events and Losses Database for the United States (SHELDUS) (Cutter *et al.*, 2008). If successful, this effort could provide standardized guidelines for loss estimation, data compilation, and metadata standards. Without these types of guidelines, a homogeneous national loss inventory will remain a vision and it will not be possible to precisely and accurately detect and assess trends in losses and quantify the value of mitigation (Figure 1.9).

To date, most efforts at quantifying trends in losses caused by impacts are based on insured loss data or on total loss (insured plus non-insured losses) estimates developed by insurers. Unfortunately, the details behind most of the insured loss data are proprietary and only aggregated loss data are available. The relationship between insured losses and total losses will likely vary as a function of extreme event and societal factors such as building codes, the extent of insurance penetration, and more complex

BOX 1.5: Heavy Precipitation and Human Health

Human-caused climate change is already affecting human health (WHO 2002, 2003, 2004; McMichael *et al.*, 2004). For the year 2000, the World Health Organization (WHO) estimated that 6% of malaria infections, 7% of dengue fever cases and 2.4% of diarrhea could be attributed to climate change (Campbell-Lendrum *et al.*, 2003). Increases in these water-borne diseases has been attributed to increases in intensity and frequency of flood events, which in turn has been linked to greenhouse-gas driven climate change (Easterling *et al.*, 2000a,b; IPCC, 2007a). Floods directly promote transmission of water-borne diseases by causing mingling of untreated or partially treated sewage with freshwater sources, as well as indirectly from the breakdown of normal infrastructure causing post-flood loss of sanitation and fresh water supplies (Atherholt *et al.*, 1998; Rose *et al.*, 2000; Curriero *et al.*, 2001; Patz *et al.*, 2003; O'Connor, 2002). Precipitation extremes also cause increases in malnutrition due to drought and flood-related crop failure. For all impacts combined, WHO estimated the total deaths

due to climate change at 150,000 people per year (WHO, 2002).

However, there is general agreement that the health sector in developed countries is strongly buffered against responses to climate change, and that a suite of more traditional factors is often responsible for both chronic and epidemic health problems. These include quality and accessibility of health care, sanitation infrastructure and practices, land-use change (particularly practices which alter timing and extent of standing water), pollution, population age structure, presence and effectiveness of vector control programs, and general socioeconomic status (Patz *et al.*, 2001; Gubler *et al.*, 2001; Campbell-Lendrum *et al.*, 2003; Wilkinson *et al.*, 2003; WHO, 2004, IPCC, 2007b). Indeed, it is generally assumed that diarrhea incidence in developed countries, which have much better sanitation infrastructure, has little or no association with climate (WHO, 2003, 2004). Yet, analyses of the U.S. indicate that the assumption that developed countries have low societal factors. The National Hurricane Center generally assumes that for the United States, total losses are twice insured loss estimates. However, this relationship will not hold for other countries or other weather phenomena.

Regardless of the uncertainties in estimating insured and total losses, it is clear that the absolute dollar value of losses from extreme events has increased dramatically over the past few decades, even after accounting for the effects of inflation (Figure 1.2). However, much of the increasing trend in losses, particularly from tropical cyclones, appears to be related to an increase in population and wealth (Pielke *et al.*, 2003; Pielke, 2005; Pielke and Landsea, 1998). The counter argument is that there is a climate change signal in recent damage trends. Damage trends have increased significantly despite ongoing adaptation efforts that have been taking place (Mills, 2005b; Stott *et al.*, 2004; Kunkel *et al.*, 1999). A number of other complicating factors also play a role in computing actual losses. For example, all other things being equal,

the losses from Hurricane Katrina would have been dramatically lower if the dikes had not failed. Looking toward the future, the potential for an increase in storm intensity (*e.g.*, tropical cyclone wind speeds and precipitation) (Chapter 3, this report) and changes in the intensity of the hydrological cycle[2] (Trenberth *et al.*, 2003) raises the possibility that changes in climate extremes will contribute to an increase in loss.

Another confounding factor in assessing extremes through their impacts is that an extreme event that lasts for a few days, or even less, can have impacts that persist for decades. For example, it will take years for Honduras and Guatemala to recover from the damage caused by Hurricane Mitch in 1998 and it seems likely that New Orleans will need years to recover from Hurricane Katrina. Furthermore, extreme events not only produce "losers" but "winners"

---

[2] The hydrologic cycle is the continuous movement of water on, above, and below the surface of the Earth where it evaporates from the surface, condenses in clouds, falls to Earth as rain or snow, flows downhill in streams and rivers, and then evaporates again.

In the U.S., 68% of water-borne disease outbreaks were preceded by downpours in the heaviest 20% of all precipitation events.

vulnerability may be premature, as independent studies have repeatedly concluded that water and food-borne pathogens (that cause diarrhea) will likely increase with projected increases in regional flooding events, primarily by contamination of main waterways (Rose *et al.*, 2000; Ebi *et al.*, 2006).

A U.S. study documented that 51% of water-borne disease outbreaks were preceded by precipitation events in the top 10% of occurrences, with 68% of outbreaks preceded by precipitation in the top 20% (Curriero *et al.*, 2001). These outbreaks comprised mainly intestinal disorders due to contaminated well water or water treatment facilities that allowed microbial pathogens, such as *E. coli*, to enter drinking water. In 1993, 54 people in Milwaukee, Wisconsin died in the largest reported flood-related disease outbreak (Curriero *et al.*, 2001). The costs associated with this one outbreak were $31.7 million in medical costs and $64.6 million in productivity losses (Corso *et al.*, 2003).

Another heavy precipitation-human health link comes from the southwestern desert of the United States. This area experienced extreme rainfalls during the intense 1992/1993 El Niño. Excess precipitation promoted lush vegetative growth, which led to population booms of deer mice (*Peromyscus maniculatus*). This wild rodent carries the hantavirus which is transmissible to humans and causes a hemorrhagic fever that is frequently lethal. The virus is normally present at moderate levels in wild mouse populations. In most years, humans in nearby settlements experienced little exposure. However, in 1993, local over-abundance of mice arising from the wet-year/population boom caused greater spillover of rodent activity. Subsequent increased contact between mice and humans and resultant higher transmission rates led to a major regional epidemic of the virus (Engelthaler *et al.*, 1999; Glass *et al.*, 2000). Similar dynamics have been shown for plague in the western United States (Parmenter *et al.*, 1999).

Weather and Climate Natural Disasters Cost

Figure 1.9 Different methodologies for collecting loss data can produce very different results. The NCDC Billion Dollar Weather Disasters loss data (Lott and Ross, 2006) assesses a subset of the largest events covered in the SHELDUS (Cutter and Emrich, 2005) loss data. SHELDUS is often less than the Billion Dollar Weather Disasters because (a) the SHELDUS event-based dataset does not fully capture drought costs and (b) SHELDUS assesses direct costs only while the Billion Dollar Weather Disasters estimates include both direct costs and indirect costs. Neither cost data set factors in the loss of life. Indeed, some extremes such as heat waves that can cause high loss of life may not show up at all in cost assessments because they cause very little property damage. Primary events contributing to peak values in the time series have been listed.

Different methodologies for collecting loss data can produce very different results.

too. Examples of two extreme-event winners are the construction industry in response to rebuilding efforts and the tourism industry at locations that receive an unexpected influx of tourists who changed plans because their first-choice destination experienced an extreme event that crippled the local tourism facilities. Even in a natural ecosystem there are winners and losers. For example, the mountain pine beetle infestation that has decimated trees in British Columbia provided an increased food source for woodpeckers.

1.5 SYSTEMS ARE ADAPTED TO THE HISTORICAL RANGE OF EXTREMES SO CHANGES IN EXTREMES POSE CHALLENGES

Over time, socioeconomic and natural systems adapt to their climate, including extremes. Snowstorms that bring traffic to a standstill in Atlanta are shrugged off in Minneapolis (WIST, 2002). Hurricane-force winds that topple tall, non-indigenous Florida trees like the Australian pine (*Casuarina equisetifolia*) may only break

a few small branches from the native live oak (*Quercus virginiana*) or gumbo-limbo (*Bursera simaruba*) trees that evolved in areas frequented by strong winds. Some species even depend on major extremes. For example, the jack pine (*Pinus banksiana*) produces very durable resin-filled cones that remain dormant until wildfire flames melt the resin. Then, the cones pop open and spread their seeds (Herring, 1999).

Therefore, it is less a question of whether extremes are good or bad, but rather, what will be the impact of their changing characteristics? For certain species and biological systems, various processes may undergo sudden shifts at specific thresholds of temperature or precipitation (Precht *et al.*, 1973; Weiser, 1973; Hoffman and Parsons, 1997), as discussed in Section 1.3. Generally, managed systems are more buffered against extreme events than natural systems, but certainly are not immune to them. The heat waves of 1995 in Chicago and 2003 in Europe caused considerable loss of life in large part because building architecture and city design were adapted for more temperate climates and not adapted for dealing with such extreme and enduring heat (Patz *et al.*, 2005). As an illustration, mortality from a future heat wave analogous to the European heat wave of 2003 is estimated to be only 2% above that of the previous hottest historical summer for Washington, D.C., while New York, with its less heat-tolerant architecture, is estimated to have mortality 155% above its previous record hot summer (Kalkstein *et al.*, 2008). On balance, because systems have adapted to their historical range of extremes, the majority of the impacts of events outside this range are negative (IPCC, 2007b).

When considering how the statistics of extreme events have changed, and may change in the future, it is important to recognize how such changes may affect efforts to adapt to them. Adaptation is important because it can reduce the extent of damage caused by extremes (*e.g.*, Mileti, 1999; Wilhite, 2005). Currently, long-term

planning uses, where possible, the longest historical climate records, including consideration of extreme events. The combined probabilities of various parameters that can occur at any given location can be considered the cumulative hazard of a place. Past observations lead to expectations of their recurrence, and these form the basis of building codes, infrastructure design and operation, land-use zoning and planning, insurance rates, and emergency response plans.

However, what would happen if statistical attributes of extreme events were to change as the climate changes? Individuals, groups, and societies would seek to adjust to changing exposure. Yet the climate may be changing in ways that pose difficulties to the historical decision-making approaches (Burton *et al.*, 1993). The solution is not just a matter of utilizing projections of future climate (usually from computer simulations). It also involves translating the projected changes in climate extremes into changes in risk.

Smit *et al.* (2000) outline an "anatomy" of adaptation to climate change and variability, consisting of four elements: a) adapt to what, b) who or what adapts, c) how does adaptation occur, and d) how good is the adaptation. Changes in the statistics of climate extremes will influence the adaptation. As noted earlier, a change in the frequency of extreme events may be relatively large, even though the change in the average is small. Increased frequencies of extreme events could lead to reduced time available for recovery, altering the feasibility and effectiveness of adaptation measures. Changes to the timing and duration of extremes, as well as the occurrence of new extreme thresholds (*e.g.*, greater precipitation intensity, stronger wind speeds), would be a challenge to both managed and unmanaged systems.

Trends in losses or productivity of climate-sensitive goods exhibit the influences of both climate variability/change and

ongoing behavioral adjustments. For example, U.S. crop yields have generally increased with the introduction of new technologies. As illustrated by Figure 1.10, climatic variability still causes short-term fluctuations in crop production, but a poor year in the 1990s tends to have better yields than a poor year (and sometimes even a good year) in the 1960s. Across the world, property losses show a substantial increase in the last 50 years, but this trend is being influenced by both increasing property development and offsetting adaptive behavior. For example, economic growth has spurred additional construction in vulnerable areas but the new construction is often better able to withstand extremes than older construction. Future changes in extreme events will be accompanied by both autonomous and planned adaptation, which will further complicate calculating losses due to extremes.

1.6 ACTIONS CAN INCREASE OR DECREASE THE IMPACT OF EXTREMES

It is important to note that most people do not use climate and weather data and forecasts directly. People who make decisions based

It is less a question of whether extremes are good or bad, but rather, what will be the impact of their changing characteristics?

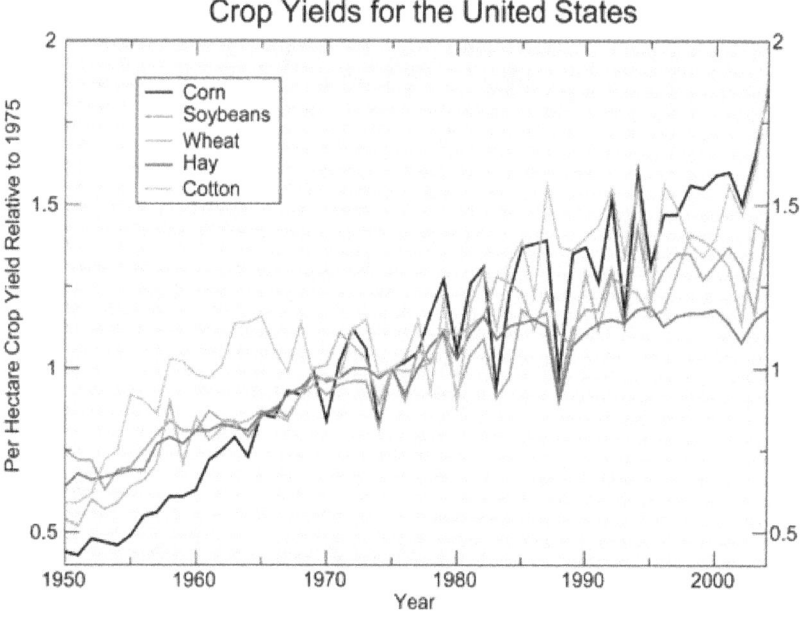

Figure 1.10 Climate variability may produce years with reduced crop yield, but because of technological improvements, a poor yield in the 1990s can still be higher than a good yield in the 1950s indicating a changing relationship between climate and agricultural yield. Data are in units of cubic meters or metric tons per unit area with the yield in 1975 assigned a value of 1. Data from USDA National Agricultural Statistics Service via update to Heinz Center (2002).

Climate Information and Decision-Making

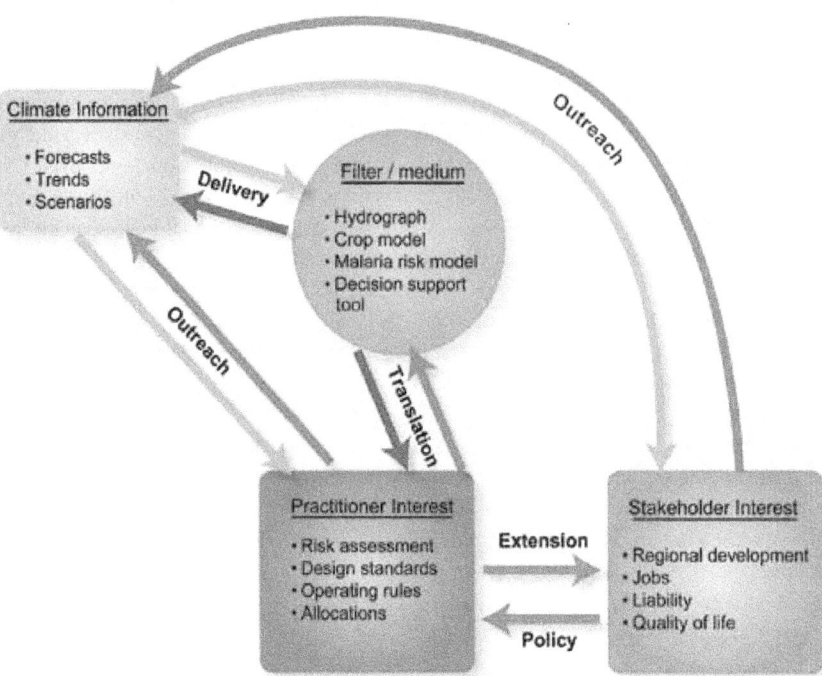

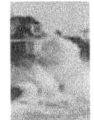

Figure 1.11 Illustration of how climate information is processed, filtered, and combined with other information in the decision process relevant to stakeholder interests (adapted from Cohen and Waddell, 2008).

on meteorological information typically base their decisions on the output of an intermediate model that translates the data into a form that is more relevant for their decision process (Figure 1.11). For example, a farmer will not use weather forecasts or climate data directly when making a decision on when to fertilize a crop or on how much pesticide to apply. Instead, the forecast is filtered through a model or mental construct that uses such information as one part of the decision process and includes other inputs such as crop type, previous pesticide application history, government regulations, market conditions, producer recommendations, and the prevalence and type of pest.

One useful decision tool is a plant hardiness zone map (Cathey, 1990). Plant hardiness zones are primarily dependent on extreme cold temperatures. Due to changing locations of plant hardiness zones, people are already planting fruit trees, such as cherries, farther north than they did 30 years ago as the probability of winterkill has diminished. This type of adaptation is common among farmers who

continually strive to plant crop species and varieties well suited to their current local climate.

To a large extent, individual losses for hazard victims have been reduced as the larger society absorbs a portion of their losses through disaster relief and insurance. Clearly relevant for settings such as New Orleans is the so-called levee effect, first discussed by Burton (1962), in which construction of levees (as well as dams, revetments, and artificially-nourished beaches) induces additional development, leading to much larger losses when the levee is eventually overtopped. A more general statement of this proposition is found in the safe development paradox in which increased perceived safety (*e.g.*, due to flood control measures), induces increased development (such as in areas considered safe due to the protection provided by levees or dams), leading to increased losses when a major event hits. The notion that cumulative reduction of smaller scale risks might increase vulnerability to large events has been referred to as the levee effect, even when the concern has nothing to do with levees (Bowden *et al.*, 1981).

After particularly severe or visible catastrophes, policy windows have been identified as windows of opportunity for creating long-term risk reduction plans that can include adaptation for climate change. A policy window opens when the opportunity arises to change policy direction and is thus an important part of agenda setting (Kingdon, 1995). Policy windows can be created by triggering or focusing events, such as disasters, as well as by changes in government and shifts in public opinion. Immediately following a disaster, the social climate may be conducive to much needed legal, economic, and social change, which can begin to reduce structural vulnerabilities. Indeed, an extreme event that is far outside normal experience can alert society to the realization that extremes are changing and that society must adapt to these changes.

The assumptions behind the utility of policy windows are that (1) new awareness of risks after a disaster leads to broad consensus, (2) agencies are reminded of disaster risks, and (3) enhanced community will and resources become available. However, during the post-emergency phase, reconstruction requires weighing, prioritizing, and sequencing of policy programming, and there are usually many diverse public and private agendas for decision makers and operational actors to incorporate, with attendant requests for resources for various actions. Thus, there is pressure to quickly return to the "normal" conditions that existed prior to the event, rather than incorporate longer-term development strategies (Berube and Katz, 2005; Christoplos, 2006). In addition, while institutional capacity for adaptation clearly matters, it is often not there in the aftermath (or even before the occurrence) of a disaster.

In contrast to the actual reconstruction plans, the *de facto* decisions and rebuilding undertaken ten months after Katrina clearly demonstrate the rush to rebuild the familiar, as found after other major disasters in other parts of the world (Kates *et al.*, 2006). This perspective helps explain the evolution of vulnerability of settings such as New Orleans, where smaller events have been mitigated, but with attendant increases in long-term vulnerability. As in diverse contexts such as El Niño-Southern Oscillation (ENSO)-related impacts in Latin America, induced development below dams or levees in the United States, and

flooding in the United Kingdom, the result is that focusing only on short-term risk reduction can actually produce greater vulnerability to future events (Pulwarty *et al.*, 2003). Thus, the evolution of responses in the short-term after each extreme event can appear logical, but might actually increase long-term risk to larger or more frequent events. Adaptation to climate change must be placed within the context of adaptation to climate across time scales (from extremes and year-to-year variability through long-term change) if it is to be embedded into effective response strategies.

Global losses from weather-related disasters amounted to a total of around $83 billion for the 1970s, increasing to a total of around $440 billion for the 1990s with the number of great natural catastrophe events increasing from 29 to 74 between those decades (MunichRe, 2004; Stern, 2006).

1.7 ASSESSING IMPACTS OF CHANGES IN EXTREMES IS DIFFICULT

As has been mentioned, assessing consequences relevant to extreme weather and climate events is not simply a function of the weather and climate phenomena but depends critically on the vulnerability of the system being impacted. Thus, the context in which these extreme events take place is crucial. This means that while the changes in extreme events are consistent with a warming climate (IPCC, 2007a), any analysis of past events or projection of future events has to carefully weigh non-climatic factors. In particular, consideration must be given to changes

There is pressure to quickly return to the "normal" conditions that existed prior to the event, rather than incorporate longer-term development strategies.

31

BOX 1.6: Tools for Assessing Impacts of Climate Extremes

There are a variety of impact tools that help translate climate information into an assessment of what the impacts will be and provide guidance on how to plan accordingly. These tools would be part of the filter/medium circle in Figure 1.11. However, as illustrated here, using the example of a catastrophe risk model, the tool has clear linkages to all the other boxes in Figure 1.11.

A catastrophe risk model can be divided into four main components, as shown in Figure Box 1.6. The hazard component provides information on the characteristics of a hazard. For probabilistic calculations, this component would include a catalog with a large number of simulated events with realistic characteristics and frequencies. Event information for each hazard would include the frequency, size, location, and other characteristics. The overall statistics should agree with an analysis of historical events.

The inventory component provides an inventory of structures that are exposed to a hazard and information on their construction. The vulnerability component simulates how structures respond to a hazard. This component requires detailed information on the statistical response of a structure to the forces produced by a hazard. This component would also account for secondary damage such as interior water damage after a structure's windows are breached. The fourth component in the risk model estimates losses produced by a hazard event and accounts for repair or replacement costs. In cases of insurance coverage, the loss component also accounts for business interruption costs and demand surge. If the model is used for emergency management purposes, the loss component also accounts for factors such as emergency supplies and shelters.

It should be noted, though, that how the loss component is treated impacts the vulnerability and inventory components, as indicated by the curved upward pointing arrows. Is a house destroyed in a flood rebuilt in the same location or on higher ground? Is a wind-damaged building repaired using materials that meet higher standards? These actions have profound effects on future catastrophe risk models for the area.

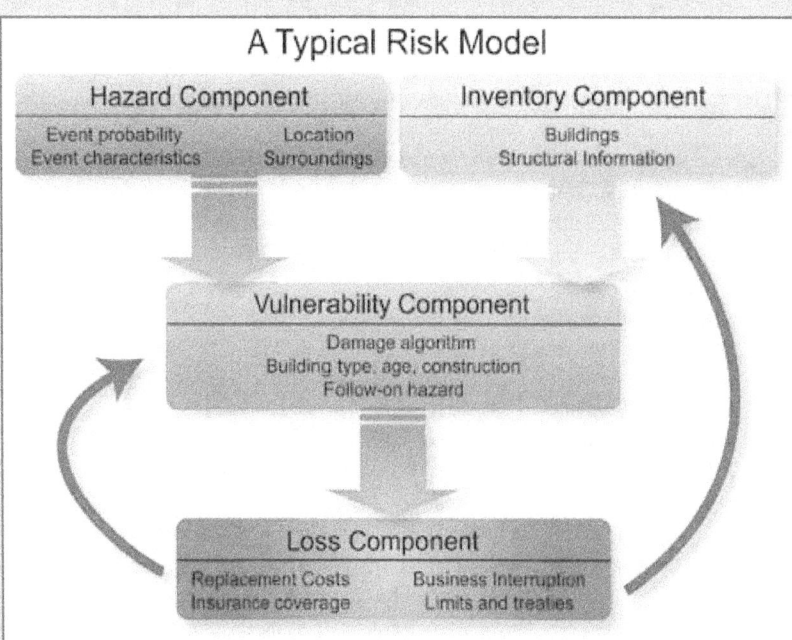

Figure Box 1.6 Schematic diagram of a typical risk model used by the insurance industry. The diagram highlights the three major components (hazard, damage, and loss) of a risk model. What happens to the loss component feeds back to the vulnerability and inventory components.

in demographic distributions and wealth, as well as the use of discount rates in assessments of future damage costs. The analysis presented by Stern (2006), regarding projected increased damage costs, has led to considerable debate on the methods for incorporating future climate and socioeconomic scenarios, including the role of adaptation, into such assessments (Pielke, 2007; Stern and Taylor, 2007; Tol and Yohe, 2006). Regarding recent trends in weather-related economic losses shown in Figure 1.2, it is likely that part of the increase in economic losses shown in Figure 1.2 has been due to increases in population in regions that are vulnerable, such as coastal communities affected by hurricanes, sea-level rise, and storm surges. In addition, property values have risen. These factors increase the sensitivity of our infrastructure to extreme events. Together with the expected increase in the frequency and severity of extreme events (IPCC, 2007a; Chapter 3 this report), our vulnerability to extreme events is very likely to increase. Unfortunately, because many extreme events occur at small temporal and spatial scales, where climate simulation skill is currently limited and local conditions are highly variable, projections of future impacts cannot always be made with a high level of confidence.

While anthropogenic climate change very likely will affect the distribution of extreme events (and is already observed to be having such effects – Chapter 2), it can be misleading to attribute any particular event solely to human causes. Nevertheless, scientifically valid statements regarding the increased risk can sometimes be made. A case in point is the 2003 heat wave in Europe, where it is very likely that human influence at least doubled the risk of such a heat wave occurring (Stott *et al.*, 2004). Furthermore, over time, there is expected to be some autonomous adaptation to experienced climate variability and other stresses. Farmers, for example, have traditionally altered their agricultural practices, such as planting different crop varieties based on experience, and water engineers have built dams and reservoirs to better manage resources during recurring floods or droughts. Such adaptation needs to be considered when assessing the importance of future extreme events.

Assessing historical extreme weather and climate events is more complicated than just the

statistical analysis of available data. Intense rain storms are often of short duration and not always captured in standard meteorological records; however, they can often do considerable damage to urban communities, especially if the infrastructure has not been enhanced as the communities have grown. Similarly, intense wind events (hurricanes are a particular example), may occur in sparsely populated areas or over the oceans, and it is only since the 1960s, with the advent of satellite observations, that a comprehensive picture can be confidently assembled. Therefore, it is important to continually update the data sets and improve the analyses. For example, probabilistic estimates of rainfall intensities for a range of event durations (from 5 minutes to 24 hours) with recurrence intervals of 20, 50, and 100 years (*e.g.*, a 10-minute rainfall that should statistically occur only once in 100 years), have long been employed by engineers when designing many types of infrastructure. In the United States, these probabilistic estimates of intense precipitation are in the process of being updated. Newer analyses based on up-to-date rainfall records often differ by more than 45% from analyses done in the 1970s (Bonnin *et al.*, 2003).

1.8 SUMMARY AND CONCLUSIONS

Weather and climate extremes have always been present. Both socioeconomic and natural systems are adapted to historical extremes. Changes from this historical range matter because people, plants, and animals tend to be more impacted by changes in extremes compared to changes in average climate. Extremes are changing, and in some cases, impacts on socioeconomic and natu-

Vulnerability is a function not only of the rate and magnitude of climate change but also of the sensitivity of the system, the extent to which it is exposed, and its adaptive capacity.

ral systems have been observed. The vulnerability of these systems is a function not only of the rate and magnitude of climate change but also of the sensitivity of the system, the extent to which it is exposed, and its adaptive capacity. Vulnerability can be exacerbated by other stresses such as social inequalities, disease, and conflict, and can be compounded by changes in other extremes events (*e.g.*, drought and heat occurring together) and by rapidly-recurring events.

Despite the widespread evidence that humans have been impacted by extreme events in the past, projecting future risk to changing climate extremes is difficult. Extreme phenomena are often more difficult to project than changes in mean climate. In addition, systems are adapting and changing their vulnerability to risk in different ways. The ability to adapt differs among systems and changes through time. Decisions to adapt to or mitigate the effect of changing extremes will be based not only on our understanding of climate processes but also on our understanding of the vulnerability of socioeconomic and natural systems.

Vulnerability to extreme events can be exacerbated by other stresses such as social inequalities, disease, and conflict.

CHAPTER 2

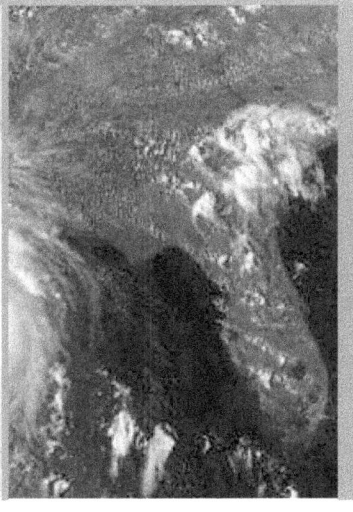

Observed Changes in Weather and Climate Extremes

Convening Lead Author: Kenneth Kunkel, Univ. Ill. Urbana-Champaign, Ill. State Water Survey

Lead Authors: Peter Bromirski, Scripps Inst. Oceanography, UCSD; Harold Brooks, NOAA; Tereza Cavazos, Centro de Investigación Científica y de Educación Superior de Ensenada, Mexico; Arthur Douglas, Creighton Univ.; David Easterling, NOAA; Kerry Emanuel, Mass. Inst. Tech.; Pavel Groisman, UCAR/NCDC; Greg Holland, NCAR; Thomas Knutson, NOAA; James Kossin, Univ. Wis., Madison, CIMSS; Paul Komar, Oreg. State Univ.; David Levinson, NOAA; Richard Smith, Univ. N.C., Chapel Hill

Contributing Authors: Jonathan Allan, Oreg. Dept. Geology and Mineral Industries; Raymond Assel, NOAA; Stanley Changnon, Univ. Ill. Urbana-Champaign, Ill. State Water Survey; Jay Lawrimore, NOAA; Kam-biu Liu, La. State Univ., Baton Rouge; Thomas Peterson, NOAA

KEY FINDINGS

Observed Changes

Long-term upward trends in the frequency of unusually warm nights, extreme precipitation episodes, and the length of the frost-free season, along with pronounced recent increases in the frequency of North Atlantic tropical cyclones (hurricanes), the length of the frost-free season, and extreme wave heights along the West Coast are notable changes in the North American climate record.

- Most of North America is experiencing more unusually hot days. The number of warm spells has been increasing since 1950. However, the heat waves of the 1930s remain the most severe in the United States historical record back to 1895.
- There are fewer unusually cold days during the last few decades. The last 10 years have seen a lower number of severe cold waves than for any other 10-year period in the historical record which dates back to 1895. There has been a decrease in the number of frost days and a lengthening of the frost-free season, particularly in the western part of North America.
- Extreme precipitation episodes (heavy downpours) have become more frequent and more intense in recent decades than at any other time in the historical record, and account for a larger percentage of total precipitation. The most significant changes have occurred in most of the United States, northern Mexico, southeastern, northern and western Canada, and southern Alaska.
- There are recent regional tendencies toward more severe droughts in the southwestern United States, parts of Canada and Alaska, and Mexico.
- For much of the continental U.S. and southern Canada, the most severe droughts in the instrumental record occurred in the 1930s. While it is more meaningful to consider drought at the regional scale, there is no indication of an overall trend at the continental scale since 1895. In Mexico, the 1950s and 1994-present were the driest periods.
- Atlantic tropical cyclone (hurricane) activity, as measured by both frequency and the Power Dissipation Index (which combines storm

intensity, duration, and frequency) has increased. The increases are substantial since about 1970, and are likely substantial since the 1950s and 60s, in association with warming Atlantic sea surface temperatures. There is less confidence in data prior to about 1950.

- There have been fluctuations in the number of tropical storms and hurricanes from decade to decade, and data uncertainty is larger in the early part of the record compared to the satellite era beginning in 1965. Even taking these factors into account, it is likely that the annual numbers of tropical storms, hurricanes, and major hurricanes in the North Atlantic have increased over the past 100 years, a time in which Atlantic sea surface temperatures also increased.

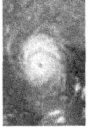

- The evidence is less compelling for significant trends beginning in the late 1800s. The existing data for hurricane counts and one adjusted record of tropical storm counts both indicate no significant linear trends beginning from the mid- to late 1800s through 2005. In general, there is increasing uncertainty in the data as one proceeds back in time.

- There is no evidence for a long-term increase in North American mainland land-falling hurricanes.

- The hurricane Power Dissipation Index shows some increasing tendency in the western north Pacific since 1980. It has decreased since 1980 in the eastern Pacific, affecting the Mexican west coast and shipping lanes, but rainfall from near-coastal hurricanes has increased since 1949.

- The balance of evidence suggests that there has been a northward shift in the tracks of strong low pressure systems (storms) in both the North Atlantic and North Pacific basins. There is a trend toward stronger intense low pressure systems in the North Pacific.

- Increases in extreme wave height characteristics have been observed along the Pacific coast of North America during recent decades based on three decades of buoy data. These increases have been greatest in the Pacific Northwest, and are likely a reflection of changes in storm tracks.

- There is evidence for an increase in extreme wave height characteristics in the Atlantic since the 1970s, associated with more frequent and more intense hurricanes.

- Over the 20th century, there was considerable decade-to-decade variability in the frequency of snowstorms of six inches or more. Regional analyses suggest that there has been a decrease in snowstorms in the South and lower Midwest of the United States, and an increase in snowstorms in the upper Midwest and Northeast. This represents a northward shift in snowstorm occurrence, and this shift, combined with higher temperatures, is consistent with a decrease in snow cover extent over the United States. In northern Canada, there has also been an observed increase in heavy snow events (top 10% of storms) over the same time period. Changes in heavy snow events in southern Canada are dominated by decade-to-decade variability.

- There is no indication of continental scale trends in episodes of freezing rain during the 20th century.

- The data used to examine changes in the frequency and severity of tornadoes and severe thunderstorms are inadequate to make definitive statements about actual changes.

- The pattern of changes of ice storms varies by region.

2.1 BACKGROUND

Weather and climate extremes exhibit substantial spatial variability. It is not unusual for severe drought and flooding to occur simultaneously in different parts of North America (*e.g.*, catastrophic flooding in the Mississippi River basin and severe drought in the southeast United States during summer 1993). These reflect temporary shifts in large-scale circulation patterns that are an integral part of the climate system. The central goal of this chapter is to identify long-term shifts/trends in extremes and to characterize the continental-scale patterns of such shifts. Such characterization requires data sets that are homogeneous, of adequate length, and with continental-scale coverage. Many data sets meet these requirements for limited periods only. For temperature and precipitation, rather high quality data are available for the conterminous United States back to the late 19th century. However, shorter data records are available for parts of Canada, Alaska, Hawaii, Mexico, the Caribbean, and U.S. territories. In practice, this limits true continental-scale analyses of temperature and precipitation extremes to the middle part of the 20th century onward. Other phenomena have similar limitations, and continental-scale characterizations are generally limited to the last 50 to 60 years or less, or must confront data homogeneity issues which add uncertainty to the analysis. We consider all studies that are available, but in many cases these studies have to be interpreted carefully because of these limitations. A variety of statistical techniques are used in the studies cited here. General information about statistical methods along with several illustrative examples are given in Appendix A.

2.2 OBSERVED CHANGES AND VARIATIONS IN WEATHER AND CLIMATE EXTREMES

2.2.1 Temperature Extremes

Extreme temperatures do not always correlate with average temperature, but they often change in tandem; thus, average temperature changes provide a context for discussion of extremes. In 2005, virtually all of North America was above to much-above average[1] (Shein, 2006) and

2006 was the second hottest year on record in the conterminous United States (Arguez, 2007). The areas experiencing the largest temperature anomalies included the higher latitudes of Canada and Alaska. Annual average temperature time series for Canada, Mexico and the United States all show substantial warming since the middle of the 20th century (Shein, 2006). Since the record hot year of 1998, six of the past ten years (1998-2007) have had annual average temperatures that fall in the hottest 10% of all years on record for the U.S.

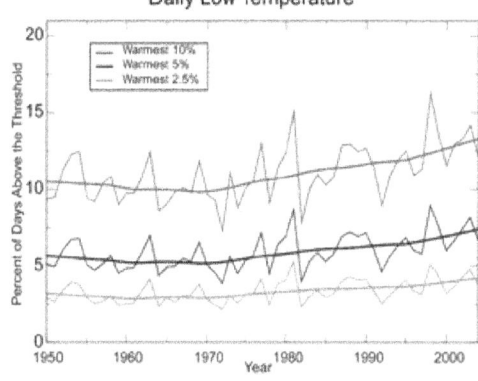

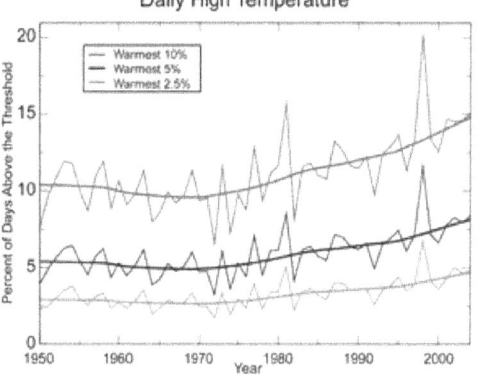

Figure 2.1 Changes in the percentage of days in a year above three thresholds for North America for daily high temperature (top) and daily low temperature (bottom) (Peterson *et al.*, 2008).

Urban warming is a potential concern when considering temperature trends. However, recent work (*e.g.*, Peterson *et al.*, 2003; Easterling *et al.*, 1997) show that urban warming is only a small part of the observed warming since the late 1800s.

Since 1950, the annual percent of days exceeding the 90th, 95th, and 97.5 percentile thresholds[2] for both maximum (hottest daytime highs) and minimum (warmest nighttime lows) temperature have increased when averaged over all of North America (Figure 2.1; Peterson *et al.*, 2008). The changes are greatest in the 90th percentile, increasing from about 10% of the

---

[1] NOAA's National Climatic Data Center uses the following terminology for classifying its monthly/ seasonal/annual U.S. temperature and precipitation rankings: "near-normal" is defined as within the middle third, "above/below normal" is within the top third/bottom third, and "much-above/much-below normal" is within the top-tenth/bottom tenth of all such periods on record.

[2] An advantage of the use of percentile, rather than absolute thresholds, is that they account for regional climate differences.

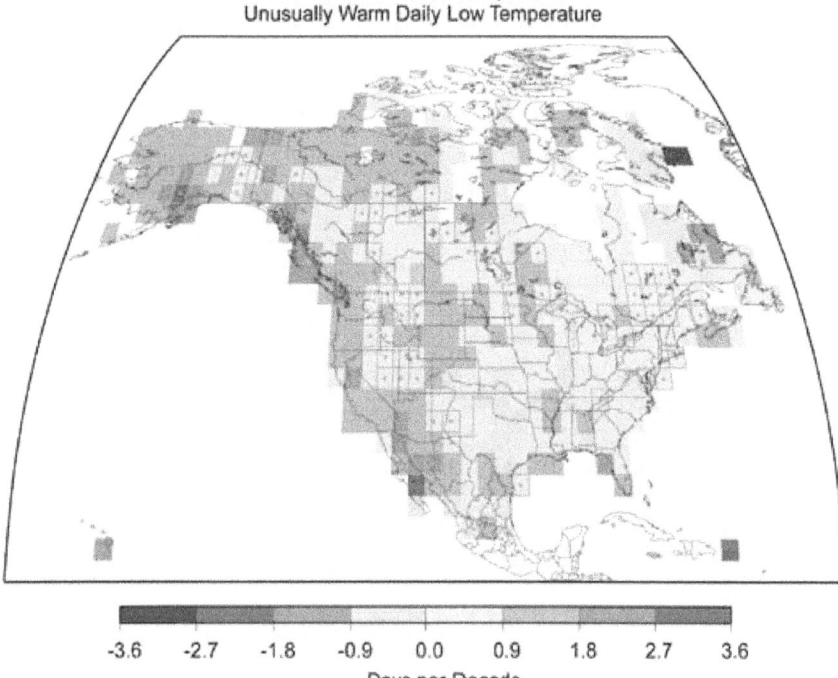

Trends in Number of Days with
Unusually Warm Daily Low Temperature

-3.6   -2.7   -1.8   -0.9   0.0   0.9   1.8   2.7   3.6
Days per Decade

Figure 2.2 Trends in the number of days in a year when the daily low is unusually warm (*i.e.*, in the top 10% of warm nights for the 1950-2004 period). Grid boxes with green squares are statistically significant at the p=0.05 level (Peterson *et al.*, 2008). A trend of 1.8 days/decade translates to a trend of 9.9 days over the entire 55 year (1950-2004) period, meaning that ten days more a year will have unusually warm nights.

There have been more rare heat events and fewer rare cold events in recent decades.

days to about 13% for maximum and almost 15% for minimum. These changes decrease as the threshold temperatures increase, indicating more rare events. The 97.5 percentage increases from about 3% of the days to 4% for maximum and 5% for minimum. The relative changes are similar. There are important regional differences in the changes. For example, the largest increases in the 90th percentile threshold temperature occur in the western part of the continent from northern Mexico through the western United States and Canada and across Alaska, while some areas, such as eastern Canada, show declines of as many as ten days per year from 1950 to 2004 (Figure 2.2).

Other regional studies have shown similar patterns of change. For the United States, the number of days exceeding the 90th, 95th and 99th percentile thresholds (defined monthly) have increased in recent years[3], but are also dominated earlier in the 20th century by the extreme

heat and drought of the 1930s[4] (DeGaetano and Allen, 2002). Changes in cold extremes (days falling below the 10th, 5th, and 1st percentile threshold temperatures) show decreases, particularly since 1960[5]. For the 1900-1998 period in Canada, there are fewer cold extremes in winter, spring and summer in most of southern Canada and more high temperature extremes in winter and spring, but little change in high temperature extremes in summer[6] (Bonsal *et al.*, 2001). However, for the more recent (1950-1998) period there are significant increases in high temperature extremes over western Canada, but decreases in eastern Canada. Similar results averaged across all of Canada are found for the longer 1900-2003 period, with 28 fewer cold nights, 10 fewer cold days, 21 more extremely warm nights, and 8 more hot days per year now than in 1900[7] (Vincent and Mekis, 2006). For the United States and Canada, the largest increases in daily maximum and minimum temperature are occurring in the colder days of each month (Robeson, 2004). For the Caribbean region, there is an 8% increase in the number of very warm nights and 6% increase in the number of hot days for the 1958-1999 period. There also has been a corresponding decrease of 7% in the number of cold days and 4% in the number of cold nights (Peterson *et al.*, 2002). The number of very warm nights has increased by 10 or more per year for Hawaii and 15 or more per year for Puerto Rico from 1950 to 2004 (Figure 2.2).

---

[3] Stations with statistically significant upward trends for 1960-1996 passed tests for field significance based on resampling.

[4] The number of stations with statistically significant negative trends for 1930-1996 was greater than the number with positive trends.

[5] Stations with statistically significant downward trends for 1960-1996 passed tests for field significance based on resampling, but not for 1930-1996.

[6] Statistical significance of trends was assessed using Kendall's tau test.

[7] These trends were statistically significant at more than 20% of the stations based on Kendall's tau test.

Analysis of multi-day very extreme heat and cold episodes[8] in the United States were updated[9] from Kunkel *et al.* (1999a) for the period 1895-2005. The most notable feature of the pattern of the annual number of extreme heat waves (Figure 2.3a) through time is the high frequency in the 1930s compared to the rest of the years in the 1895-2005 period. This was followed by a decrease to a minimum in the 1960s and 1970s and then an increasing trend since then. There is no trend over the entire period, but a highly statistically significant upward trend since 1960. The heat waves during the 1930s were characterized by extremely high daytime temperatures while nighttime temperatures were not as unusual (Figure 2.3b,c). An extended multi-year period of intense drought undoubtedly played a large role in the extreme heat of this period, particularly the daytime temperatures, by depleting soil moisture and reducing the moderating effects of evaporation. By contrast, the recent period of increasing heat wave index is distinguished by the dominant contribution of a rise in extremely high nighttime temperatures (Figure 2.3c). Cold waves show a decline in the first half of the 20th century, then a large spike of events during the mid-1980s, then a decline[10]. The last 10 years have seen a lower number of severe cold waves in the United States than in any other 10-year period since record-keeping began in 1895, consistent with observed impacts such as increasing insect populations (Chapter 1, Box 1.2). Decreases in the frequency of extremely low nighttime temperatures have

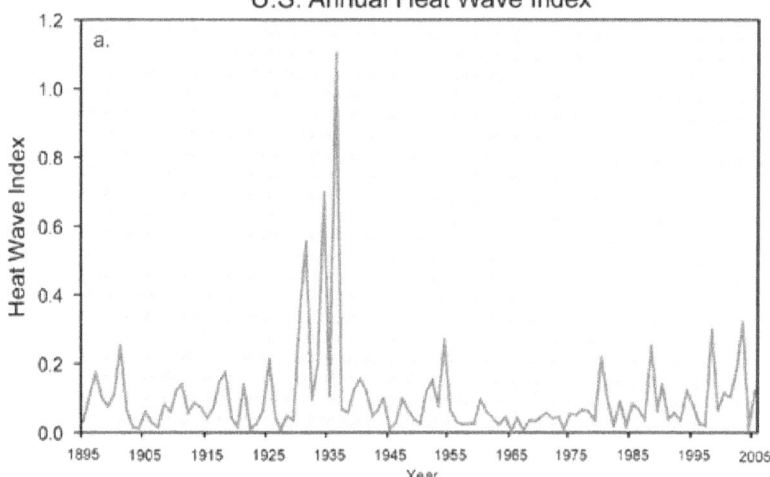

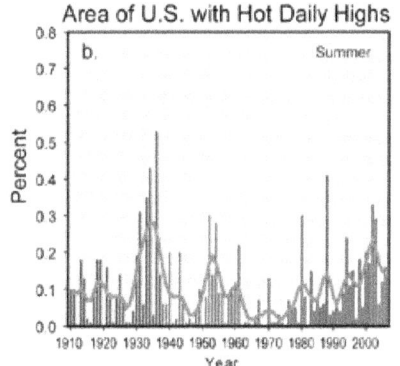 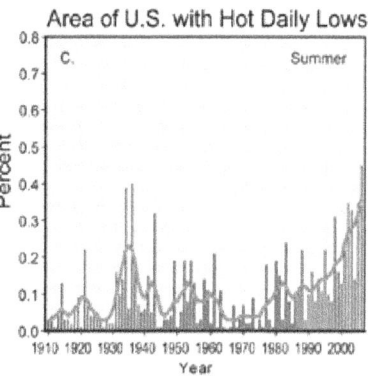 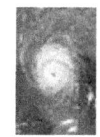

Figure 2.3 Time series of (a) annual values of a U.S. national average "heat wave" index. Heat waves are defined as warm spells of 4 days in duration with mean temperature exceeding the threshold for a 1 in 10 year event. (updated from Kunkel *et al.*, 1999); (b)Area of the United States (in percent) with much above normal daily high temperatures in summer; (c) Area of the United States (in percent) with much above normal daily low temperatures in summer. Blue vertical bars give values for individual seasons while red lines are smoothed (9-year running) averages. The data used in (b) and (c) were adjusted to remove urban warming bias.

made a somewhat greater contribution than extremely low daytime temperatures to this recent low period of cold waves. Over the entire period there is a downward trend, but it is not statistically significant at the p=0.05 level.

The annual number of warm spells[11] averaged over North America has increased since 1950 (Peterson *et al.*, 2008). The frequency and extent of hot summers[12] was highest in the 1930s, 1950s, and 1995-2003; the geographic

In contrast to the 1930s, the recent period of increasing heat wave index is distinguished by the dominant contribution of a rise in extremely high nighttime temperatures.

---

[8] The threshold is approximately the 99.9 percentile.
[9] The data were first transformed to create near-normal distributions using a log transformation for the heat wave index and a cube root transformation for the cold wave index. The transformed data were then subjected to least squares regression. Details are given in Appendix A, Example 2.
[10] Details of this analysis are given in Appendix A, Example 1.

---

[11] Defined as at least three consecutive days above the 90th percentile threshold done separately for maximum and minimum temperature.
[12] Based on percentage of North American grid points with summer temperatures above the 90th or below the 10th percentiles of the 1950-1999 summer climatology.

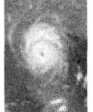

pattern of hot summers during 1995-2003 was similar to that of the 1930s (Gershunov and Douville, 2008).

The occurrence of temperatures below the biologically and societally important freezing threshold (0°C, 32°F) is an important aspect of the cold season climatology. Studies have typically characterized this either in terms of the number of frost days (days with the minimum temperature below freezing) or the length of the frost-free season[13]. The number of frost days decreased by four days per year in the United States during the 1948-1999 period, with the largest decreases, as many as 13 days per year, occurring in the western United States[14] (Easterling, 2002). In Canada, there have been significant decreases in frost day occurrence over the entire country from 1950 to 2003, with the largest decreases in extreme western Canada where there have been decreases of up to 40 or more frost days per year, and slightly smaller decreases in eastern Canada (Vincent and Mekis, 2006). The start of the frost-free season in the northeastern United States occurred 11 days earlier in the 1990s than in the

1950s (Cooter and LeDuc, 1995). For the U.S. as a whole, the average length of the frost-free season over the 1895-2000 period increased by almost two weeks[15] (Figure 2.4; Kunkel *et al.*, 2004). The change is characterized by four distinct regimes, with decreasing frost-free season length from 1895 to 1910, an increase in length of about one week from 1910 to 1930, little change during 1930-1980, and large increases since 1980. The frost-free season length has increased more in the western United States than in the eastern United States (Easterling, 2002; Kunkel *et al.*, 2004), which is consistent with the finding that the spring pulse of snow melt water in the Western United States now comes as much as 7-10 days earlier than in the late 1950s (Cayan *et al.*, 2001).

Ice cover on lakes and the oceans is a direct reflection of the number and intensity of cold, below freezing days. Ice cover on the Laurentian Great Lakes of North American usually forms along the shore and in shallow areas in December and January, and in deeper mid-lake areas in February due to their large depth and heat storage capacity. Ice loss usually starts in early to mid-March and lasts through mid- to late April (Assel, 2003).

Annual maximum ice cover on the Great Lakes has been monitored since 1963. The maximum extent of ice cover over the past four decades varied from less than 10% to over 90%. The winters of 1977-1982 were characterized by a higher ice cover regime relative to the prior 14 winters (1963-1976) and the following 24 winters (1983-2006) (Assel *et al.*, 2003; Assel, 2005a; Assel personal communication for winter 2006). A majority of the mildest winters with lowest seasonal average ice cover (Assel, 2005b) over the past four decades occurred during the most recent 10-year period (1997-2006). Analysis of ice breakup dates on other smaller lakes in North America with at least 100 years of data (Magnuson *et al.*, 2000) show a uniform trend toward earlier breakup dates (up to 13 days earlier per 100 years)[16].

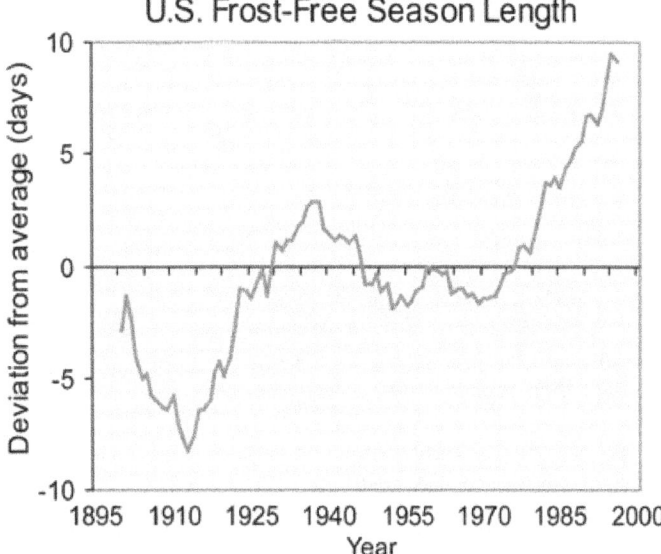

U.S. Frost-Free Season Length

Figure 2.4 Change in the length of the frost-free season averaged over the United States (from Kunkel *et al.*, 2003). The frost-free season is at least ten days longer on average than the long-term average.

---

[13] The difference between the date of the last spring frost and the first fall frost.
[14] Trends in the western half of the United States were statistically significant based on simple linear regression.

---

[15] Statistically significant based on least-squares linear regression.
[16] Statistically significant trends were found for 16 of 24 lakes.

Reductions in Arctic sea ice, especially near-shore sea ice, allow strong storm and wave activity to produce extensive coastal erosion resulting in extreme impacts. Observations from satellites starting in 1978 show that there has been a substantial decline in Arctic sea ice, with a statistically significant decreasing trend in annual Arctic sea ice extent of -33,000 (± 8,800) km² per year (equivalent to approximately -7% ± 2% since 1978). Seasonally the largest changes in Arctic sea ice have been observed in the ice that survives the summer, where the trend in the minimum Arctic sea ice extent, between 1979 and 2005, was -60,000 ± 24,000 km² per year (–20% ± 8%) (Lemke *et al.*, 2007). The 2007 summer sea-ice minimum was dramatically lower than the previous record low year of 2005.

Rising sea surface temperatures have led to an increase in the frequency of extreme high SST events causing coral bleaching (Box 1.1, Chapter 1). Mass bleaching events were not observed prior to 1980. However, since then, there have been six major global cycles of mass bleaching, with increasing frequency and intensity (Hoegh-Guldberg, 2005). Almost 30% of the world's coral reefs have died in that time.

Less scrutiny has been focused on Mexico temperature extremes, in part, because much of the country can be classified as a "tropical climate" where temperature changes are presumed fairly small, or semi-arid to arid climate where moisture availability exerts a far greater influence on human activities than does temperature.

Most of the sites in Mexico's oldest temperature observing network are located in major metropolitan areas, and there is considerable evidence to indicate that trends at least partly reflect urbanization and urban heat island influences (Englehart and Douglas, 2003). To avoid such issues in analysis, a monthly rural temperature data set has recently been developed[17]. Exam-

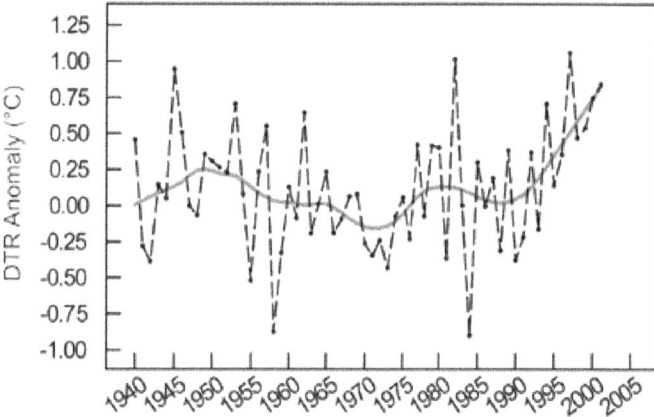

Figure 2.5 Change in the daily range of temperature (difference between the daily low and the daily high temperature) during the warm season (June-Sept) for Mexico. This difference is known as a Diurnal Temperature Range (DTR). The recent rise in the daily temperature range reflects hotter daily summer highs. The time series represents the average DTR taken over the four temperature regions of Mexico as defined in Englehart and Douglas (2004). Trend line (red) based on LOWESS smoothing (*n*=30).

ined in broad terms as a national aggregate, a couple of basic behaviors emerge. First, long period temperature trends over Mexico are generally compatible with continental-scale trends which indicate a cooling trend over North America from about the mid-1940s to the mid-1970s, with a warming trend thereafter.

The rural gridded data set indicates that much of Mexico experienced decreases in both maximum daily temperature and minimum daily temperature during the 1941-1970 period (-0.8°C for maximum daily temperature and -0.6°C for minimum daily temperature), while the later period of 1971-2001 is dominated by upward trends that are most strongly evident in maximum daily temperature (1.1°C for maximum daily temperature and 0.3°C for minimum daily temperature). Based on these results, it appears very likely that much of Mexico has experienced an increase in average temperature driven in large measure by increases in maximum daily temperature. The diurnal temperature range (the difference between the daily high and the daily low temperature) for the warm season (June-September) averaged over all of Mexico has increased by 0.8°C since 1970, with particularly rapid rises since 1990 (Figure

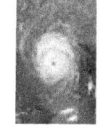

---

[17] It consists of monthly historical surface air temperature observations (1940-2001) compiled from stations (n=103) located in places with population <10,000 (2000 Census) and outside the immediate environment of large metropolitan areas. About 50% of the stations are located in places with <1000 inhabitants, and fewer than 10% of the stations are in places with >5000 population. To accommodate variable station record lengths and missing monthly observations, the

data set is formatted as a grid-type (2.5° x 2.5° lat.-long.) based on the climate anomaly method (Jones and Moberg, 2003).

2.5), reflecting a comparatively rapid rise in maximum daily temperature with respect to minimum daily temperature (Englehart and Douglas, 2005)[18]. This behavior departs from the general picture for many regions of the world, where warming is attributable mainly to a faster rise in minimum daily temperature than in maximum daily temperature (*e.g.,* Easterling *et al.*, 1997). Englehart and Douglas (2005) indicate that the upward trend in diurnal temperature range for Mexico is in part a response to land use changes as reflected in population trends, animal numbers and rates of change in soil erosion.

2.2.2 Precipitation Extremes
2.2.2.1 DROUGHT

Droughts are one of the most costly natural disasters (Chapter 1, Box 1.4), with estimated annual U.S. losses of $6–8 billion (Federal Emergency Management Agency, 1995). An extended period of deficient precipitation is the root cause of a drought episode, but the intensity can be exacerbated by high evaporation rates arising from excessive temperatures, high winds, lack of cloudiness, and/or low humidity. Drought can be defined in many ways, from acute short-term to chronic long-term hydrological drought, agricultural drought, meteorological drought, and so on. The assessment in this report focuses mainly on meteorological droughts based on the Palmer (1965) Drought Severity Index (PDSI), though other indices are also documented in the report (Chapter 2, Box 2.1).

Individual droughts can occur on a range of geographic scales, but they often affect large

Droughts are one of the most costly natural disasters, with estimated annual U.S. losses of $6–8 billion.

areas, and can persist for many months and even years. Thus, the aggregate impacts can be very large. For the United States, the percentage area affected by severe to extreme drought (Figure 2.6) highlights some major episodes of extended drought. The most widespread and severe drought conditions occurred in the 1930s and 1950s (Andreadis *et al.*, 2005). The early 2000s were also characterized by severe droughts in some areas, notably in the western United States. When averaged across the entire United States (Figure 2.6), there is no clear tendency for a trend based on the PDSI. Similarly, long-term trends (1925-2003) of hydrologic droughts based on model derived soil moisture and runoff show that droughts have, for the most part, become shorter, less frequent, and cover a smaller portion of the U. S. over the last century (Andreadis and Lettenmaier, 2006). The main exception is the Southwest and parts of the interior of the West, where increased temperature has led to rising drought trends (Groisman *et al.*, 2004; Andreadis and Lettenmaier, 2006). The trends averaged over all of North America since 1950 (Figure 2.6) are similar to U.S. trends for the same period, indicating no overall trend.

Since the contiguous United States has experienced an increase in both temperature and precipitation during the 20th century, one question is whether these increases are impacting the occurrence of drought. Easterling *et al.* (2007) examined this possibility by looking at drought, as defined by the PDSI, for the United States using detrended temperature and precipitation. Results indicate that without the upward trend

---

[18] Statistically significant trends were found in the northwest, central, and south, but not the northeast regions.

in precipitation, the increase in temperatures would have led to an increase in the area of the United States in severe-extreme drought of up to 30% in some months. However, it is most useful to look at drought in a regional context because as one area of the country is dry, often another is wet.

Summer conditions, which relate to fire danger, have trended toward lesser drought in the upper Mississippi, Midwest, and Northwest, but the fire danger has increased in the Southwest, in California in the spring season (not shown), and, surprisingly, over the Northeast, despite the fact that annual precipitation here has increased. A century-long warming in this region is quite significant in summer, which attenuates the precipitation contribution to soil wetness (Groisman *et al.*, 2004). Westerling *et al.* (2006) document that large wildfire activity in the western United States increased suddenly and markedly in the mid-1980s, with higher large-wildfire frequency, longer wildfire durations, and longer wildfire seasons. The greatest increases occurred in mid-elevation Northern Rockies forests, where land-use histories have relatively little effect on fire risks, and are strongly associated with increased spring and summer temperatures and an earlier spring snowmelt.

For the entire North American continent, there is a north-south pattern in drought trends (Dai *et al.*, 2004). Since 1950, there is a trend toward wetter conditions over much of the conterminous United States, but a trend toward drier conditions over southern and western Canada, Alaska, and Mexico. The summer PDSI averaged for Canada indicates dry conditions during the 1940s and 1950s, generally wet conditions from the 1960s to 1995, but much drier after 1995 (Shabbar and Skinner, 2004). In Alaska and Canada, the upward trend in temperature, resulting in increased evaporation rates, has made a substantial contribution to the upward trend in drought (Dai *et al.*, 2004). In agreement with this drought index analysis, the area of forest fires in Canada

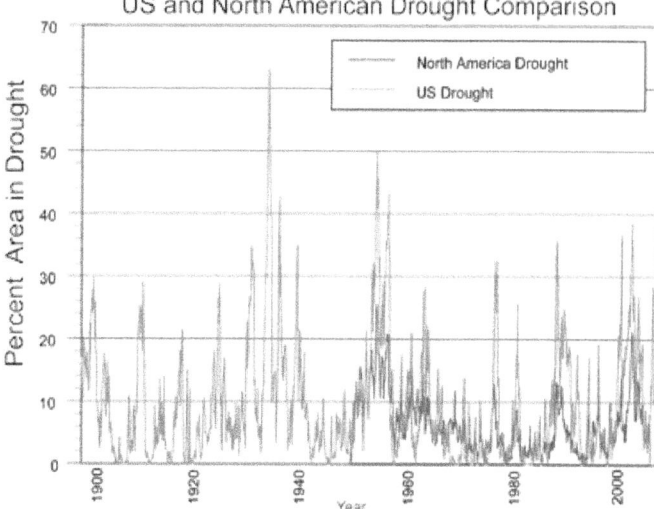

Figure 2.6 The area (in percent) of area in severe to extreme drought as measured by the Palmer Drought Severity Index for the United States (red) from 1900 to present and for North America (blue) from 1950 to present.

has been quite high since 1980 compared to the previous 30 years, and Alaska experienced a record high year for forest fires in 2004 followed by the third highest in 2005 (Soja *et al.*, 2007). During the mid-1990s and early 2000s, central and western Mexico (Kim *et al.*, 2002; Nicholas and Battisti, 2008; Hallack-Alegria and Watkins, 2007) experienced continuous cool-season droughts having major impacts in agriculture, forestry, and ranching, especially during the warm summer season. In 1998, "El Niño" caused one of the most severe droughts in Mexico since the 1950s (Ropelewski, 1999), creating the worst wildfire season in Mexico's history. Mexico had 14,445 wildfires affecting 849,632 hectares–the largest area ever burned in Mexico in a single season (SEMARNAT, 2000).

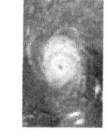

The greatest increases in wildfire activity have occurred in mid-elevation Northern Rockies forests, and are strongly associated with increased spring and summer temperatures and earlier spring snowmelt.

Reconstructions of drought prior to the instrumental record based on tree-ring chronologies indicate that the 1930s may have been the worst drought since 1700 (Cook *et al.*, 1999). There were three major multiyear droughts in the United States during the latter half of the 1800s: 1856-1865, 1870-1877, and 1890-1896 (Herweijer *et al.*, 2006). Similar droughts have been reconstructed for northern Mexico (Therrell *et al.*, 2002). There is evidence of earlier, even more intense drought episodes (Woodhouse and Overpeck, 1998). A period in the mid- to late 1500s has been termed a "mega-drought" and was longer-lasting and more widespread than the 1930s Dust Bowl (Stahle *et al.*, 2000). Several additional mega-droughts occurred during the years 1000-1470 (Herweijer *et al.*, 2007). These droughts were about as severe as the 1930s Dust

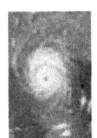

BOX 2.1: "Measuring" Drought

Drought is complex and can be "measured" in a variety of ways. The following list of drought indicators are all based on commonly-observed weather variables and fixed values of soil and vegetation properties. Their typical application has been for characterization of past and present drought intensity. For use in future projections of drought, the same set of vegetation properties is normally used. However, some properties may change. For example, it is known that plant water use efficiency increases with increased CO_2 concentrations. There have been few studies of the magnitude of this effect in realistic field conditions. A recent field study (Bernacchi *et al.*, 2007) of the effect of CO_2 enrichment on evapotranspiration (ET) from unirrigated soybeans indicated an ET decrease in the range of 9-16% for CO_2 concentrations of 550 parts per million (ppm) compared to present-day CO_2 levels. Studies in native grassland indicated plant water use efficiency can increase from 33% - 69% under 700-720 ppm CO_2 (depending on species). The studies also show a small increase in soil water content of less than 20% early on (Nelson *et al.*, 2004), which disappeared after 50 days in one experiment (Morgan *et al.*, 1998), and after 3 years in the other (Ferretti *et al.*, 2003). These experiments over soybeans and grassland suggest that the use of existing drought indicators without any adjustment for consideration of CO_2 enrichment would tend to have a small overestimate of drought intensity and frequency, but present estimates indicate that it is an order of magnitude smaller effect, compared to traditional weather and climate variations. Another potentially important effect is the shift of whole biomes from one vegetation type to another. This could have substantial effects on ET over large spatial scales (Chapin *et al.*, 1997).

- **Palmer Drought Severity Index (PDSI; Palmer, 1965)** – meteorological drought. The PDSI is a commonly used drought index that measures intensity, duration, and spatial extent of drought. It is derived from measurements of precipitation, air temperature, and local estimated soil moisture content. Categories range from less than -4 (extreme drought) to more than +4 (extreme wet conditions), and have been standardized to facilitate comparisons from region to region. Alley (1984) identified some positive characteristics of the PDSI that contribute to its popularity: (1) it is an internationally recognized index; (2) it provides decision makers with a measurement of the abnormality of recent weather for a region; (3) it provides an opportunity to place current conditions in historical perspective; and (4) it provides spatial and temporal representations of historical droughts. However, the PDSI has some limitations: (1) it may lag emerging droughts by several months; (2) it is less well suited for mountainous land or areas of frequent climatic extremes; (3) it does not take into account streamflow, lake and reservoir levels, and other long-term hydrologic impacts (Karl and Knight, 1985), such as snowfall and snow cover; (4) the use of temperature alone to estimate potential evapotranspiration (PET) can introduce biases in trend estimates because humidity, wind, and radiation also affect PET, and changes in these elements are not accounted for.

- **Crop Moisture Index (CMI; Palmer, 1968)** – short-term meteorological drought. Whereas the PDSI monitors long-term meteorological wet and dry spells, the CMI was designed to evaluate short-term moisture conditions across major crop-producing regions. It is based on the mean temperature and total precipitation for each week, as well as the CMI value from the previous week. Categories range from less than -3 (severely dry) to more than +3 (excessively wet). The CMI responds rapidly to changing conditions, and it is weighted by location and time so that maps, which commonly display the weekly CMI across the United States, can be used to compare moisture conditions at different

Bowl episode but much longer, lasting 20 to 40 years. In the western United States, the period of 900-1300 was characterized by widespread drought conditions (Figure 2.7; Cook *et al.*, 2004). In Mexico, reconstructions of seasonal precipitation (Stahle *et al.*, 2000, Acuña-Soto *et al.*, 2002, Cleaveland *et al.*, 2004) indicate that there have been droughts more severe than the 1950s drought, *e.g.*, the mega-drought in the

mid- to late-16th century, which appears as a continental-scale drought.

During the summer months, excessive heat and drought often occur simultaneously because the meteorological conditions typically causing drought are also conducive to high temperatures. The impacts of the Dust Bowl droughts and the 1988 drought were compounded by

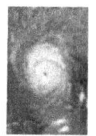

locations. Weekly maps of the CMI are available as part of the USDA/JAWF Weekly Weather and Crop Bulletin.

- **Standardized Precipitation Index (SPI; McKee et al., 1993)** – precipitation-based drought. The SPI was developed to categorize rainfall as a standardized departure with respect to a rainfall probability distribution function; categories range from less than -3 (extremely dry) to more than +3 (extremely wet). The SPI is calculated on the basis of selected periods of time (typically from 1 to 48 months of total precipitation), and it indicates how the precipitation for a specific period compares with the long-term record at a given location (Edwards and McKee, 1997). The index correlates well with other drought indices. Sims *et al.* (2002) suggested that the SPI was more representative of short-term precipitation and a better indicator of soil wetness than the PDSI. The 9-month SPI corresponds closely to the PDSI (Heim, 2002; Guttman, 1998).

- **Keetch-Byram Drought Index (KBDI; Keetch and Byram, 1968)** – meteorological drought and wildfire potential index. This was developed to characterize the level of potential fire danger. It uses daily temperature and precipitation information and estimates soil moisture deficiency. High values of KBDI are indicative of favorable conditions for wildfires. However, the index needs to be regionalized, as values are not comparable among regions (Groisman *et al.*, 2004, 2007).

- **No-rain episodes** – meteorological drought. Groisman and Knight (2007, 2008) proposed to directly monitor frequency and intensity of prolonged no-rain episodes (greater than 20, 30, 60, *etc.* days) during the warm season, when evaporation and transpiration are highest and the absence of rain may affect natural ecosystems and agriculture. They found that during the past four decades the duration of prolonged dry episodes has significantly increased over the eastern and southwestern United States

and adjacent areas of northern Mexico and southeastern Canada.

- **Soil Moisture and Runoff Index (SMRI; Andreadis and Lettenmaier, 2006)**–hydrologic and agricultural droughts. The SMRI is based on model-derived soil moisture and runoff as drought indicators; it uses percentiles and the values are normalized from 0 (dry) to 1 (wet conditions). The limitation of this index is that it is based on land-surface model-derived soil moisture. However, long-term records of soil moisture – a key variable related to drought – are essentially nonexistent (Andreadis and Lettenmaier, 2006). Thus, the advantage of the SMRI is that it is physically based and with the current sophisticated land-surface models it is easy to produce multimodel average climatologies and century-long reconstructions of land surface conditions, which could be compared under drought conditions.

Resources: A list of these and other drought indicators, data availability, and current drought conditions based on observational data can be found at NOAA's National Climatic Data Center (NCDC, http://www.ncdc.noaa.gov). The North American Drought Monitor at NCDC monitors current drought conditions in Canada, the United States, and Mexico. Tree-ring reconstruction of PDSI across North America over the last 2000 years can be also found at NCDC.

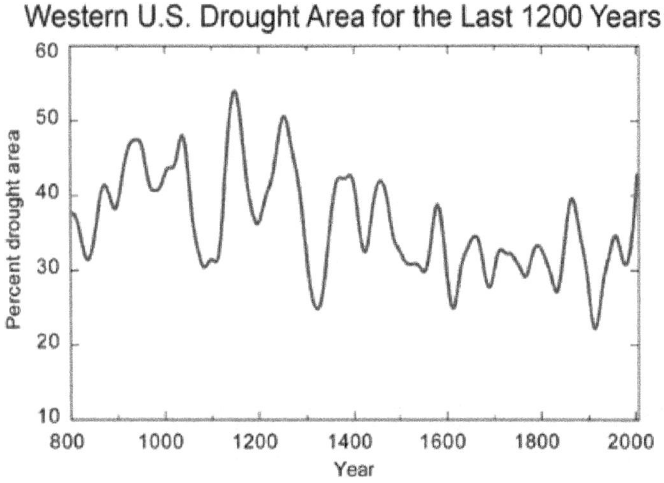

Western U.S. Drought Area for the Last 1200 Years

Figure 2.7 Area of drought in the western United States as reconstructed from tree rings (Cook *et al.*, 2004).

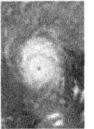

United States, regional average values of the 99.9 percentile threshold for daily precipitation are lowest in the Northwest and Southwest (average of 55 mm) and highest in the South (average of 130mm)[19].

As noted above, spatial patterns of precipitation have smaller spatial correlation scales (for example, compared to temperature and atmospheric pressure) which means that a denser network is required in order to achieve a given uncertainty level. While monthly precipitation time series for flat terrain have typical radii of correlation[20] (ρ) of approximately 300 km or even more, daily precipitation may have ρ less than 100 km with typical values for convective rainfall in isolated thunderstorms of approximately 15 to 30 km (Gandin and Kagan, 1976). Values of ρ can be very small for extreme rainfall events, and sparse networks may not be adequate to detect a desired minimum magnitude of change that can result in societally-important impacts and can indicate important changes in the climate system.

2.2.2.2.2 United States

One of the clearest trends in the United States observational record is an increasing frequency and intensity of heavy precipitation events (Karl and Knight, 1998; Groisman *et al.*, 1999, 2001, 2004, 2005; Kunkel *et al.*, 1999b; Easterling *et al.*, 2000; IPCC, 2001; Semenov and Bengtsson, 2002; Kunkel, 2003). One measure of this is how much of the annual precipitation at a location comes from days with precipitation exceeding 50.8 mm (2 inches) (Karl and Knight, 1998). The area of the United States affected by a much above normal contribution from these heavy precipitation days increased by a statistically significant amount, from about 9% in the 1910s to about 11% in the 1980s and 1990s (Karl and Knight, 1998). Total precipitation also increased during this time, due in large part to increases in the intensity of heavy precipitation events (Karl and Knight, 1998). In

episodes of extremely high temperatures. The month of July 1936 in the central United States is a notable example. To illustrate, Lincoln, NE received only 0.05" of precipitation that month (after receiving less than 1 inch the previous month) while experiencing temperatures reaching or exceeding 110°F on 10 days, including 117°F on July 24. Although no studies of trends in such "compound" extreme events have been performed, they represent a significant societal risk.

2.2.2.2 SHORT DURATION HEAVY PRECIPITATION

2.2.2.2.1 Data Considerations and Terms

Intense precipitation often exhibits higher geographic variability than many other extreme phenomena. This poses challenges for the analysis of observed data since the heaviest area of precipitation in many events may fall between stations. This adds uncertainty to estimates of regional trends based on the climate network. The uncertainty issue is explicitly addressed in some recent studies.

Precipitation extremes are typically defined based on the frequency of occurrence (by percentile [*e.g.*, upper 5%, 1%, 0.1%, *etc.*] or by return period [*e.g.*, an average occurrence of once every 5 years, once every 20 years, *etc.*]), and/or their absolute values (*e.g.*, above 50 mm, 100 mm, 150 mm, or more). Values of percentile or return period thresholds vary considerably across North America. For example, in the

One of the clearest trends in the U.S. observational record is an increasing frequency and intensity of heavy precipitation events.

---

[19] The large magnitude of these differences is a major motivation for the use of regionally-varying thresholds based on percentiles.

[20] Spatial correlation decay with distance, r, for many meteorological variables, X, can be approximated by an exponential function of distance: Corr (X(A), X(B)) ~ e$^{-r/\rho}$ where r is a distance between point A and B and ρ is a radius of correlation, which is a distance where the correlation between the points is reduced to 1/e compared to an initial "zero" distance.

fact, there has been little change or decrease in the frequency of light and average precipitation days (Easterling *et al.*, 2000; Groisman *et al.*, 2004, 2005) during the last 30 years, while heavy precipitation frequencies have increased (Sun and Groisman, 2004). For example, the amount of precipitation falling in the heaviest 1% of rain events increased by 20% during the 20th century, while total precipitation increased by 7% (Groisman *et al.*, 2004). Although the exact character of those changes has been questioned (*e.g.*, Michaels *et al.*, 2004), it is highly likely that in recent decades extreme precipitation events have increased more than light to medium events.

Over the last century there was a 50% increase in the frequency of days with precipitation over 101.6 mm (four inches) in the upper midwestern U.S.; this trend is statistically significant (Groisman *et al.*, 2001). Upward trends in the amount of precipitation occurring in the upper 0.3% of daily precipitation events are statistically significant for the period of 1908-2002 within three major regions (the South, Midwest, and Upper Mississippi; Figure 2.8) of the central United States (Groisman *et al.*, 2004, 2005). The upward trends are primarily a warm season phenomenon when the most intense rainfall events typically occur. A time series of the frequency of events in the upper 0.3% averaged for these 3 regions (Fig 2.8) shows a 20% increase over the period of 1893-2002 with all of this increase occurring over the last third of the 1900s (Groisman *et al.*, 2005).

Examination of intense precipitation events defined by return period, covering the period

The amount of precipitation falling in the heaviest 1% of rain events increased by 20% during the 20th century, while total precipitation increased by 7%.

Regions of N. America where Intense Precipitation has Increased

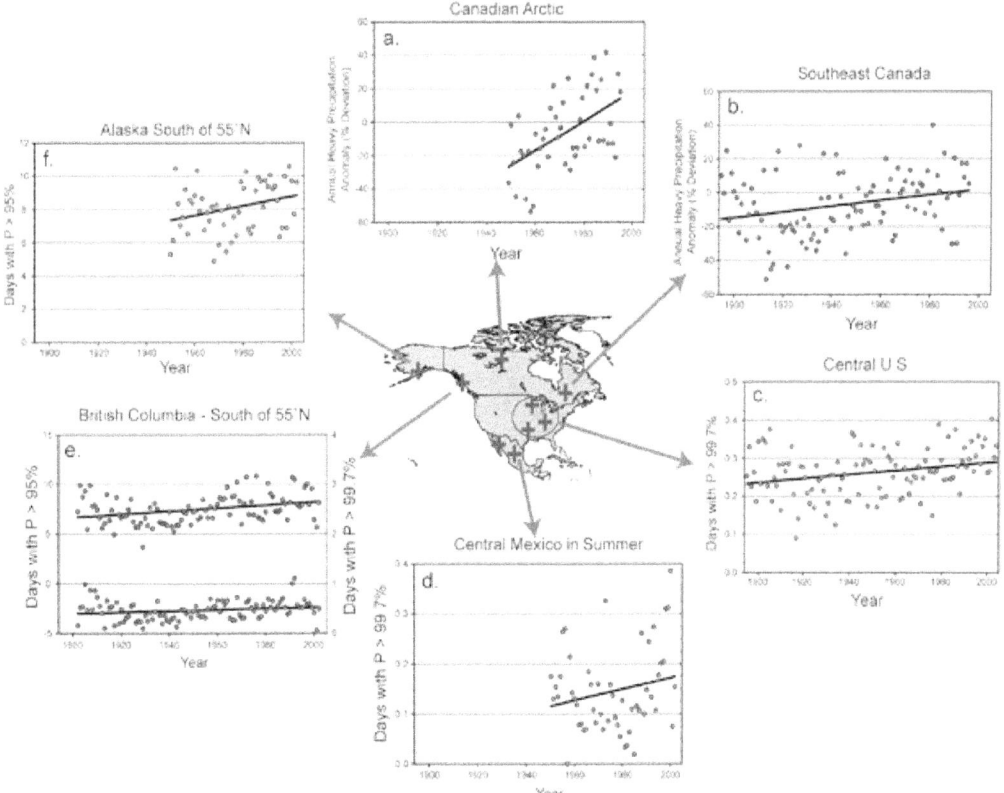

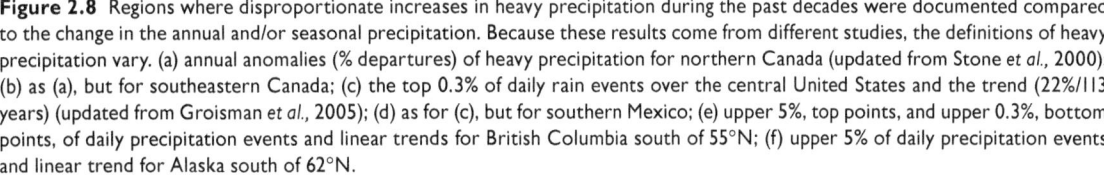

Figure 2.8 Regions where disproportionate increases in heavy precipitation during the past decades were documented compared to the change in the annual and/or seasonal precipitation. Because these results come from different studies, the definitions of heavy precipitation vary. (a) annual anomalies (% departures) of heavy precipitation for northern Canada (updated from Stone *et al.*, 2000); (b) as (a), but for southeastern Canada; (c) the top 0.3% of daily rain events over the central United States and the trend (22%/113 years) (updated from Groisman *et al.*, 2005); (d) as for (c), but for southern Mexico; (e) upper 5%, top points, and upper 0.3%, bottom points, of daily precipitation events and linear trends for British Columbia south of 55°N; (f) upper 5% of daily precipitation events and linear trend for Alaska south of 62°N.

of 1895-2000, indicates that the frequencies of extreme precipitation events before 1920 were generally above the long-term averages for durations of 1 to 30 days and return periods 1 to 20 years, and only slightly lower than values during the 1980s and 1990s (Kunkel *et al.*, 2003). The highest values occur after about 1980, but the elevated levels prior to about 1920 are an interesting feature suggesting that there is considerable variability in the occurrence of extreme precipitation on decade-to-decade time scales

There is a seeming discrepancy between the results for the 99.7th percentile (which do not show high values early in the record in the analysis of Groisman *et al.*, 2004), and for 1 to 20-year return periods (which do show high values in the analysis of Kunkel *et al.*, 2003). The number of stations with available data is only about half (about 400) in the late 1800s of what is available in most of the 1900s (800-900). Furthermore, the geographic distribution of stations throughout the record is not uniform; the density in the western United States is relatively lower than in the central and eastern United States. It is possible that the resulting uncertainties in heavy precipitation estimates are too large to make unambiguous statements about the recent high frequencies.

Recently, this question was addressed (Kunkel *et al.*, 2007a) by analyzing the modern dense network to determine how the density of stations affects the uncertainty, and then to estimate the level of uncertainty in the estimates of frequencies in the actual (sparse) network used in the long-term studies. The results were unambiguous. For all combinations of three precipitation durations (1-day, 5-day and 10-day) and three return periods (1-year, 5-year, and 20-year), the frequencies for 1983-2004 were significantly higher than those

It is highly likely that the recent elevated frequencies in heavy precipitation in the United States are the highest on record.

for 1895-1916 at a high level of confidence. In addition, the observed linear trends were all found to be upward, again with a high level of confidence. Based on these results, it is highly likely that the recent elevated frequencies in heavy precipitation in the United States are the highest on record.

2.2.2.2.3 Alaska and Canada
The sparse network of long-term stations in Canada increases the uncertainty in estimates of extremes. Changes in the frequency of heavy precipitation events exhibit considerable decade-to-decade variability since 1900, but no long-term trend for the century as a whole (Zhang *et al.*, 2001). However, according to Zhang *et al.* (2001), there are not sufficient instrumental data to discuss the nationwide trends in precipitation extremes over Canada prior to 1950. Nevertheless, there are changes that are noteworthy. For example, the frequency of the upper 0.3% of events exhibits a statistically significant upward trend of 35% in British Columbia since 1910 (Figure 2.8; Groisman *et al.*, 2005). For Canada, increases in precipitation intensity during the second half of the 1900s are concentrated in heavy and intermediate events, with the largest changes occurring in Arctic areas (Stone *et al.*, 2000). The tendency for increases in the frequency of intense precipitation, while the frequency of days with average and light precipitation does not change or decreases, has also been observed in Canada over the last 30 years (Stone *et al.*, 2000),

mirroring United States changes. Recently, Vincent and Mekis (2006) repeated analyses of precipitation extremes for the second half of the 1900s (1950-2003 period). They reported a statistically significant increase of 1.8 days over the period in heavy precipitation days (defined as the days with precipitation above 10 mm) and statistically insignificant increases in the maximum 5-day precipitation (by approximately 5%) and in the number of "very wet days" (defined as days with precipitation above the upper 5th percentiles of local daily precipitation [by 0.4 days]).

There is an upward trend of 39% in southern Alaska since 1950, although this trend is not statistically significant (Figure 2.8; Groisman et al., 2005).

2.2.2.2.4 Mexico

On an annual basis, the number of heavy precipitation (P > 10 mm) days has increased in northern Mexico and the Sierra Madre Occidental and decreased in the south-central part of the country (Alexander et al., 2006). The percent contribution to total precipitation from heavy precipitation events exceeding the 95th percentile threshold has increased in the monsoon region (Alexander et al., 2006) and along the southern Pacific coast (Aguilar et al., 2005), while some decreases are documented for south-central Mexico (Aguilar et al., 2005).

On a seasonal basis, the maximum precipitation reported in five consecutive days during winter and spring has increased in northern Mexico and decreased in south-central Mexico (Alexander et al., 2006). Northern Baja California, the only region in Mexico characterized by a Mediterranean climate, has experienced an increasing trend in winter precipitation exceeding the 90th percentile, especially after 1977 (Cavazos and Rivas, 2004). Heavy winter precipitation in this region is significantly correlated with El Niño events (Pavia and Badan, 1998; Cavazos and Rivas, 2004); similar results have been documented for California (e.g., Gershunov and Cayan, 2003). During the summer, there has been a general increase of 2.5 mm in the maximum five-consecutive-day precipitation in most of the country, and an upward trend in the intensity of events exceeding the 99th and 99.7th percentiles in the high plains of northern

Mexico during the summer season (Groisman et al., 2005).

During the monsoon season (June-September) in northwestern Mexico, the frequency of heavy events does not show a significant trend (Englehart and Douglas, 2001; Neelin et al., 2006). Similarly, Groisman et al. (2005) report that the frequency of very heavy summer precipitation events (above the 99th percentile) in the high plains of Northern Mexico (east of the core monsoon) has not increased, whereas their intensity has increased significantly.

The increase in the mean intensity of heavy summer precipitation events in the core monsoon region during the 1977-2003 period are significantly correlated with the Oceanic El Niño Index[21] conditions during the cool season. El Niño SST anomalies antecedent to the monsoon season are associated with less frequent, but more intense, heavy precipitation events[22] (exceeding the 95th percentile threshold), and vice versa.

There has been an insignificant decrease in the number of consecutive dry days in northern Mexico, while an increase is reported for south-central Mexico (Alexander et al., 2006), and the southern Pacific coast (Aguilar et al., 2005).

2.2.2.2.5 Summary

All studies indicate that changes in heavy precipitation frequencies are *always* higher than changes in precipitation totals and, in some regions, an *increase* in heavy and/or very heavy precipitation occurred while no change or even a decrease in precipitation totals was observed (*e.g.*, in the summer season in central Mexico). There are regional variations in which these changes are statistically significant (Figure 2.8). The most significant changes occur in the central United States; central Mexico; southeastern, northern, and western Canada; and

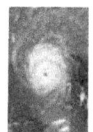

Changes in heavy precipitation frequencies are always higher than changes in precipitation totals, and in some regions, an increase in heavy and/or very heavy precipitation occurred while no change or even a decrease in precipitation totals was observed.

---

[21] Oceanic El Niño Index:http://www.cpc.ncep.noaa.gov/products/analysis_monitoring/ensostuff/ensoyears.shtml
Warm and cold episodes based on a threshold of +/- 0.5°C for the oceanic El Niño Index [3 month running mean of ERSST.v2 SST anomalies in the Niño 3.4 region (5°N-5°S, 120°-170°W)], based on the 1971-2000 base period.

[22] The correlation coefficient between oceanic El Niño Index and heavy precipitation frequency (intensity) is -0.37 (+0.46).

Increase in the Occurence of Periods of Heavy Rainfall Lasting at Least 90 Days

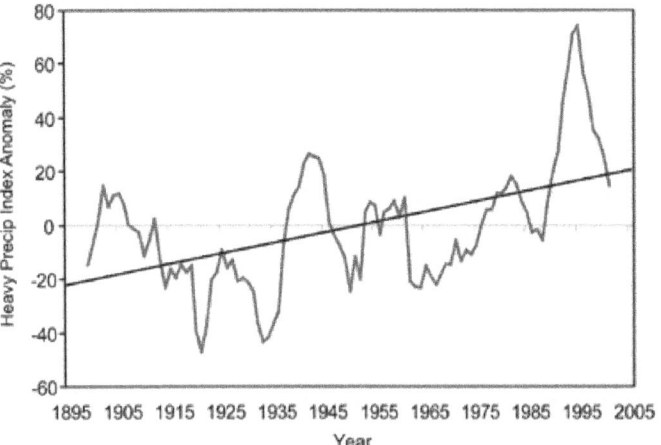

Figure 2.9 Frequency (expressed as a percentage anomaly from the period of record average) of excessive precipitation periods of 90 day duration exceeding a 1-in-20-year event threshold for the U.S. The periods are identified from a time series of 90-day running means of daily precipitation totals. The largest 90-day running means were identified and these events were counted in the year of the first day of the 90-day period. Annual frequency values have been smoothed with a 9-yr running average filter. The black line shows the trend (a linear fit) for the annual values.

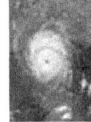

southern Alaska. These changes have resulted in a wide range of impacts, including human health impacts (Chapter 1, Box 1.3).

2.2.2.3 MONTHLY TO SEASONAL HEAVY PRECIPITATION

On the main stems of large river basins, significant flooding will not occur from short duration extreme precipitation episodes alone. Rather, excessive precipitation must be sustained for weeks to months. The 1993 Mississippi River flood, which resulted in an estimated $17 billion

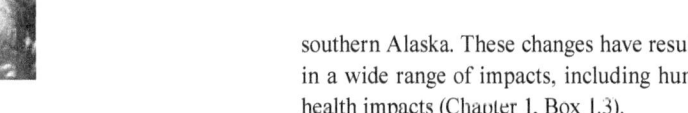

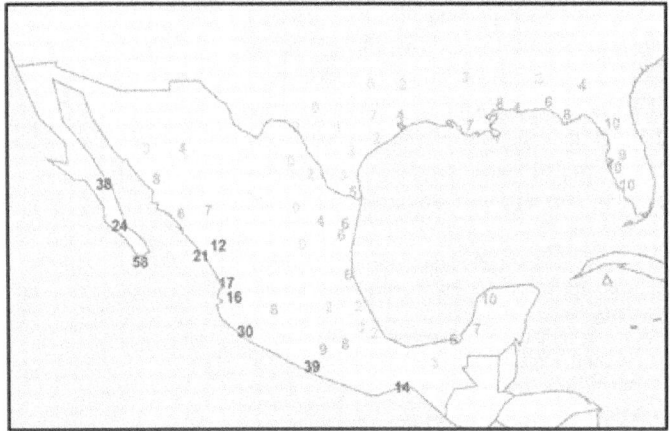

Figure 2.10 Average (median) percentage of warm season rainfall (May-November) from hurricanes and tropical storms affecting Mexico and the Gulf Coast of the United States. Figure updated from Englehart and Douglas (2001).

in damages, was caused by several months of anomalously high precipitation (Kunkel *et al.*, 1994).

A time series of the frequency of 90-day precipitation totals exceeding the 20-year return period (a simple extension of the approach of Kunkel *et al.*, 2003) indicates a statistically significant upward trend (Figure 2.9). The frequency of such events during the last 25 years is 20% higher than during any earlier 25-year period. Even though the causes of multi-month excessive precipitation are not necessarily the same as for short duration extremes, both show moderately high frequencies in the early 20th century, low values in the 1920s and 1930s, and the highest values in the past two to three decades. The trend[23] over the entire period is highly statistically significant.

2.2.2.4 NORTH AMERICAN MONSOON

Much of Mexico is dominated by a monsoon type climate with a pronounced peak in rainfall during the summer (June through September) when up to 60% to 80% of the annual rainfall is received (Douglas *et al.*, 1993; Higgins *et al.*, 1999; Cavazos *et al.*, 2002). Monsoon rainfall in southwest Mexico is often supplemented by tropical cyclones moving along the coast. Farther removed from the tracks of Pacific tropical cyclones, interior and northwest sections of Mexico receive less than 10% of the summer rainfall from passing tropical cyclones (Figure 2.10; Englehart and Douglas, 2001). The main influences on total monsoon rainfall in these regions rests in the behavior of the monsoon as defined by its start and end date, rainfall intensity, and duration of wet and dry spells (Englehart and Douglas, 2006). Extremes in any one of these parameters can have a strong effect on the total monsoon rainfall.

The monsoon in northwest Mexico has been studied in detail because of its singular importance to that region, and because summer rainfall from this core monsoon region spills over into the United States Desert Southwest (Douglas *et al.*, 1993; Higgins *et al.*, 1999; Cavazos *et al.*, 2002). Based on long term data

---

[23] The data were first subjected to a square root transformation to produce a data set with an approximate normal distribution, then least squares regression was applied. Details can be found in Appendix A, Example 4.

from eight stations in southern Sonora, the summer rains have become increasingly late in arriving (Englehart and Douglas, 2006), and this has had strong hydrologic and ecologic repercussions for this northwest core region of the monsoon. Based on linear trend, the mean start date for the monsoon has been delayed almost 10 days (9.89 days with a significant trend of 1.57 days per decade) over the past 63 years (Figure 2.11a). Because extended periods of intense heat and desiccation typically precede the arrival of the monsoon, the trend toward later starts to the monsoon will place additional stress on the water resources and ecology of the region if continued into the future.

Accompanying the tendency for later monsoon starts, there also has been a notable change in the "consistency" of the monsoon as indicated by the average duration of wet spells in southern Sonora (Figure 2.11b). Based on a linear trend, the average wet spell[24] has decreased by almost one day (0.88 days with a significant trend of -0.14 days per decade) from nearly four days in the early 1940s to slightly more than three days in recent years. The decrease in wet spell length indicates a more erratic monsoon is now being observed. Extended periods of consecutive days with rainfall are now becoming less common during the monsoon. These changes can have profound influences on surface soil moisture levels which affect both plant growth and runoff in the region.

A final measure of long-term change in monsoon activity is associated with the change in rainfall intensity over the past 63 years (Figure 2.11c). Based on linear trend, rainfall intensity[25] in the 1940s was roughly 5.6 mm per rain day, but in recent years has risen to nearly 7.5 mm per rain day[26]. Thus, while the summer monsoon has become increasingly late in arriving and wet spells have become shorter, the average rainfall during rain events has actually increased very significantly by 17% or 1.89 mm over the 63 year period (0.3 mm per decade) as

well as the intensity of heavy precipitation events (Figure 2.9). Taken together, these statistics indicate that rainfall in the core region of the monsoon (i.e., northwest Mexico) has become more erratic with a tendency towards high intensity rainfall events, a shorter monsoon, and shorter wet spells.

Variability in Mexican monsoon rainfall shows modulation by large-scale climate modes. Englehart and Douglas (2002) demonstrate that a well-developed inverse relationship exists between ENSO and total seasonal rainfall (June-September) over much of Mexico, but the relationship is only operable in the positive phase of the PDO. Evaluating monsoon rainfall behavior on intraseasonal time scales, Englehart and Douglas (2006) demonstrate that rainfall intensity (mm per rain day) in the core region of the

monsoon is related to PDO phase, with the positive (negative) phase favoring relatively high (low) intensity rainfall events. Analysis indicates that other rainfall characteristics of

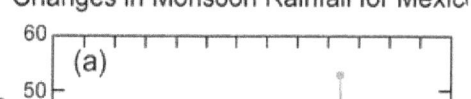

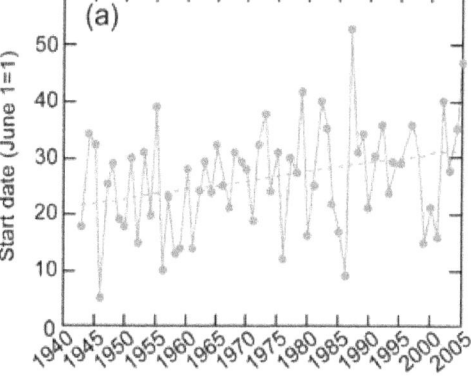

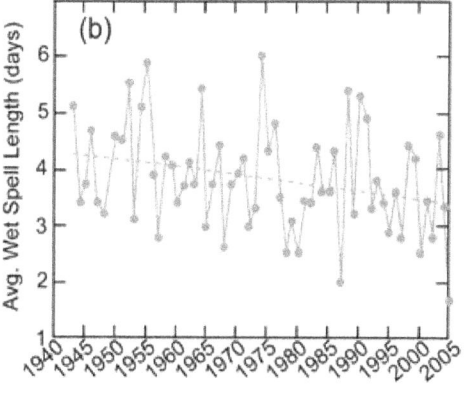

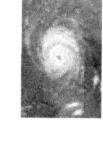

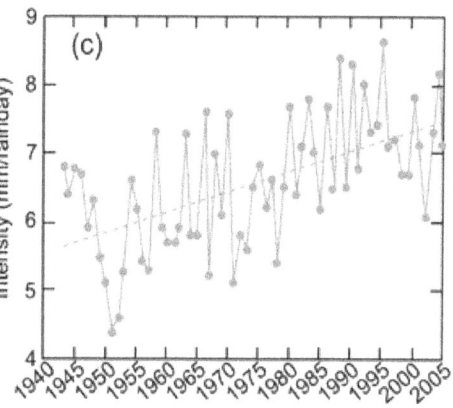

Figure 2.11 Variations and linear trend in various characteristics of the summer monsoon in southern Sonora, Mexico; including: (a) the mean start date June 1 = Day 1 on the graph; (b) the mean wet spell length defined as the mean number of consecutive days with mean regional precipitation >1 mm; and (c) the mean daily rainfall intensity for wet days defined as the regional average rainfall for all days with rainfall > 1 mm.

---

[24] For southern Sonora, Mexico, wet spells are defined as the mean number of consecutive days with mean regional precipitation ≥1 mm.

[25] Daily rainfall intensity during the monsoon is defined as the regional average rainfall for all days with rainfall ≥ 1 mm.

[26] The linear trend in this time series is significant at the p=0.01 level.

Hurricane/Tropical Storm Rainfall for Manzanillo, Mexico

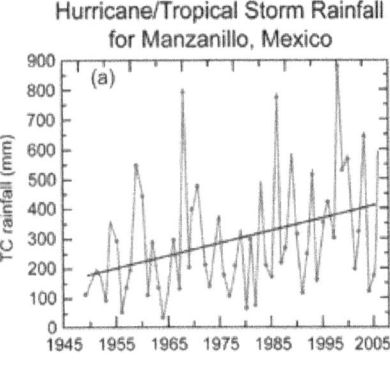

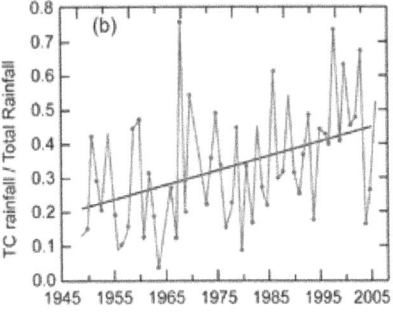

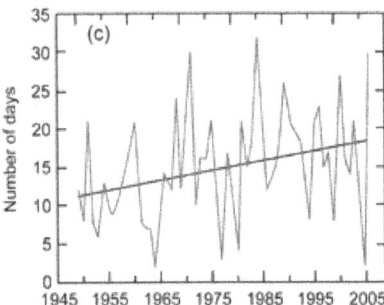

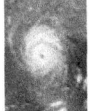

Figure 2.12 Trends in hurricane/tropical storm rainfall statistics at Manzanillo, Mexico, including: (a) the total warm season rainfall from hurricanes/tropical storms; (b) the ratio of hurricane/tropical storm rainfall to total summer rainfall; and (c) the number of days each summer with a hurricane or tropical storm within 550km of the stations.

the monsoon respond to ENSO, with warm events favoring later starts to the monsoon and shorter length wet spells (days) with cold events favoring opposite behavior (Englehart and Douglas, 2006).

2.2.2.5 TROPICAL STORM RAINFALL IN WESTERN MEXICO

Across southern Baja California and along the southwest coast of Mexico, 30% to 50% of warm season rainfall (May to November) is attributed to tropical cyclones (Figure 2.10), and in years heavily affected by tropical cyclones (upper 95th percentile), 50% to 100% of the summer rainfall comes from tropical cyclones. In this region of Mexico, there is a long-term, upward trend in tropical cyclone-derived rainfall at both Manzanillo (41.8 mm/decade; Figure 2.12a) and Cabo San Lucas (20.5 mm/decade)[27]. This upward trend in tropical cyclone rainfall has led to an increase in the importance of tropical cyclone rainfall in the total warm season rainfall for southwest Mexico (Figure 2.12b), and this has resulted in a higher ratio of tropical cyclone rainfall to total warm season rainfall. Since these two stations are separated by more than 700 km, these significant trends in tropical cyclone rainfall imply large scale shifts in the summer climate of Mexico.

This recent shift in emphasis on tropical cyclone warm season rainfall in western Mexico has

strong repercussions as rainfall becomes less reliable from the monsoon and becomes more dependent on heavy rainfall events associated with passing tropical cyclones. Based on the large scale and heavy rainfall characteristics associated with tropical cyclones, reservoirs in the mountainous regions of western Mexico are often recharged by strong tropical cyclone events which therefore have benefits for Mexico despite any attendant damage due to high winds or flooding.

This trend in tropical cyclone-derived rainfall is consistent with a long term analysis of near-shore tropical storm tracks along the west coast of Mexico (storms passing within 5° of the coast) which indicates an upward trend in the number of near-shore storms over the past 50 years (Figure 2.12c, see Englehart et al., 2008). While the number of tropical cyclones occurring in the entire east Pacific basin is uncertain prior to the advent of satellite tracking in about 1967, it should be noted that the long term data sets for near shore storm activity (within 5° of the coast) are considered to be much more reliable due to coastal observatories and heavy ship traffic to and from the Panama Canal to Pacific ports in Mexico and the United States. The number of near shore storm days (storms less than 550 km from the station) has increased by 1.3 days/decade in Manzanillo and about 0.7days/decade in Cabo San Lucas (1949-2006)[28]. The long term correlation between tropical cyclone days at each station and total tropical cyclone rainfall is $r = 0.61$ for Manzanillo and $r = 0.37$ for Cabo San Lucas, illustrating the strong tie between passing tropical cyclones and the rain that they provide to coastal areas of Mexico.

Interestingly, the correlations between tropical cyclone days and total tropical cyclone rainfall actually drop slightly when based only on the satellite era (1967-2006) ($r = 0.54$ for Manzanillo and $r = 0.31$ for Cabo San Lucas). The fact that the longer time series has the higher set of correlations shows no reason to suggest problems with near shore tropical cyclone tracking in the pre-satellite era. The lower correlations in the more recent period between tropical cyclone days and total tropical cyclone

[27] The linear trends in tropical cyclone rainfall at these two stations are significant at the p=0.01 and p=0.05 level, respectively.

[28] The linear trends in near shore storm days are significant at the p=0.05 level and p=0.10 level, respectively.

rainfall may be tied to tropical cyclone derived rainfall rising at a faster pace compared to the rise in tropical cyclone days. In other words, tropical cyclones are producing more rain per event than in the earlier 1949-1975 period when SSTs were colder.

2.2.2.6 TROPICAL STORM RAINFALL IN THE SOUTHEASTERN UNITED STATES

Tropical cyclone-derived rainfall along the southeastern coast of the United States on a century time scale has changed insignificantly in summer (when no century-long trends in precipitation was observed) as well as in autumn (when the total precipitation increased by more than 20% since the 1900s; Groisman *et al.*, 2004).

2.2.2.7 STREAMFLOW

The flooding in streams and rivers resulting from precipitation extremes can have devastating impacts. Annual average flooding losses rank behind only those of hurricanes. Assessing whether the observed changes in precipitation extremes has caused similar changes in streamflow extremes is difficult for a variety of reasons. First, the lengths of records of stream gages are generally shorter than neighboring precipitation stations. Second, there are many human influences on streamflow that mask the climatic influences. Foremost among these is the widespread use of dams to control streamflow. Vörösmarty *et al.*, (2004) showed that the influence of dams in the United States increased from minor areal coverage in 1900 to a large majority of the U.S. area in 2000. Therefore, long-term trend studies of climatic influences on streamflow have necessarily been

very restricted in their areal coverage. Even on streams without dams, other human effects such as land-use changes and stream channelization may influence the streamflow data.

A series of studies by two research groups (Lins and Slack, 1999, 2005; Groisman *et al.*, 2001, 2004) utilized the same set of streamgages not affected by dams. This set of gages represents streamflow for approximately 20% of the contiguous U.S. area. The initial studies both examined the period 1939-1999. Differences in definitions and methodology resulted in opposite judgments about trends in high streamflow. Lins and Slack (1999, 2005) reported no significant changes in high flow above the 90th percentile. On the other hand, Groisman *et al.* (2001) showed that for the same gauges, period, and territory, there were statistically significant regional average increases in the uppermost fractions of total streamflow. However, these trends became statistically insignificant after Groisman *et al.* (2004) updated the analysis to include the years 2000 through 2003, all of which happened to be dry years over most of the eastern United States. They concluded that "… during the past four dry years the contribution of the upper two 5-percentile classes to annual precipitation remains high or (at least) above the average while the similar contribution to annual streamflow sharply declined. This could be anticipated due to the accumulative character of high flow in large and medium rivers; it builds upon the base flow that remains low during dry years…" All trend estimates are sensitive to the values at the edges of the time series, but for high streamflow, these estimates are also sensitive to the mean values of the flow.

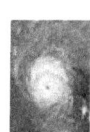

Tropical cyclones rank with flash floods as the most lethal and expensive natural catastrophes.

2.2.3 Storm Extremes
2.2.3.1 TROPICAL CYCLONES
2.2.3.1.1 Introduction

Each year, about 90 tropical cyclones develop over the world's oceans, and some of these make landfall in populous regions, exacting heavy tolls in life and property. The global number has been quite stable since 1970, when global satellite coverage began in earnest, having a standard deviation of 10 and no evidence of any substantial trend (*e.g.,*

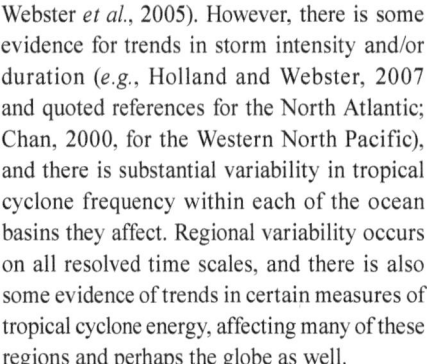

Webster *et al.*, 2005). However, there is some evidence for trends in storm intensity and/or duration (*e.g.*, Holland and Webster, 2007 and quoted references for the North Atlantic; Chan, 2000, for the Western North Pacific), and there is substantial variability in tropical cyclone frequency within each of the ocean basins they affect. Regional variability occurs on all resolved time scales, and there is also some evidence of trends in certain measures of tropical cyclone energy, affecting many of these regions and perhaps the globe as well.

There are at least two reasons to be concerned with such variability. The first and most obvious is that tropical cyclones rank with flash floods as the most lethal and expensive natural catastrophes, greatly exceeding other phenomena such as earthquakes. In developed countries, such as the United States, they are enormously costly: Hurricane Katrina is estimated to have caused in excess of $80 billion 2005 dollars in damage and killed more than 1,500 people. Death and injury from tropical cyclones is yet higher in developing nations; for example, Hurricane Mitch of 1998 took more than 11,000 lives in Central America. Any variation or trend in tropical cyclone activity is thus of concern to coastal residents in affected areas, compounding trends related to societal factors such as changing coastal population.

A second, less obvious and more debatable issue is the possible feedback of tropical cyclone activity on the climate system itself. The inner cores of tropical cyclones have the highest specific entropy content of any air at sea level, and for this reason such air penetrates higher into the stratosphere than is the case with other storm systems. Thus tropical cyclones may play a role in injecting water, trace gases, and microscopic airborne particles into the upper

troposphere and lower stratosphere, though this idea remains largely unexamined. There is also considerable evidence that tropical cyclones vigorously mix the upper ocean, affecting its circulation and biogeochemistry, perhaps to the point of having a significant effect on the climate system. Since the current generation of coupled climate models greatly underresolves tropical cyclones, such feedbacks are badly underrepresented, if they are represented at all. For these reasons, it is important to quantify, understand, and predict variations in tropical cyclone activity. The following sections review current knowledge of these variations on various time scales.

2.2.3.1.2 Data Issues
Quantifying tropical cyclone variability is limited, sometimes seriously, by a large suite of problems with the historical record of tropical cyclone activity. In the North Atlantic and eastern North Pacific regions, responsibility for the tropical cyclone database rests with NOAA's National Hurricane Center (NHC), while in other regions, archives of hurricane activity are maintained by several organizations, including the U.S. Joint Typhoon Warning Center (JTWC), the Japan Meteorological Agency (JMA), the Hong Kong Observatory (HKO) and the Australian Bureau of Meteorology (BMRC). The data, known as "best track" data (Jarvinen *et al.*, 1984; Chu *et al.*, 2002), comprise a global historical record of tropical cyclone position and intensity, along with more recent structural information. Initially completed in real time, the best tracks are finalized by teams of forecasters who update the best track data at the end of the hurricane season in each ocean basin using data collected during and after each hurricane's lifetime.

It should first be recognized that the primary motivation for collecting data on tropical cyclones was initially to support real-time forecasts, and this remains the case in many regions today. From the 1970s onward, increasing emphasis has been placed on improving the archive for climate purposes and on extending the record back to include historical systems (*e.g.*, Lourensz, 1981; Neumann, 1993; Landsea *et al.*, 2004). Unfortunately, improvements in measurement and estimation techniques have often been implemented with little or no effort

to calibrate against existing techniques, and with poor documentation where such calibrations were done. Thus the available tropical cyclone data contain an inhomogeneous mix of changes in quality of observing systems, reporting policies, and the methods utilized to analyze the data. As one example, the Dvorak technique – a subjective method that is routinely applied in all ocean basins to estimate tropical cyclone intensity using satellite imagery – was not introduced until the early 1970s and has evolved markedly since then. The technique originally utilized visible satellite imagery and was based on subjective pattern recognition. At that time, intensity estimates could only be made during daylight hours. In the early- to mid 1980s, the Dvorak technique was significantly modified to include digital infrared satellite imagery (which is available 24 hours per day), and has become the *de facto* method for estimating intensity in the absence of aircraft reconnaissance. (See Chapter 4 for suggested measures to improve consistency).

Insufficient efforts in re-examining and quality controlling the tropical cyclone record on a year to year basis, particularly outside the Atlantic and eastern North Pacific regions, have resulted in substantial uncertainties when using best track data to calculate changes over time. Efforts are ongoing to reanalyze the historic best track data, but such a *posteriori* reanalyses are less than optimal because not all of the original data that the best track was based on are readily available.

Compared to earlier periods, tropical cyclonecounts are acceptable for application to long-term trend studies back to about 1945 in the Atlantic and 1970 in the Eastern Pacific (*e.g.,* Holland and Webster, 2007 and references therein), and back to about 1975 for the Western and Southern Pacific basins, thanks to earth-orbiting satellites (*e.g.,* Holland, 1981). Until the launch of MeteoSat-7 in 1998, the Indian Oceans were seen only obliquely, but storm counts may still be expected to be accurate after 1977. In earlier periods, it is more likely that storms could be missed entirely, especially if they did not pass near ships at sea or land masses. For the North Atlantic, it is likely that up to 3 storms per year were missing before 1900, dropping to zero by the early 1960s

(Holland and Webster, 2007; Chang and Guo, 2007). Estimates of the duration of storms are considered to be less reliable prior to the 1970s due particularly to a lack of good information on their time of genesis. Since the 1970s, storms were more accurately tracked throughout their lifetimes by geostationary satellites.

Estimates of storm intensity are far less reliable, and this remains true for large portions of the globe even today. Airborne hurricane reconnaissance flight became increasingly routine in the North Atlantic and western North Pacific regions after 1945, but was discontinued in the western North Pacific region in 1987. Some missions are today being conducted under the auspices of the government of Taiwan. However, airborne reconnaissance only samples a small fraction of storms, and then only over a fraction of their lifetimes; moreover, good, quantitative estimates of wind speeds from aircraft did not become available until the late 1950s. Beginning in the mid-1970s, tropical cyclone intensity has been estimated from satellite imagery. Until relatively recently, techniques for doing so were largely subjective, and the known lack of homogeneity in both the data and techniques applied in the post-analyses has resulted in significant skepticism regarding the consistency of the intensity estimates in the data set. This lack of temporal consistency renders the data suspect for identifying trends, particularly in metrics related to intensity.

Recent studies have addressed these known data issues. Kossin *et al.* (2007a) constructed a more homogeneous record of hurricane activity, and found remarkably good agreement in

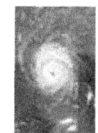

Compared to earlier periods, tropical cyclone counts are acceptable for application to long-term trend studies back to about 1945 in the Atlantic and 1970 in the Eastern Pacific.

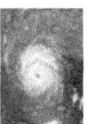

both variability and trends between their new record and the best track data in the North Atlantic and Eastern Pacific basins during the period 1983–2005. They concluded that the best track maintained by the NHC does not appear to suffer from data quality issues during this period. On the other hand, they were not able to corroborate the presence of upward intensity trends in any of the remaining tropical cyclone-prone ocean basins. This could be due to inaccuracies in the satellite best tracks, or could be due to the training of the Kossin *et al.* technique on North Atlantic data. The results of Kossin *et al.* (2007a) are supported by Wu *et al.* (2006), who considered Western Pacific best track data constructed by other agencies (HKMO and JMA) that construct best track data for the western North Pacific. Harper and Callaghan (2006) report on reanalyzed data from the Southeastern Indian Ocean and showed some biases, but an upward intensity trend remains. These studies underscores the need for improved care in analyzing tropical cyclones and in obtaining better understanding of the climatic controls of tropical cyclone activity beyond SST-based arguments alone.

The standard tropical cyclone databases do not usually contain information pertaining to the geometric size of tropical cyclones. Exceptions include the Australian region and the enhanced database for the North Atlantic over the last few decades. A measure of size of a tropical cyclone is a crucial complement to estimates of intensity as it relates directly to storm surge and damage area associated with landfalling storms. Such size measures can be inferred from aircraft measurements and surface pressure distributions, and can now be estimated from satellite imagery (*e.g.*, Mueller *et al.*, 2006; Kossin *et al.*, 2007b).

2.2.3.1.3 Low-frequency Variability and Trends of Tropical Cyclone Activity Indices

"Low frequency" variability is here defined as variations on time scales greater than those associated with ENSO (*i.e.*, more than three to four years). Several papers in recent years have quantified interdecadal variability of tropical cyclones in the Atlantic (Goldenberg *et al.*, 2001; Bell and Chelliah, 2006) and the western North Pacific (Chan and Shi, 1996), attributing most of the variability to natural decade-to-decade variability of regional climates in the Atlantic and Pacific, respectively. In the last few years, however, several papers have attributed both low frequency variability and trends in tropical cyclone activity to changing radiative forcing owing to human-caused particulates (sulfate aerosols) and greenhouse gases.

Emanuel (2005a) developed a "Power Dissipation Index" (PDI) of tropical cyclones, defined as the sum of the cubed estimated maximum sustained surface wind speeds at 6-hour intervals accumulated over each Atlantic tropical cyclone from the late 1940s to 2003. Landsea (2005) commented on the quality of data comprising the index, arguing that the PDI from the 1940s to the mid-1960s was likely underestimated due to limited coverage of the basin by aircraft reconnaissance in that era. An updated version of this analysis (Emanuel 2007), shown in Figure 2.13, confirms that there has been a substantial increase in tropical cyclone activity since about 1970, and indicates that the low-frequency Atlantic PDI variations are strongly correlated with low-frequency variations in tropical Atlantic SSTs. PDI, which integrates over time, is relatively insensitive to

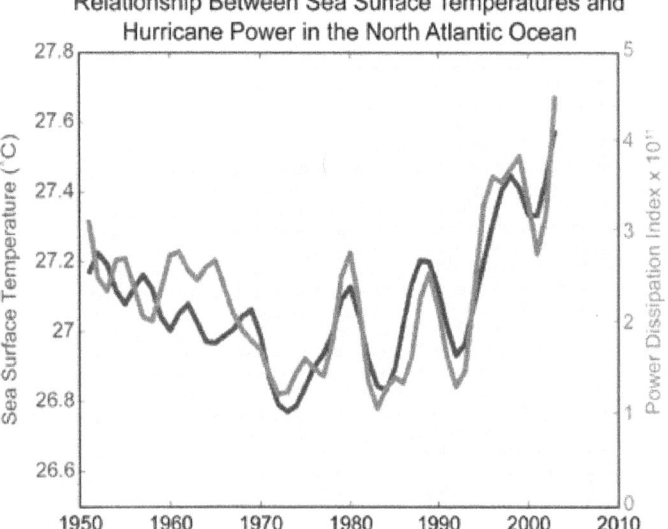

Figure 2.13 Sea surface temperatures (blue) correlated with the Power Dissipation Index for North Atlantic hurricanes (Emanuel, 2007). Sea Surface Temperature is from the Hadley Centre data set and is for the Main Development Region for tropical cyclones in the Atlantic, defined as 6-18°N, 20-60°W. The time series have been smoothed using a 1-3-4-3-1 filter to reduce the effect of interannual variability and highlight fluctuations on time scales of three years and longer.

random errors in intensity. Taking into account limitations in data coverage from aircraft reconnaissance and other issues, we conclude that it is likely that hurricane activity, as measured by the Power Dissipation Index (PDI), has increased substantially since the 1950s and '60s in association with warmer Atlantic SSTs. The magnitude of this increase depends on the adjustment to the wind speed data from the 1950s and '60s (Landsea 2005; Emanuel 2007). It is very likely that PDI has generally tracked SST variations on decadal time scales in the tropical Atlantic since 1950, and likely that it also generally tracked the general increase of SST. Confidence in these statistics prior to the late 1940s is low, due mainly to the decreasing confidence in hurricane duration and intensity observations. While there is a global increase in PDI over the last few decades it is not spatially uniform. For example, the PDI in the eastern Pacific has decreased since the early 1980s in contrast to the increase in the tropical Atlantic and some increasing tendency in the western Pacific. (Kossin *et al.*, 2007a; Emanuel *et al.*, 2008).

The Power Dissipation Index for U.S. landfalling tropical cyclones has not increased since the late 1800s (Landsea 2005). Pielke (2005) noted that there are no evident trends in observed damage in the North Atlantic region, after accounting for population increases and coastal development. However, Emanuel (2005b) notes that a PDI series such as Landsea's (2005), based on only U.S. landfalling data, contains only about 1 percent of the data that Emanuel's (2005a) basin-wide PDI contains, which is based on all storms over their entire lifetimes. Thus a trend in basin-wide PDI may not be detectable in U.S. landfalling PDI since the former index has a factor of 10 advantage in detecting a signal in a variable record (the signal-to-noise ratio).

Figure 2.14 (from Holland and Webster, 2007) indicates that there has been no distinct trend in the mean intensity of all Atlantic storms, hurricanes, and major hurricanes. A distinct increase in the most intense storms occurred around the time of onset of aircraft reconnaissance, but this is considered to be largely due to better observing methods. Holland and Webster also found that the overall proportion of hurricanes in the North Atlantic has remained remarkably

constant during the 20th century at around 50%, and there has been a marked oscillation in major hurricane proportions, which has no observable trend. Webster *et al.* (2005) reported that the number of category 4 and 5 hurricanes has almost doubled globally over the past three decades. The recent reanalysis of satellite data beginning in the early 1980s by Kossin *et al.* (2007a) support these results in the Atlantic, although the results in the remaining basins were not corroborated.

The recent Emanuel and Webster *et al.* studies have generated much debate in the hurricane research community, particularly with regard to homogeneity of the tropical cyclone data over time and the required adjustments (*e.g.,* Landsea 2005; Knaff and Sampson, 2006; Chan, 2006; Hoyos *et al.,* 2006; Landsea *et al.,* 2006; Sriver and Huber, 2006; Klotzbach, 2006; Elsner *et al.,* 2006; Maue and Hart, 2007; Manning and Hart,2007; Holland and Webster, 2007; Landsea, 2007; Mann *et al.,* 2007; Holland, 2007). Several of these studies argue that data problems preclude determination of significant trends in various tropical cyclone measures, while others provide further evidence in support of reported trends. In some cases, differences between existing historical data sets maintained by different nations can yield strongly contrasting results (*e.g.,* Kamahori *et al.,* 2006).

There is evidence that Atlantic tropical cyclone formation regions have undergone systematic long-term shifts to more eastward developments. These shifts affect track and duration, which subsequently affect intensity.

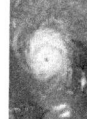

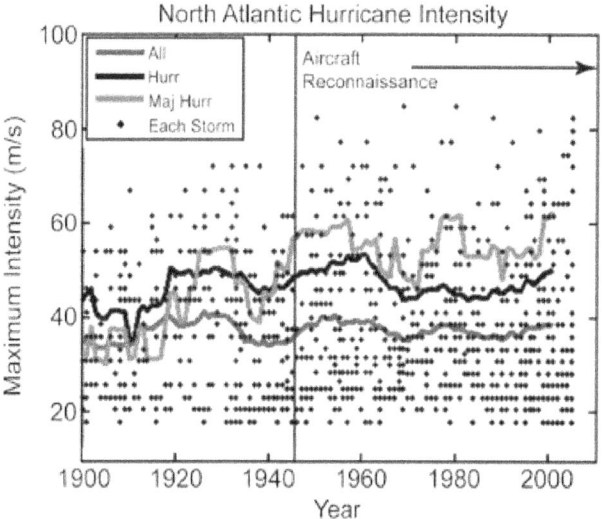

Figure 2.14 Century changes in the intensity of North Atlantic tropical cyclones, hurricanes, and major hurricanes. Also shown are all individual tropical cyclone intensities. (From Holland and Webster, 2007).

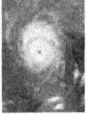

Atlantic tropical
cyclone counts
closely track
low-frequency
variations in tropical
Atlantic sea surface
temperatures,
including a long-
term increase since
the late 1800s and
early 1900s.

Several studies have examined past regional variability in tropical cyclone tracks (Wu *et al.,* 2005; Xie *et al.,* 2005; Vimont and Kossin, 2007; Kossin and Vimont, 2007). Thus far, no clear long-term trends in tracks have been reported, but there is evidence that Atlantic tropical cyclone formation regions have undergone systematic long-term shifts to more eastward developments (Holland, 2007). These shifts affect track and duration, which subsequently affects intensity. The modulation of the Atlantic tropical cyclone genesis region occurs through systematic changes of the regional SST and circulation patterns. Thus SST affects intensity not just through thermodynamic pathways that are local to the storms, but also through changes in basinwide circulation patterns (Kossin and Vimont, 2007).

In summary, we conclude that Atlantic tropical storm and hurricane destructive potential as measured by the Power Dissipation Index (which combines storm intensity, duration, and frequency) has increased. This increase is substantial since about 1970, and is likely substantial since the 1950s and '60s, in association with warming Atlantic sea surface temperatures.

2.2.3.1.4 Low-frequency Variability and Trends of Tropical Cyclone Numbers

Mann and Emanuel (2006) reported that Atlantic tropical cyclone counts closely track low-frequency variations in tropical Atlantic SSTs, including a long-term increase since the late 1800s and early 1900s (see also Figure 2.15 from Holland and Webster, 2007). There is currently debate on the relative roles of internal climate variability (*e.g.,* Goldenberg *et al.,* 2001) versus radiative forcing, including greenhouse gases, and sulfate aerosols (Mann and Emanuel, 2006; Santer *et al.,* 2006) in producing the multidecadal cooling of the tropical North Atlantic. This SST variation is correlated with reduced hurricane activity during the 1970s and '80s relative to the 1950s and '60s or to the period since 1995 (see also Zhang *et al.,* 2007).

On a century time scale, time series of tropical cyclone frequency in the Atlantic (Figure 2.15) show substantial interannual variability and a marked increase (of over 100%) since about 1900. This increase occurred in two sharp jumps of around 50%, one in the 1930s and another that commenced in 1995 and has not yet stabilized. Holland and Webster (2007) have

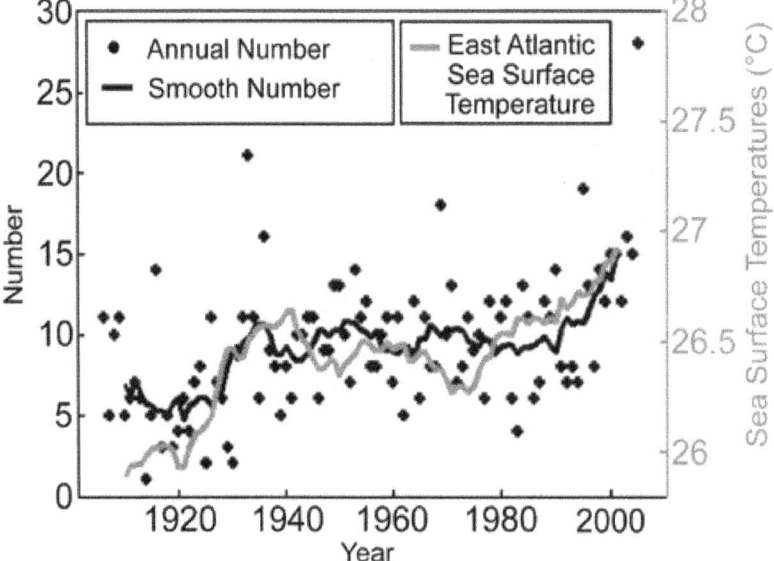

Figure 2.15 Combined annual numbers of hurricanes and tropical storms for the North Atlantic (black dots), together with a 9-year running mean filter (black line) and the 9-year smoothed sea surface temperature in the eastern North Atlantic (red line). Adapted from Holland and Webster (2007).

suggested that these sharp jumps are transition periods between relatively stable climatic periods of tropical cyclone frequency (Figure 2.15). Figure 2.15 uses unadjusted storm data—an issue which will be addressed further below.

For tropical cyclone frequency, the finding that the largest recorded increases over the past century have been in the eastern North Atlantic (*e.g.*, see recent analysis in Vecchi and Knutson, 2008; Holland, 2007), which historically has been the least well observed, has led to questions of whether this may be due to data issues (Landsea *et al.*, 2004; Landsea, 2007). The major observing system change points over the past century have been:

- The implementation of routine aircraft reconnaissance in 1944-45;
- The use of satellite observations and related analysis procedures from the late 1960s onwards; and
- A change in analysis practice by the National Hurricane Center from 1970 to include more mid-latitude systems.

In addition, there have been steady improvements in techniques and instrumentation, which may also introduce some spurious trends.

Landsea (2007) has used the fraction of tropical cyclones striking land in the satellite and pre-satellite era to estimate the number of missing tropical cyclones in the pre-satellite era (1900 to 1965) to be about 3.2 per year. He argued that since about 2002, an additional one tropical cyclone per year is being detected due to improved measurement tools and methods. His first adjustment (2.2 per year) assumes that the fraction of all tropical cyclones that strike land in the real world has been relatively constant over time, which has been disputed by Holland (2007). Holland also shows that the smaller fraction of tropical cyclones that made landfall during the past fifty years (1956-2005) compared to the previous fifty years (1906-1955) is related to changes in the main formation

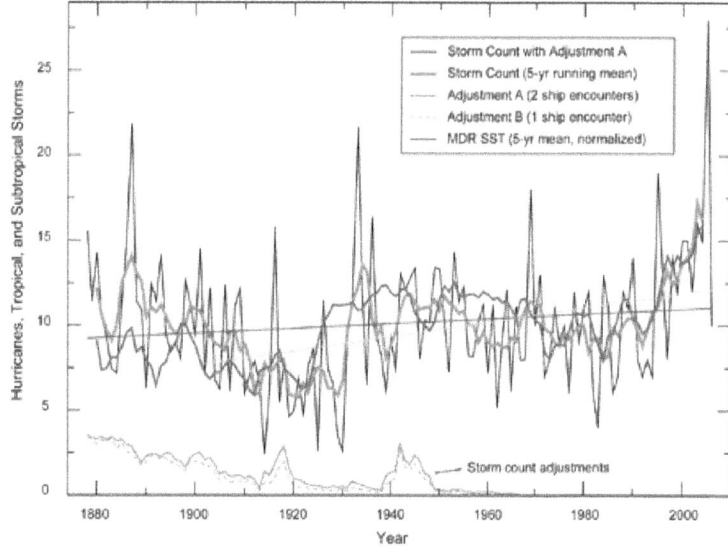

Atlantic Hurricanes/Tropical Storms (Adjusted for Estimated Missing Storms)

Figure 2.16 Atlantic hurricanes and tropical storms for 1878-2006, using the adjustment method for missing storms described in the text. Black curve is the adjusted annual storm count, red is the 5-year running mean, and solid blue curve is a normalized (same mean and variance) 5-year running mean sea surface temperature index for the Main Development Region of the tropical Atlantic (HadISST, 80-20°W, 10-20°N; Aug.-Oct.). Green curve shows the adjustment that has been added for missing storms to obtain the black curve, assuming two simulated ship-storm "encounters" are required for a modern-day storm to be "detected" by historical ship traffic for a given year. Straight lines are least squares trend lines for the adjusted storm counts. (Adapted from Vecchi and Knutson, 2008).

location regions, with a decrease in western Caribbean and Gulf of Mexico developments and an increase in the eastern Atlantic.

Alternative approaches to estimating the earlier data deficiencies have been used by Chang and Guo (2007), Vecchi and Knutson (2008), and Mann *et al.* (2007). The first two studies use historical ship tracks from the pre-satellite era, combined with storm track information from the satellite era, to infer an estimated adjustment for missing storms in the pre-satellite era (assumed as all years prior to 1965). Mann *et al.* used statistical climate relationships to estimate potential errors. Vecchi and Knutson found 2 to 3 storms per year were missing prior to 1900, decreasing to near zero by 1960. Chang and Guo found 1.2 storms missing around 1910, also decreasing to near zero by 1960. Mann *et al.* estimated a more modest undercount bias of 1 per year prior to 1970. The adjusted time series by Vecchi and Knutson (Figure 2.16) suggest a statistically significant (p=0.002 or less) positive linear trend in adjusted storm counts of 55% per century since 1900. However, beginning the trend from 1878, the trend through 2006 is

There has been an increase in both overall storm frequency and the proportion of major hurricanes since 1995. Taken together, these result in a very sharp increase in major hurricane numbers, which can be associated with changes of sea surface temperature.

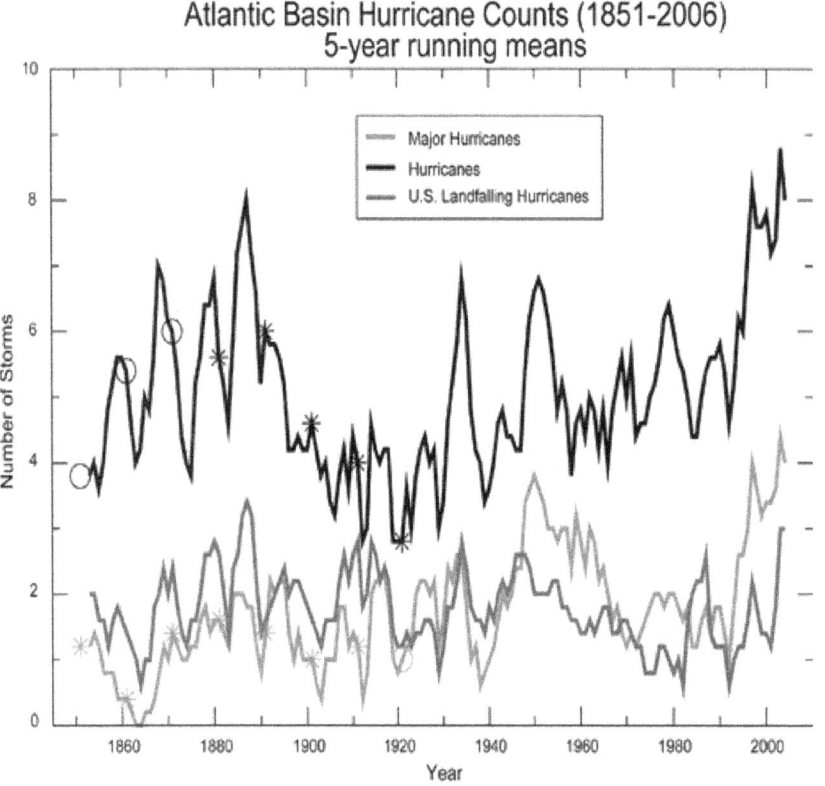

Figure 2.17 Counts of total North Atlantic basin hurricanes (black), major hurricanes (red), and U.S. landfalling hurricanes (blue) based on annual data from 1851 to 2006 and smoothed (using a 5-year running mean). Asterisks on the time series indicate years where trends beginning in that year and extending through 2005 are statistically significant (p=0.05) based on annual data; circles indicate non-significant trend results (data obtained from NOAA's Oceanographic and Meteorological Laboratory: http://www.aoml.noaa.gov/hrd/hurdat/ushurrlist18512005-gt.txt).

uncertainty in the late 1800s storm counts is greater than that during the 1900s.

Hurricane frequency closely follows the total tropical cyclone variability, with a stable 50% of all cyclones developing to hurricane strength over much of the past century (Holland and Webster, 2007). However, there has been a concomitant increase in both overall storm frequency and the proportion of major hurricanes since 1995. Taken together, these result in a very sharp increase in major hurricane numbers, which can be associated with changes of SST (Holland and Webster, 2007; Webster *et al.,* 2005). The PDI trend reported by Emanuel (2007) is largely due to this increase in major hurricane numbers.

Atlantic basin total hurricane counts, major hurricane counts, and U.S. landfalling hurricane counts as recorded in the HURDAT data base for the period 1851-2006 are shown in Figure 2.17. These have not been adjusted for missing storms, as there was likely less of a tendency to miss both hurricanes and major hurricanes in earlier years compared to tropical storms, largely because of their intensity and damage potential. However, even though intense storms were less likely than weaker systems to be missed entirely, lack of satellite coverage and other data issues imply that it would have been much more difficult to measure their maximum intensity accurately, leading to a potential undercount in the hurricane and major hurricane numbers. Using the unadjusted data, hurricane counts ending in 2005 and beginning in 1881 through 1921 increased and are statistically significant (p=0.05), whereas trends beginning in 1851 through 1871 are not statistically significant, owing to the high counts reported from 1851 to 1871. For major hurricanes, trends to 2005 beginning in 1851 through 1911 show

smaller (+15% per century) and not statistically significant at the p=0.05 level (p-value of about 0.2)[29]. It is notable that the degree of increase over the past century depends on the analysis methodology. When using a linear trend, as above, the increase from 1900 to 2005 is around 55% in the adjusted storm counts. However, using the essentially non-linear approach by Holland and Webster (2007) of separate climatic regimes, the increase in adjusted storm counts from the 1900-1920 regime to the 1995-2006 regime is 85%. The trend from 1900 begins near a local minimum in the time series and ends with the recent high activity, perhaps exaggerating the significance of the trend due to multidecadal variability. On the other hand, high levels of activity during the late 1800s, which lead to the insignificant trend result, are indirectly inferred in large part from lack of ship track data, and the

---

[29] Details of the statistical analysis are given in the Appendix, Example 5.

an increase and are statistically significant, whereas the trend beginning from 1921 also shows an increase but is not statistically significant[30]. The significant increase since 1900 in hurricane and major hurricane counts is supported by the significant upward trends in tropical storm counts since 1900 and the observation that hurricane and major hurricane counts as a proportion of total tropical cyclone counts are relatively constant over the long term (Holland and Webster, 2007). Regarding the trends from the 1800s, the lack of significant trend in hurricane counts from earlier periods is qualitatively consistent with the lack of significant trend in adjusted tropical storm counts from 1878 (Figure 2.16). For major hurricanes, the counts from the late 1800s, and thus the significant positive trends from that period, are considered less reliable, as the proportion of storms that reached major hurricane intensity, though relatively constant over the long-term in the 20th century, decreases strongly prior to the early 1900s, suggestive of strong data inhomogeneities. There is no evidence for a significant trend in U.S. landfalling hurricane frequency.

Regional storm track reconstructions for the basin (Vecchi and Knutson, 2008; Holland and Webster, 2007) indicate a decrease in tropical storm occurrence in the western part of the basin, consistent with the minimal change or slight decrease in U.S. landfalling tropical storm or hurricane counts. These analyses further suggest that—after adjustment for missing storms—an increase in basin-wide Atlantic tropical cyclone occurrence has occurred since 1900, with increases mainly in the central and eastern parts of the basin (also consistent with Chang and Guo, 2007). From a climate variability perspective, Kossin and Vimont (2007) have argued that the Atlantic Meridional Mode is correlated to a systematic eastward extension of the genesis region in the Atlantic. Kimberlain and Elsner (1998) and Holland and Webster (2007) have shown that the increasing frequency over the past 30 years is associated with a changeover to equatorial storm developments, and particularly to developments in the eastern equatorial region. Over the past century, the relative contributions of data fidelity issues

---

[30] Further details of the statistical analysis are given in Appendix A, Example 6.

(Landsea, 2007) versus real climatic modulation (Landsea *et al.*, 1999; Kimberlain and Elsner, 1998; Kossin and Vimont, 2007; Holland and Webster, 2007) to the apparent long-term increase in eastern Atlantic tropical storm activity is presently an open question.

In summary, we conclude that there have been fluctuations in the number of tropical storms and hurricanes from decade to decade, and data uncertainty is larger in the early part of the record compared to the satellite era beginning in 1965. Even taking these factors into account, it is likely that the annual numbers of tropical storms, hurricanes, and major hurricanes in the North Atlantic have increased over the past 100 years, a time in which Atlantic sea surface temperatures also increased. The evidence is less compelling for significant trends beginning in the late 1800s. The existing data for hurricane counts and one adjusted record of tropical storm counts both indicate no significant linear trends beginning from the mid- to late 1800s through 2005. In general, there is increasing uncertainty in the data as one proceeds back in time. There is no evidence for a long-term increase in North American mainland land-falling hurricanes.

2.2.3.1.5 Paleoclimate Proxy Studies of Past Tropical Cyclone Activity

Paleotempestology is an emerging field of science that attempts to reconstruct past tropical cyclone activity using geological proxy evidence or historical documents. This work attempts to expand knowledge about hurricane occurrence back in time beyond the limits

It is likely that the annual numbers of tropical storms, hurricanes, and major hurricanes in the North Atlantic have increased over the past 100 years, a time in which Atlantic sea surface temperatures also increased.

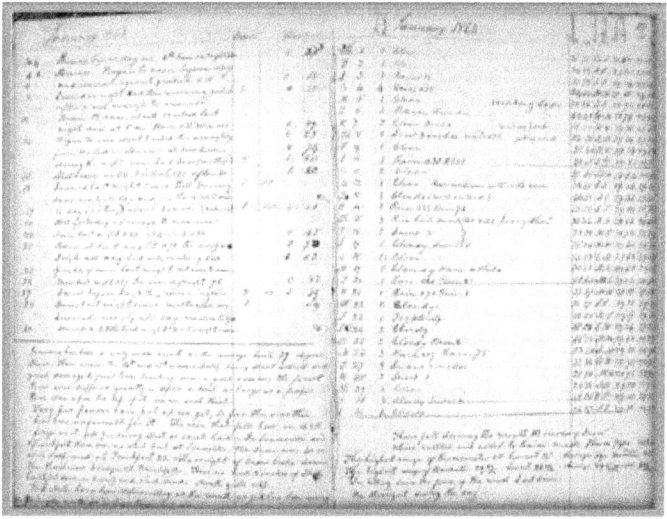

of conventional instrumental records, which cover roughly the last 150 years. A broader goal of paleotempestology is to help researchers explore physically based linkages between prehistoric tropical cyclone activity and other aspects of past climate.

Among the geologically-based proxies, overwash sand layers deposited in coastal lakes and marshes have proven to be quite useful (Liu and Fearn, 1993, 2000; Liu, 2004; Donnelly and Webb, 2004). Similar methods have been used to produce proxy records of hurricane strikes from back-barrier marshes in Rhode Island and New Jersey extending back about 700 years (Donnelly et al., 2001a, 2001b; Donnelly et al., 2004; Donnelly and Webb, 2004), and more recently in the Caribbean (Donnelly, 2005). Stable isotope signals in tree rings (Miller et al., 2006), cave deposits (Frappier et al., 2007) and coral reef materials are also being actively explored for their utility in providing paleoclimate information on tropical cyclone activity. Historical documents apart from traditional weather service records (newspapers, plantation diaries, Spanish and British historical archives, etc.) can also be used to reconstruct some aspects of past tropical cyclone activity (Ludlum, 1963; Millás, 1968; Fernández-Partagás and Diaz, 1996; Chenoweth, 2003; Mock, 2004; García Herrera et al., 2004; 2005; Liu et al., 2001; Louie and Liu, 2003, 2004).

Donnelly and Woodruff's (2007) proxy reconstruction of the past 5,000 years of intense hurricane activity in the western North Atlantic suggests that hurricane variability has been strongly modulated by El Niño during this time, and that the past 250 years has been relatively active in the context of the past 5,000 years. Nyberg et al. (2007) suggest that major hurricane activity in the Atlantic was anomalously low in the 1970s and 1980s relative to the past 270 years. As with Donnelly and Woodruff, their proxy measures were located in the western part of the basin (near Puerto Rico), and in their study, hurricane activity was inferred indirectly through statistical associations with proxies for vertical wind shear and SSTs.

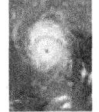

Combined with non-tropical storms, rising sea level extends the zone of impact from storm surge and waves farther inland, and will likely result in increasingly greater coastal erosion and damage.

2.2.3.2 STRONG EXTRATROPICAL CYCLONES OVERVIEW

Extra-tropical cyclone (ETC)[31] is a generic term for any non-tropical, large-scale low pressure storm system that develops along a boundary between warm and cold air masses. These types of cyclonic[32] disturbances are the dominant weather phenomenon occurring in the mid- and high latitudes during the cold season because they are typically large and often have associated severe weather. The mid-latitude North Pacific and North Atlantic basins, between approximately 30°N-60°N, are regions where large-numbers of ETCs develop and propagate across the ocean basins each year. Over land or near populous coastlines, strong or extreme ETC events generate some of the most devastating impacts associated with extreme weather and climate, and have the potential to affect large areas and dense population centers. A notable example was the blizzard of 12-14 March 1993 along the East Coast of the United States that is often referred to as the "super-storm" or "storm of the century"[33] (e.g., Kocin et al., 1995). Over the ocean, strong ETCs generate high waves that can cause extensive coastal erosion when combined with storm surge as they reach the shore, resulting in significant economic impact. Rising sea level extends the zone of impact

---

[31] The fundamental difference between the characteristics of extra-tropical and tropical cyclones is that ETCs have a cold core and their energy is derived from baroclinic instability, while tropical cyclones have a warm core and derive their energy from barotropic instability (Holton, 1979).

[32] A term applied to systems rotating in the counterclockwise direction in the Northern Hemisphere.

[33] The phrase "Storm of the Century" is also frequently used to refer to the 1991 Halloween ETC along the Northeast US coast, immortalized in the book and-movie The Perfect Storm (Junger, 1997).

from storm surge and waves farther inland, and will likely result in increasingly greater coastal erosion and damage from storms of equal intensity.

Studies of changes in strong ETCs and associated frontal systems have focused on locations where ETCs form, and the resulting storm tracks, frequencies, and intensities[34]. The primary constraint on these studies has been the limited period of record available that has the best observation coverage for analysis and verification of results, with most research focused on the latter half of the 20th century. Model reanalysis data is used in the majority of studies, either NCEP-NCAR (Kalnay *et al.*, 1996) or ERA-40 (Upalla *et al.*, 2005) data sets, although prior to 1965, data quality has been shown to be less reliable.

It is important to stress that any observed changes in ETC storm tracks, frequencies, or intensities are highly dependent on broad-scale atmospheric modes of variability, and the noise associated with this variability is large in relation to any observed linear trend. Therefore, detection and attribution of long-term (decade-to-century-scale) changes in ETC activity is extremely difficult, especially when considering the relatively short length of most observational records.

2.2.3.2.1 Variability of Extra-Tropical Cyclone Activity

Inter-annual and inter-decadal variability of ETCs is primarily driven by the location and other characteristics associated with the Polar jet stream. The mean location of the Polar jet stream is often referred to as the "storm track." The large-scale circulation is governed by the equator-to-pole temperature gradient, which is strongly modulated by SSTs over the oceans. The magnitude of the equator-to-pole temperature gradient is an important factor in determining the intensity of storms: the smaller (larger) the gradient in temperature, the smaller (larger) the potential energy available for extra-tropical cyclone formation. The observed intensity of ETCs at the surface is related to the amplitude of the large-scale circulation pattern,

with high-amplitude, negatively tilted troughs favoring stronger development of ETCs at the surface (Sanders and Gyakum, 1980).

From a seasonal perspective, the strongest ETCs are temporally out of phase in the Pacific and Atlantic basins, since the baroclinic wave energy climatologically reaches a peak in late fall in the North Pacific and in January in the North Atlantic (Nakamura, 1992; Eichler and Higgins, 2006). While it remains unclear what the physical basis is for the offset in peak storm activity between the two basins, Nakamura (1992) showed statistically that when the Pacific jet exceeds 45 m per second, there is a suppression of baroclinic wave energy, even though the low-level regional baroclinicity and strength of the Pacific jet are at a maximum. (This effect is not evident in the Atlantic basin, since the peak strength of the jet across the basin rarely exceeds 45 m per second). Despite the observed seasonal difference in the peak of ETC activity, Chang and Fu (2002) found a strong positive correlation between the Pacific and Atlantic storm tracks using monthly mean reanalysis data covering 51 winters (1949 to 1999). They found the correlations between the two basins remained positive and robust over individual months during winter (DJF) or over the entire season (Chang and Fu, 2002).

It has been widely documented that the track position, intensity, and frequency of ETCs is strongly modulated on inter-annual time-scales by different modes of variability, such as the El Niño/Southern Oscillation (ENSO) phenomenon (Gershunov and Barnett, 1998; An *et al.*, 2007). In a recent study, Eichler and Higgins (2006) used both NCEP-NCAR and ERA-40 reanalysis data to diagnose the behavior of ETC activity during different ENSO phases. Their results showed that during El Niño events, there is an equatorward shift in storm tracks in the North Pacific basin, as well as an increase of storm track activity along the United States East Coast. However, they found significant variability related to the magnitude of the El Niño event. During strong El Niños, ETC frequencies peak over the North Pacific and along the eastern United States, from the southeast coast to the Maritime Provinces of Canada (Eichler and Higgins, 2006), with a secondary track across the Midwest from the

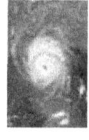

During El Niño events, there is an equatorward shift in storm tracks in the North Pacific basin, as well as an increase of storm track activity along the United States East Coast.

---

[34] These studies use *in situ* observations (both surface and upper-air), reanalysis fields, and Atmospheric-Ocean Global Climate Model (GCM) hind casts.

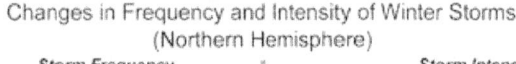

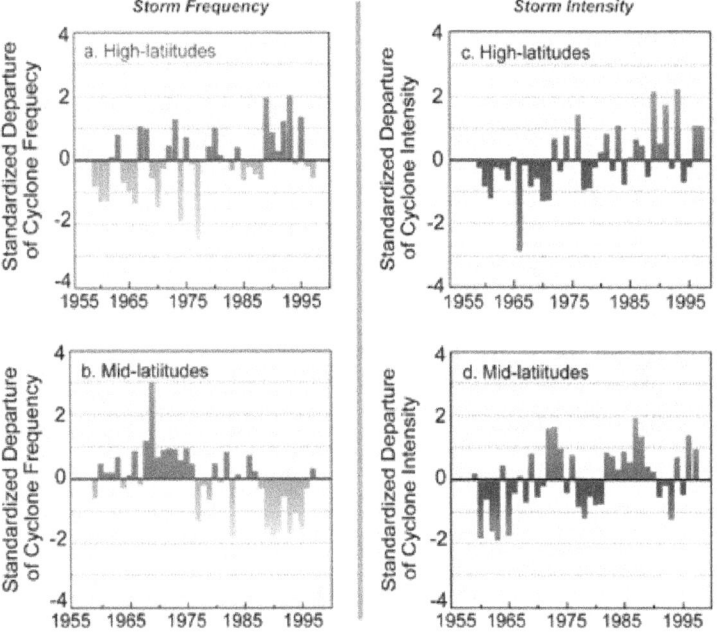

Figure 2.18 Changes from average (1959-1997) in the number of winter (Nov-Mar) storms each year in the Northern Hemisphere for: (a) high latitudes (60°-90°N), and (b) mid-latitudes (30°-60°N), and the change from average of winter storm intensity in the Northern Hemisphere each year for (c) high latitudes (60°-90°N), and (d) mid-latitudes (30°-60°N). (Adapted from McCabe *et al.*, 2001).

tive AO conditions, the polar vortex is weaker and cyclone activity shifts southward. Since the North Atlantic Oscillation (NAO) represents the primary component of the AO, it has a similar effect on storm track position, especially over the eastern North Atlantic basin (McCabe *et al.*, 2001).

2.2.3.2.2 Changes in Storm Tracks and Extra-Tropical Cyclone Characteristics

Many studies have documented changes in storm track activity. Specifically, a significant northward shift of the storm track in both the Pacific and Atlantic ocean basins has been verified by a number of recent studies that have shown a decrease in ETC frequency in mid-latitudes, and a corresponding increase in ETC activity in high latitudes (Wang *et al.*, 2006a; Simmonds and Keay, 2002; Paciorek *et al.*, 2002; Graham and Diaz, 2001; Geng and Sugi, 2001; McCabe *et al.*, 2001; Key and Chan, 1999; Serreze *et al.*, 1997). Several of these studies have examined changes in storm tracks over the entire Northern Hemisphere (McCabe *et al.*, 2001; Paciorek *et al.*, 2002; Key and Chan, 1999), while several others have focused on the storm track changes over the Pacific (Graham and Diaz, 2001) and Atlantic basins (Geng and Sugi, 2001), or both (Wang and Swail, 2001). Most of these studies focused on changes in frequency and intensity observed during winter (Dec., Jan., Feb.) or the entire cold season (Oct.-Mar.). However, for spring, summer, and autumn, Key and Chan (1999) found opposite trends in 1000-hPa and 500-hPa cyclone frequencies for both the mid- and high latitudes of the Northern Hemisphere. The standardized annual departures[35] of ETC frequency for the entire Northern Hemisphere over the period 1959-1997 (Figure 2.18a,b; McCabe *et al.*, 2001) shows that cyclone frequency has decreased for the mid-latitudes (30°-60°N) and increased for the high latitudes (60°-90°N).

lee of the Rocky Mountains to the Great Lakes. During weak to moderate El Niños, the storm tracks are similar to the strong El Niños, except there is a slight increase in the number of ETCs over the northern Plains, and the frequency of ETC activity decreases over the mid-Atlantic region. Similar to other previous studies (*e. g.* Hirsch *et al.*, 2001; Noel and Changnon, 1998), an inverse relationship typically exists during La Niñas; as the strength of La Niña increases, the frequency maxima of East Coast storms shifts poleward, the North Pacific storm track extends eastward toward the Pacific Northwest, and the frequency of cyclones increases across the Great Lakes region (Eichler and Higgins, (2006).

In addition to ENSO, studies have shown that the Arctic Oscillation (AO) can strongly influence the position of storm tracks and the intensity of ETCs. Previous studies have shown that during positive AO conditions, Northern Hemisphere cyclone activity shifts poleward (Serreze *et al.*, 1997; Clark *et al.*, 1999). Inversely, during nega-

A significant northward shift of the storm track in both the Pacific and Atlantic ocean basins has been verified by a number of recent studies.

---

[35] Standardized departures (*z* scores) were computed for each 5° latitudinal band by subtracting the respective 1959-1997 mean from each value and dividing by the respective 1959-1997 standard deviation (McCabe *et al.*, 2001).

For the 55-year period of 1948-2002, a metric called the Cyclone Activity Index (CAI)[36] was developed by Zhang *et al.* (2004) to document the variability of Northern Hemisphere cyclone activity. The CAI has increased in the Arctic Ocean (70°-90°N) during the latter half of the 20th century, while it has decreased in mid-latitudes (30°-60°N) from 1960 to 1993, which is evidence of a poleward shift in the average storm track position. Interestingly, the number and intensity of cyclones entering the Arctic from the mid-latitudes has increased, particularly during summer (Zhang *et al.*, 2004). The increasing activity in the Arctic was more recently verified by Wang *et al.* (2006a), who analyzed ETC counts by applying two separate cyclone detection thresholds to ERA-40 reanalysis of mean sea level pressure data. Their results showed an increase in high latitude storm counts, and a decrease in ETC counts in the mid-latitudes during the latter half of the 20th century.

Northern Hemisphere ETC intensity has increased over the period 1959-1997 across both mid- and high latitudes cyclone intensity (McCabe *et al.*, 2001; Figure 2.18c,d), with the upward trend more significant for the high latitudes (0.01 level) than for the mid-latitudes (0.10 level). From an ocean basin perspective, the observed increase in intense ETCs appears to be more robust across the Pacific than the Atlantic. Using reanalysis data covering the period 1949-1999, Paciorek *et al.* (2002) found that extreme wind speeds have increased significantly in both basins (Figure 2.19a,d). Their results also showed that the observed upward trend in the frequency of intense cyclones has been more pronounced in the Pacific basin (Figure 2.19c), although the inter-annual variability is much less in the Atlantic (Figure 2.19f). Surprisingly, they found that the overall counts of ETCs showed either no long-term change, or a decrease in the total number of cyclones (Figure 2.19b,e). However, this may be a result of the large latitudinal domain used in their

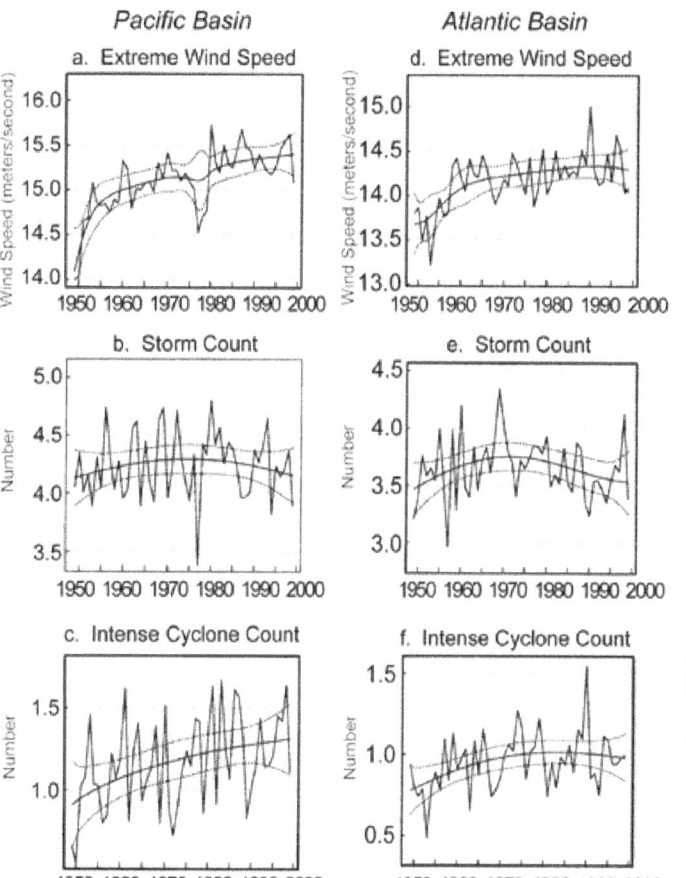

Winter Storm Characteristics for the Pacific and Atlantic

Figure 2.19 Extreme wind speed (meters per second), number of winter storms, and number of intense (≤980 hPa) winter storms for the Pacific region (20°-70°N, 130°E-112.5°W; panels a-b-c) and the Atlantic region (20°-70°N, 7.5°E-110°W; panels d-e-f). The thick smooth lines are the trends determined using a Bayesian spline model, and the thin dashed lines denote the 95% confidence intervals. (Adapted from Paciorek *et al.*, 2002).

study (20°-70°N), which included parts of the tropics, sub-tropics, mid- and high latitudes.

On a regional scale, ETC activity has increased in frequency, duration, and intensity in the lower Canadian Arctic during 1953-2002, with the most statistically significant trends during winter[37] (p=0.05 level; Wang *et al.*, 2006b). In contrast to the Arctic region, cyclone activity was less frequent and weaker along the southeast and southwest coasts of Canada. Winter cyclone deepening rates (*i.e.*, rates of intensification) have increased in the zone around 60°N, but decreased further south in the Great Lakes area and southern Prairies-British Columbia region

Extreme wind speeds in storms outside the tropics have increased significantly in both the Atlantic and Pacific basins

---

[36] The CAI integrates information on cyclone intensity, frequency, and duration into a comprehensive index of cyclone activity. The CAI is defined as the sum over all cyclone centers, at a 6-hourly resolution, of the differences between the cyclone central SLP and the climatological monthly mean SLP at corresponding grid points in a particular region during the month (Zhang *et al.*, 2004).

[37] Results based on hourly average sea level pressure data observed at 83 stations.

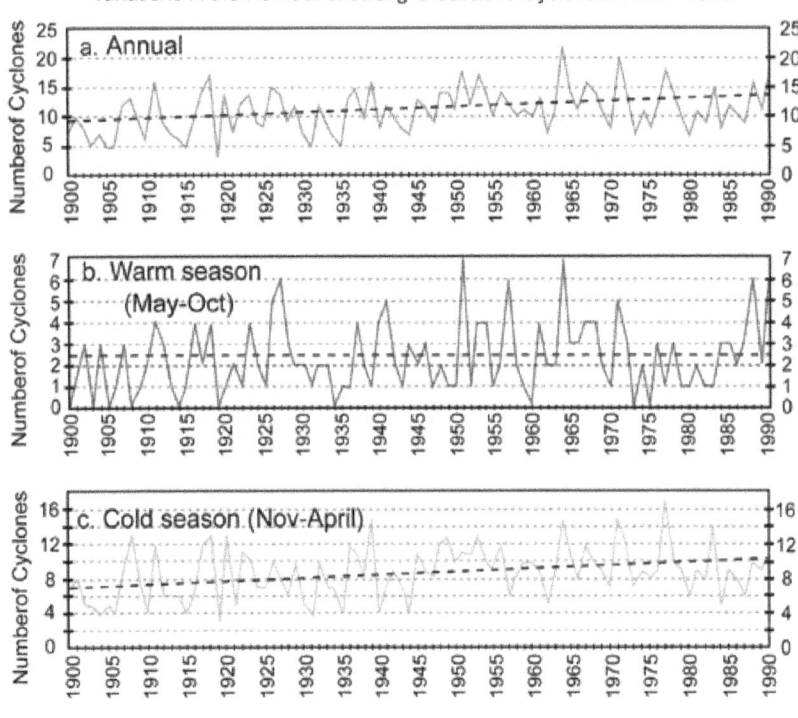

Figure 2.20 Time series of the number of strong (≤992 hPa) cyclones across the Great Lakes region (40°-50°N, 75°-93°W) over the period 1900-1990 for (a) Annual, (b) Warm season (May-Oct), and (c) Cold season (Nov-Apr). All trends were significant at the 95% level. (Adapted from Angel and Isard, 1998).

(*e.g.,* Lewis, 1987; Harman *et al.,* 1980; Garriott, 1903). Over the period 1900 to 1990, the number of strong cyclones (≤992 mb) increased significantly across the Great Lakes (Angel and Isard, 1998). This increasing trend was evident (at the p=0.05 level) both annually and during the cold season (Figure 2.20). In fact, over the 91-year period analyzed, they found that the number of strong cyclones per year more than doubled during both November and December.

In addition to studies using reanalysis data, which have limited record lengths, other longer-term studies of the variability of *storminess* typically use wave or water level measurements as proxies for storm frequency and intensity. Along the United States West Coast, one of the longest continuous climate-related instrumental time series in existence is the hourly tide gauge record at San Francisco that dates back to 1858. A derived metric called non-tide residuals (NTR)[40], which are related to broad-scale atmospheric circulation patterns across the eastern North Pacific that affect storm track location, provides a measure of *storminess* variability along the California coast (Bromirski *et al.,* 2003). Average monthly variations in NTR, which are associated with the numbers and intensities of all ETCs over the eastern North Pacific, did not change substantially over the period 1858-2000 or over the period covered by most ETC reanalysis studies (1951-2000). However, the highest 2% of extreme winter NTR (Figure 2.21), which are related to the intensity of the most extreme ETCs, had a significant upward trend since approximately 1950, with a pronounced quasi-periodic decadal-scale variability that is relatively consistent over the last 140 years. Changes in storm inten-

of Canada. This is also indicative of a pole-ward shift in ETC activity, and corresponding weakening of ETCs in the mid latitudes and an increase in observed intensities in the high latitudes. For the period of 1949-1999, the intensity of Atlantic ETCs increased from the 1960s to the 1990s during the winter season[38] (Harnik and Chang, 2003). Their results showed no significant trend in the Pacific region, but this is a limited finding because of a lack of upper-air (*i.e.,* radiosonde) data over the central North Pacific[39] in the region of the storm track peak (Harnik and Chang, 2003).

There have been very few studies that have analyzed the climatological frequencies and intensities of ETCs across the central United States, specifically in the Great Lakes region

> Over the period 1900 to 1990, the number of strong storms increased significantly across the Great Lakes, more than doubling for both November and December.

---

[38] Results based on gridded rawinsonde observations covering the Northern Hemisphere.

[39] Besides the few radiosonde sites located on islands (*i.e.,* Midway or the Azores), most upper-air observations over the vast expanses of the North Pacific and Atlantic are from automated pilot reports (pireps) that measure temperature, wind speed, and barometric pressure onboard commercial aircraft traveling at or near jet stream level (between 200-300 hPa).

---

[40] Non-tide residuals are obtained by first removing the known tidal component from the water level variations using a spectral method; then, variations longer than 30 days and shorter than 2.5 days are removed with a bandpass filter.

sity from the mid-1970s to early 1980s are also suggested by substantial pressure decreases at an elevation above sea level of about 3000 m over the eastern North Pacific and North America (Graham, 1994), indicating that the pattern of variability of extreme storm conditions observed at San Francisco (as shown in Figure 2.21) likely extends over much of the North Pacific basin and the United States. The oscillatory pattern of variability is thought to be

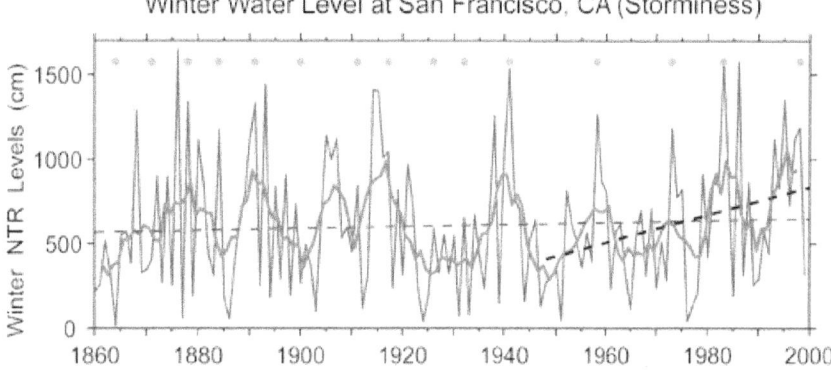

Winter Water Level at San Francisco, CA (Storminess)

Figure 2.21 Cumulative extreme Non-Tide Residuals (NTR) water levels exceeding the hourly 98th percentile NTR levels at San Francisco, during winter months (Dec-Mar), with the 5-yr running mean (red line). Least squares trend estimates for the entire winter record (light dashed line; not significant) and since 1948 (bold dashed line; significant at the 97.5% level), the period covered by NCEP reanalysis and ERA-40 data used in most ETC studies. (Adapted from Bromirski *et al.*, 2003).

influenced by teleconnections from the tropics, predominately during ENSO events (Trenberth and Hurrell, 1994), resulting in a deepened Aleutian low shifted to the south and east that causes both ETC intensification and a shift in storm track. It is interesting to note that peaks in the 5-year moving average in Figure 2.21 generally correspond to peaks in extreme rainfall in Figure 2.10, suggesting that the influence of El Niño and broad-scale atmospheric circulation patterns across the Pacific that affect sea level variability along the West Coast are associated with storm systems that affect rainfall variability across the United States.

The amplitude and distribution of ocean wave energy measured by ocean buoys is determined by ETC intensity and track location. Changes in long period (>12 sec), intermediate period, and short period (<6 sec) components in the wave-energy spectra permit inferences regarding the changes over time of the paths of the storms, as well as their intensities and resulting wave energies (Bromirski *et al.*, 2005). Analysis of the combination of observations from several buoys in the eastern North Pacific supports a progressive northward shift of the dominant Pacific storm tracks to the central latitudes (section 2.2.3.3).

2.2.3.2.3 Nor'easters

Those ETCs that develop and propagate along the East Coast of the United States and southeast Canada are often termed colloquially as

Nor'easters[41]. In terms of their climatology and any long-term changes associated with this subclass of ETCs, there are only a handful of studies in the scientific literature that have analyzed their climatological frequency and intensity (Jones and Davis 1995), likely due to a lack of any formal objective definition of this important atmospheric phenomenon (Hirsch *et al.*, 2001).

Because waves generated by ETCs are a function of storm size and the duration and area over which high winds persist, changes in significant wave heights can also be used as a proxy for changes in Nor'easters. Using hindcast wave heights and assigning a minimum criterion of open ocean waves greater than 1.6 m in height (a commonly used threshold for storms that caused some degree of beach erosion along the mid-Atlantic coast) to qualify as a nor'easter, the frequency of nor'easters along the Atlantic coast peaked in the 1950s, declined to a minimum in the 1970s, and then increased again to the mid-1980s (Dolan *et al.*, 1988; Davis *et al.*, 1993).

An alternate approach (Hirsch *et al.*, 2001) uses the NCEP-NCAR reanalysis pressure field to determine the direction of movement and wind speed to identify storm systems, and generically names them as East Coast Winter

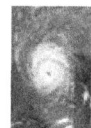

The pattern of variability of extreme storm conditions observed at San Francisco (as shown in Figure 2.21) likely extends over much of the North Pacific basin and the United States.

---

[41] According to the *Glossary of Meteorology* (Huschke, 1959), a *nor'easter* is any cyclone forming within 167 km of the East Coast between 30°-40°N and tracking to the north-northeast.

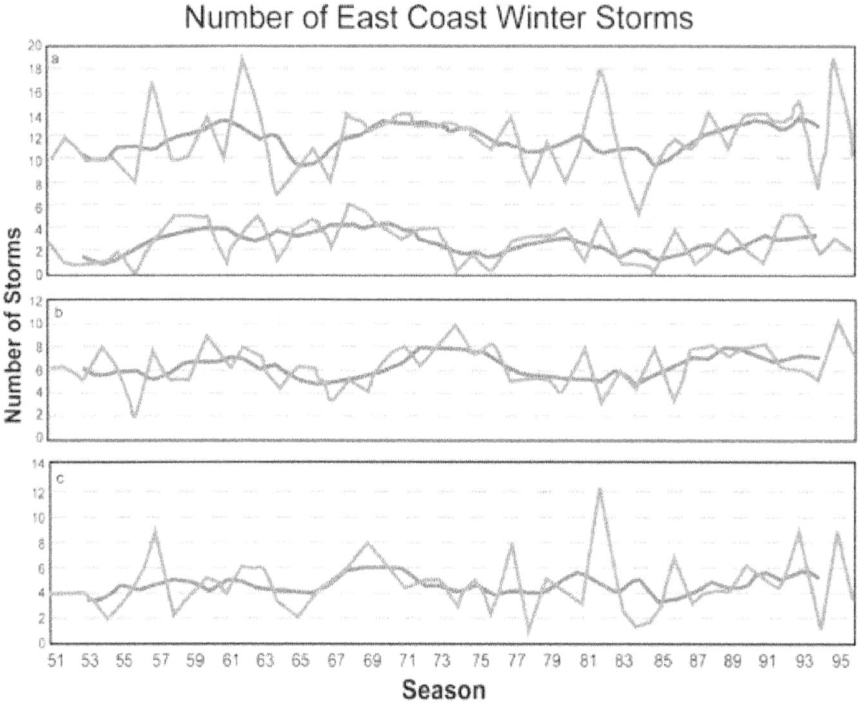

Number of East Coast Winter Storms

Figure 2.22 Seasonal totals (gray line) covering the period of 1951-1997 for: (a) all East Coast Winter Storms (ECWS; top curve) and strong ECWS (bottom curve); (b) northern ECWS (>35°N); and (c) those ECWS tracking along the full coast. Data points along the 5-year moving average (black) correspond to the middle year. (Adapted from Hirsch *et al.*, 2001).

The damage from a storm's waves depends to a large extent on whether it makes landfall, when the elevated water levels of its surge combine with the high waves to produce coastal erosion and flooding.

Storms (ECWS)[42]. An ECWS is determined to be "strong" if the maximum wind speed is greater than 23.2 m s^{-1} (45 kt). During the period of 1951-1997, their analysis showed that there were an average of 12 ECWS events occurring each winter (October-April), with a maximum in January, and an average of three strong events (Figure 2.22a). They found a general tendency toward weaker systems over the past few decades, based on a marginally significant (at the p=0.1 level) increase in average storm minimum pressure (not shown). However, their analysis found no statistically significant trends in ECWS frequency for all nor'easters identified in their analysis, specifically for those storms that occurred over the northern portion of the domain (>35°N), or those that traversed full coast (Figure 2.22b, c) during the 46-year period of record used in this study.

---

[42] According to Hirsch *et al.* (2001), in order to be classified as an ECWS, an area of low pressure is required to (1) have a closed circulation; (2) be located along the east coast of the United States, within the box bounded at 45°N by 65° and 70°W and at 30°N by 75° and 85°W; (3) show general movement from the south-southwest to the north-northeast; and (4) contain winds greater than 10.3 meters per second (20 kt) during at least one 6-hour period.

2.2.3.3 COASTAL WAVES: TRENDS OF INCREASING HEIGHTS AND THEIR EXTREMES

The high wind speeds of hurricanes and extratropical storms generate extremes in the heights and energies of ocean waves. Seasonal and long-term changes in storm intensities and their tracks produce corresponding variations in wave heights and periods along coasts, defining their wave climates (Komar *et al.*, 2008). Waves generated by extratropical storms dominate the oceans at higher latitudes, including the Northeast Pacific along the shores of Canada and the west coast of the United States, and along the Atlantic shores of North America where they occur as destructive Nor'easters. Tropical cyclones dominate the coastal wave climatologies at lower latitudes during the warm season (June-September), including the southeast Atlantic coast of the United States, the Gulf of Mexico, and the Caribbean, while tropical cyclones in the East Pacific generate waves along the western shores of Mexico and Central America. In addition, from mid- to late autumn, tropical cyclones sometimes combine with extratropical storms to generate extreme waves

BOX 2.2: Extreme Coastal Storm Impacts: "The Perfect Storm" as a True Nor'easter

From a coastal impacts perspective, damage is greatest when large storms are propagating *towards* the coast, which generally results in both a larger storm surge and more long period wave energy (resulting in greater run-up, causing more beach/coastal erosion/damage). Storm intensity (wind speed) is usually greatest in the right-front quadrant of the storm (based on the cyclone's forward movement), so the typical track of east coast winter storms propagating parallel to the coast leaves the most intense part of the storm out to sea. In contrast to storms propagating parallel to the coast, Nor'easters (such as "the Perfect Storm") that propagate from east-to-west in a retrograde track at some point in their lifetime (Figure Box 2.2) can generate much greater surge and greater long period wave energy, and also potentially have the most intense associated winds making landfall along the coast.

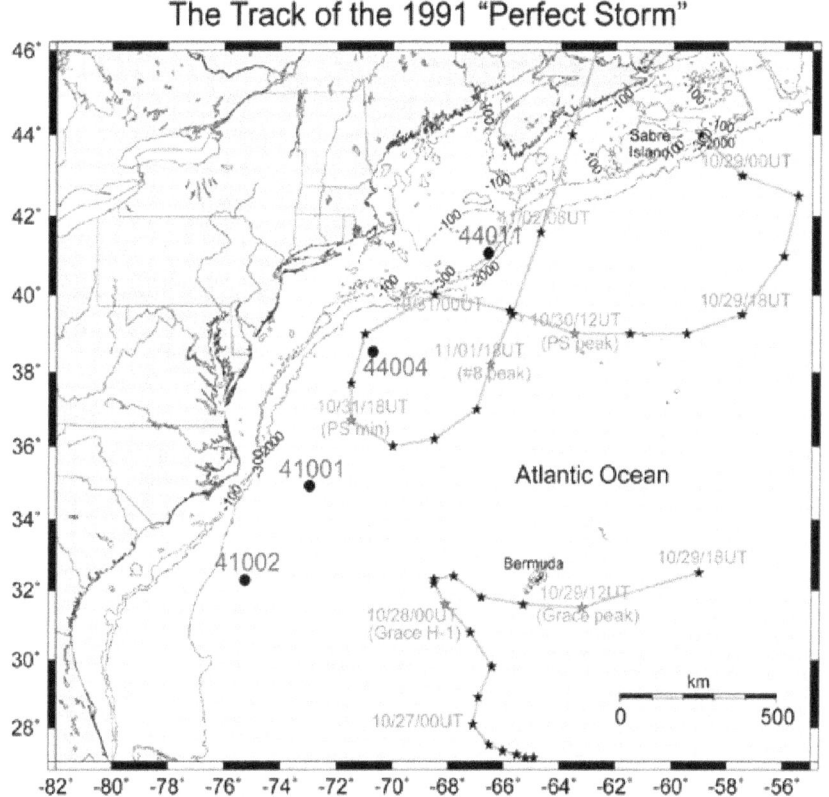

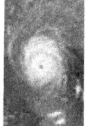

Figure Box 2.2 Track of the October 1991 "Perfect Storm" (PS) center showing the east-to-west retrograde propagation of a non-typical Nor'easter. The massive ETC was reenergized as it moved southward by absorbing northward propagating remnants of Hurricane Grace, becoming unnamed Hurricane #8 and giving rise to the name "Perfect Storm" for this composite storm. Storm center locations with date/hour time stamps at 6-hour intervals are indicated by stars. Also shown are locations of open ocean NOAA buoys that measured the extreme waves generated by these storms. (Adapted from Bromirski, 2001).

(Bromirski, 2001). The damage from a storm's waves depends to a large extent on whether it makes landfall, when the elevated water levels of its surge combine with the high waves to produce coastal erosion and flooding.

2.2.3.3.1 The Waves of Extratropical Storms and Hurricanes

The heights of waves generated by a storm depend on its wind speeds, the area over which the winds blow (the storm's fetch), and on the duration of the storm, factors that govern the amount of energy transferred to the waves. The resulting wave energy and related heights have been estimated from: (1) direct measurements

by buoys; (2) visual observations from ships; (3) hindcast analyses[43]; and (4) in recent years from satellite altimetry. The reliability of the data ranges widely for these different sources, depending on the collection methodology and processing techniques. However, multi-decadal records from these sources has made it possible to estimate progressive trends and episodic occurrences of extreme wave heights, and in some cases to establish their underlying causes.

In the Northern Hemisphere the hurricane winds are strongest on the right-hand side of the storm relative to its track, where the highest winds and rate of storm advance coincide, producing the highest waves within the cyclonic storm. Extreme wave heights are closely associated with the Saffir-Simpson hurricane classification, where the central atmospheric pressures are lower and the associated wind speeds are higher for the more-intense hurricane categories. Correlations have established that on average the measured wave heights and the central atmospheric pressures by Hsu *et al.* (2000) allows the magnitude of the significant wave height[44], H_S to be related to the hurricane category[45]. Estimates of the maximum H_S generated close to the wall of the hurricane's eye in the storm's leading right quadrant, where the wind speeds are greatest, range from about 6 to 7 m for Category 1 storms to 20 m and greater for Category 5. In response to the decreasing wind speeds outward from the eye of the storm,

Trends of increasing wave heights have been found along the U.S. Atlantic coast for waves generated by hurricanes.

H_S decreases by about 50% at a distance of approximately five times the radius of the eye, typically occurring at about 250 km from the storm's center (Hsu *et al.*, 2000).

This empirically-based model may under-predict the highest waves of Category 4 and 5 storms. Measurements of waves were obtained by six wave gauges deployed at depths of 60 to 90 m in the Gulf of Mexico, when the Category 4 Hurricane Ivan passed directly over the array on 15 September 2004 (Wang *et al.*, 2005). The empirical relationship of Hsu *et al.* (2000) yields a maximum H_S of 15.6 m for Ivan's 935-mb central pressure, seemingly in agreement with the 16 m waves measured by the gauges, but they were positioned about 30 km outward from the zone of strongest winds toward the forward face of Ivan rather than in its right-hand quadrant. Wang *et al.* (2005) therefore concluded it is likely that the maximum significant wave height was greater than 21 m for the Category 4 storm, with the largest individual wave heights having been greater than 40 m, indicating that the Hsu *et al.* (2000) empirical formula somewhat under-predicts the waves generated by high-intensity hurricanes. However, the study by Moon *et al.* (2003) has found that hurricane waves assessed from more complex models that use spatially distributed surface wind measurements compare well with satellite and buoy observations, both in deep water and in shallow water as hurricanes make landfall.

Increasing intensities of hurricanes or of extra-tropical storms should therefore on average be reflected in upward trends in their generated wave heights. This is evident in the empirical relationship of Hsu *et al.* (2000), which indicates that the significant wave heights should increase from on the order of 5 m to more than 20 m for the range of hurricane categories. Therefore, even a relatively modest increase in average hurricane intensities over the decades might be expected to be detected in measurements of their generated waves.

2.2.3.3.2 Atlantic Coast Waves

Trends of increasing wave heights have been found in wave-buoy measurements along the United States Atlantic coast for waves generated by hurricanes. Komar and Allan

---

[43] Hindcasts are model estimates of waves using forecast models that are run retrospectively using observed meteorological data.

[44] The "significant wave height" is a commonly used statistical measure for the waves generated by a storm, defined as the average of the highest one-third of the measured wave heights.

[45] Hsu *et al.* (2000) give the empirical formula $H_{smax}=0.2(P_n-P_c)$ where P_c and $P_n \sim 1013$ mbar are respectively the atmospheric pressures at the center and edge of the tropical cyclone, and H_{smax} is the maximum value of the significant wave height.

(2007a, 2008) analyzed data from three NOAA buoys located in deep water to the east of Cape May, New Jersey; Cape Hatteras, North Carolina; and offshore from Charleston, South Carolina. These buoys were selected due to their record lengths that began in the mid-1970s, and because the sites represent a range of latitudes where the wave climate is affected by both tropical hurricanes and extratropical storms. Separate analyses were undertaken for the winter season dominated by extratropical storms and the summer season of hurricanes[46], with only the latter having increased. Part of the difficulty in assessing trends in extreme wave heights stems from changes in temporal sampling of the buoy data and the sparse spatial coverage, so it is important to examine the consistency of the resulting buoy analyses with other related measures of wave heights, *e.g.*, hindcasts, changes in storm frequency, tracks, and intensity.

The analyses of the hurricane waves included trends in the annual averages of measured significant wave heights greater than 3 m, as in almost all occurrences they could be identified as having been generated by specific hurricanes (Komar and Allan, 2008). Of the three Atlantic buoys analyzed, the net rate of increase was greatest for that offshore from Charleston (0.59 m/decade, 1.8 m in 30 years). There were several years not included in these analyses due to application of the standard criterion, where a year is excluded, if more than 20% of its potential measurements are missing, the resulting gaps add some uncertainty to the analysis trends. However, the missing years, primarily occurring before 1990, are a mix of very quiet and moderately active hurricane

Number of Significant Wave Height Occurences

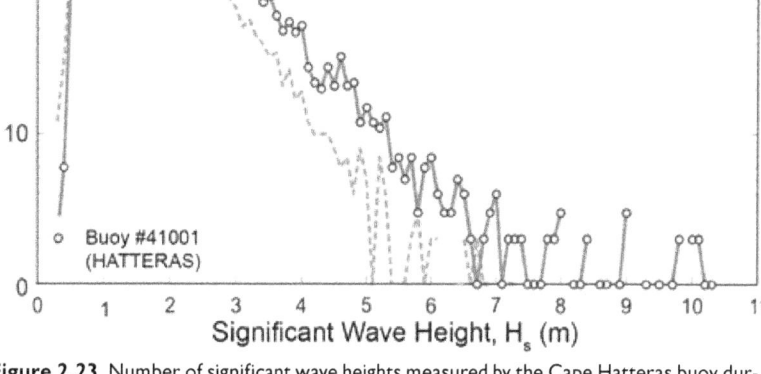

Figure 2.23 Number of significant wave heights measured by the Cape Hatteras buoy during the July-September season, early in its record 1976-1991 and during the recent decade, 1996-2005 (from Komar and Allan, 2007a,b).

years, thus the finding of an overall increase in significant wave heights is unlikely to have been affected by the missing data. Furthermore, the results are generally consistent with the increase in frequency of hurricane landfalls since the 1970s along the U.S. coasts (Figure 2.17) and the overall increase in the PDI in the Atlantic (Emanuel, 2005a).

The analyses by Komar and Allan (2007a, 2008) also included histograms of all measured significant wave heights for each buoy. Figure 2.23 shows the histograms for Cape Hatteras, one based on data from early in the buoy's record (1977-1990), the second from 1996-2005 (a period of increased hurricane activity in the Atlantic) to document the changes, including the most extreme measured waves[47]. Increases in numbers of occurrences are evident for significant wave heights greater than 2 to 3 m, with regularity in the increases extending up to about 6 m heights, whereas the histograms become irregular for the highest measured waves, the extreme but rare occurrences. Nearly all hurricanes passing through the north Atlantic (not having entered the Gulf of Mexico

The observed increase in the hurricane Power Dissipation Index since the 1970s is consistent with the measured increasing wave heights.

---

[46] The hurricane waves were analyzed for the months of July through September, expected to be dominated by tropical cyclones, while the waves of extratropical storms were based on the records from November through March; important transitional months such as October were not included, when both types of storms could be expected to be important in wave generation. Also, strict missing data criteria eliminated some years from the analysis that were both quiet and moderately active hurricane years.

[47] Traditionally a wave histogram is graphed as the percentages of occurrences, but in Figure 2.23 the actual numbers of occurrences for the range of wave heights have been plotted, using a log scale that emphasizes the most-extreme heights (Komar and Allan, 2007b)

The intensity of storms outside the tropics has increased in the North Pacific.

or Caribbean) produce measured significant wave heights in the range 3 to 6 m, and this accounts for the decadal trends of increasing annual averages. In contrast, the highest measured significant wave heights are sensitively dependent on the buoy's proximity to the path of the hurricane, being recorded only when the storm passes close to the buoy, with the measured wave heights tending to produce scatter in the data and occasional outliers in the annual averages. This scatter does not lend itself to any statistical conclusions about trends in these most extreme significant wave heights.

Increases in significant wave heights along the United States Atlantic coast therefore depend on changes from year to year in the numbers and intensities of hurricanes, and their tracks, *i.e.*, how closely the storms approached the buoys. Analyses by Komar and Allan (2008) indicate that all of these factors have been important to the observed wave-height increases. The observed increase in the PDI since the 1970s (Figure 2.13) and increases in land-falling hurricane frequency since the 1970s (Figure 2.17) are consistent with the measured increasing wave heights (Figure 2.23).

In contrast to the changes in hurricane waves, analyses of the winter wave heights generated by extratropical storms and recorded since the mid-1970s by the three buoys along the central United States Atlantic shore have shown little change (Komar and Allan, 2007a,b, 2008). This result is in agreement with the hindcast analyses by Wang and Swail (2001) based on the meteorological records of extratropical storms, analyzed with respect to changes in the 90th and 99th percentiles of the significant wave heights, thereby representing the trends for the more extreme wave conditions. A histogram of significant wave heights generated by extratropical storms, measured by the Cape Hatteras buoy during the winter, is similar to the 1996-2005 histogram in Figure 2.23 for the summer hurricane waves, with about the same 10 m extreme measurement (Komar and Allan, 2007b). Therefore, the summer and winter wave climates are now similar despite having different origins, although 30 years ago at the beginning of buoy measurements, the winter waves from extratropical storms were systematically higher than those generated by hurricanes, with the subsequent increase in the latter accounting for their present similarity.

2.2.3.3.3 Pacific Coast Waves
As reviewed earlier there is evidence that the intensities of extratropical storms have increased in the North Pacific. During strong El Niños, the storms intensify and follow tracks at more southerly latitudes as the Aleutian Low also intensifies and shifts southward (Mo and Livezey, 1986), in contrast to La Niña conditions when tracks are anomalously farther north (Bromirski *et al.*, 2005). These factors have been found to govern the heights of waves measured along the United States West Coast (Seymour *et al.*, 1984; Bromirski *et al.*, 2005; Allan and Komar, 2000). Allan and Komar (2006) analyzed the data from six buoys, from south-central California northward to the coast of Washington, including graphs for the decadal increases in the annual averages of the winter wave heights[48]. The highest rate of increase was found to have occurred on the Washington coast, slightly less offshore from Oregon, with northern to central California being a zone of transition having still lower rates of wave-height increases, until off the coast of south-central California there has not been a statistically significant change. It was further established that some of the "scatter" in the data above and below these linear regressions correlated with the Multivariate ENSO Index, showing that increased wave heights occurred at all latitudes along the U.S. Pacific Coast during major El Niños, but with

---

[48] "Winter" was taken as the months of October through March, the dominant season of significant storms and of relevance to coastal erosion.

the greatest increases along the shore of southern California where this cycle between El Niños and La Niñas exerts the primary climate control on the storm wave heights and their extremes (Allan and Komar, 2006; Cayan *et al.*, 2008).

More detailed analyses of the extreme wave heights measured off the Washington coast have been undertaken (important with respect to erosion), with example trends graphed in Figure 2.24 and a complete listing of the regression results in Table 2.1 (Allan and Komar, 2006). The annual averages of the winter wave heights and the averages of the five largest significant wave heights measured each winter have been increasing, the latter showing a higher rate of increase (0.95 m/decade, a 2.85 m increase in the significant wave heights in 30 years). The series of analyses demonstrate that the higher waves are increasing faster, at a rate of about 1 m/decade for the single highest measured significant wave height each year. This pattern of increase is reflected in progressive shifts in the histograms of measured significant wave heights, like that seen in Figure 2.23 for the East Coast hurricane-generated waves, with the orderly progression in Table 2.1 reflecting the increasing skewness of the histograms for the ranges of most extreme measured waves (Komar *et al.*, 2008).

While the linear trends for averages of the largest five storm-wave occurrences each year are statistically significant at the 0.05 significance level (Figure 2.24), the linear trends for the more extreme waves are not statistically significant (Table 2.1). However, in analyses using other statistical models, changes in the most extreme measured wave heights are also statistically significant (Komar *et al.*, 2008).

The identification of varying wave heights in the North Pacific affected by climate variations

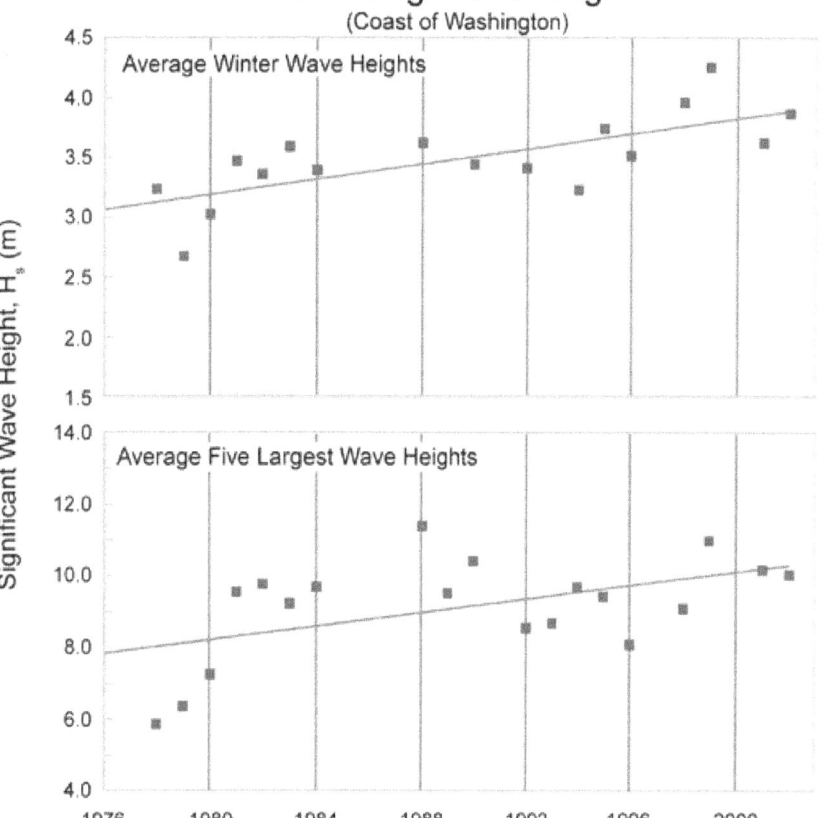

Figure 2.24 The trends of increasing wave heights measured by NOAA's National Data Buoy Center (NDBC) buoy #46005 off the coast of Washington (after Allan and Komar, 2006).

and changes is limited by the relatively short record length from the buoys, extending back only to the 1970s. Visual observations from ships in transit provide longer time series of wave height estimates, but of questionable quality. Gulev and Grigorieva (2004) examined this source of wave data for the North Pacific, finding that there has been a general increase in the significant wave heights beginning in about 1960, in agreement with that found by the wave buoy measurements (Figure 2.24). The wave hindcasts by Wang and Swail (2001) also show a general increase in the 90th and 99th percentile wave heights throughout the central to eastern North Pacific.

2.2.3.4 WINTER STORMS
2.2.3.4.1 Snowstorms
The amount of snow that causes serious impacts varies depending on a given location's usual snow conditions. A snowstorm is defined here as an event in which more than 15 cm of snow

There has been a northward shift in heavy snowstorm occurrence for the U.S. and Canada.

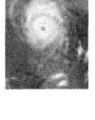

falls in 24 hours or less at some location in the United States. This is an amount sufficient to cause societally-important impacts in most locations. During the 1901-2001 period, 2,257 snowstorms occurred (Changnon *et al.*, 2006). Temporal assessment of the snowstorm incidences during 1901-2000 revealed major regional differences. Comparison of the storm occurrences in 1901-1950 against those in 1951-2000 revealed that much of the eastern United States had more storms in the early half of the 20th century, whereas in the West and New England, the last half of the century had more storms. Nationally, 53% of the weather stations had their peaks in 1901-1950 and 47% peaked in 1951-2000.

The South and lower Midwest had distinct statistically significant downward trends in snowstorm frequency from 1901 to 2000. In direct contrast, the Northeast and upper Midwest had statistically significant upward linear trends, although with considerable decade-to-decade variability. These contrasting regional trends suggest a net northward shift in snowstorm occurrence over the 20th century. Nationally, the regionally varying up and down trends resulted in a national storm trend that was not statistically significant for 1901-2000. Research has shown that cyclonic activity was low during 1931-1950, a period of few snowstorms in the United States.

In the United States, 39 of 231 stations with long-term records had their lowest frequencies of storms during 1931-1940, whereas 29 others had their peak of incidences then. The second ranked decade with numerous stations having low snowstorm frequencies was 1981-1990. Very few low storm occurrences were found during 1911-1920 and in the 1961-1980 period, times when storms were quite frequent. The 1911-1920 decade had the greatest number of high station values with 38 stations. The fewest peak values occurred in the next decade, 1921-1930. Comparison of the decades of high and low frequencies of snowstorms reveals, as expected, an inverse relationship. That is, when many high storm values occurred, there are few low storm frequencies.

Changes in heavy snowfall frequency have been observed in Canada (Zhang *et al.*, 2001). In southern Canada, the number of heavy snowfall events increased from the beginning of the 20th century until the late 1950s to the 1970s, then decreased to the present. By contrast, heavy snowfall events in northern Canada have been increasing with marked decade-to-decade variation.

The analyses of United States and Canadian heavy snowfall trends both suggest a northward shift in the tracks of heavy snowfall storms, but it is unclear why southern Canadian data reveal a decrease in heavy snowstorms since 1970. Specifically, the upward trend in heavy snowstorms in the United States over the past several decades is absent in adjacent southern Canada. To date, no analyses have been completed to better understand these differences.

In the United States, the decades with high snowstorm frequencies were characterized by cold winters and springs, especially in the West. The three highest decades for snowstorms (1911-1920, 1961-1970, and 1971-1980) were ranked 1st, 4th, and 3rd coldest, respectively, while the two lowest decades (1921-1930 and 1931-1940) were ranked as 3rd and 4th warmest. One exception to this general relationship is the warmest decade (1991-2000), which experienced a moderately high number of snowstorms across the United States.

Very snowy seasons (those with seasonal snowfall totals exceeding the 90th percentile threshold) were infrequent in the 1920s and 1930s and have also been rare since the mid-1980s (Kunkel *et al.*, 2007b). There is a high correlation with average winter temperature. Warm winters tend to have few stations with high snowfall totals, and most of the snowy seasons have also been cold.

Some of the snowiest regions in North America are the southern and eastern shores of the Great Lakes, where cold northwesterly winds flowing over the warmer lakes pick up moisture and deposit it on the shoreline areas. There is evidence of upward trends in snowfall since 1951 in these regions even while locations away from the snowy shoreline areas have not experienced increases (Burnett *et al.*, 2003). An analysis of historical heavy lake-effect snowstorms identified several weather conditions to be closely related to heavy lake-effect snowstorm occurrence including moderately high surface wind speed, wind direction promoting a long fetch over the lakes, surface air temperature in the range of -10 to 0°C, lake surface to air temperature difference of at least 7°C, and an unstable lower troposphere (Kunkel *et al.*, 2002). It is also necessary that the lakes be mostly ice-free.

lowest in 1973-1976. The 52-year linear trends for all three regions were downward over time. The time distributions for the Central, West North Central, and East North Central regions are alike, all showing that high values occurred early (1949-1956). All climate regions had their lowest FZRA during 1965-1976. The East North Central, Central, Northwest, and Northeast regions, which embrace the northern half of the conterminous United States, all had statistically significant downward linear trends. This is in contrast to trends in snowstorm incidences.

Snow cover extent for North America based on satellite data (Robinson *et al.*, 1993) abruptly decreased in the mid-1980s and generally has remained low since then (http://climate.rutgers.edu/snowcover/chart_anom.php?ui_set=0&ui_region=nam&ui_month=6).

2.2.3.4.2 Ice Storms
Freezing rain is a phenomenon where even light amounts can have substantial impacts. All days with freezing rain (ZR) were determined during the 1948-2000 period based on data from 988 stations across the United States (Changnon and Karl, 2003). The national frequency of freezing rain days (FZRA) exhibited a downward trend, being higher during 1948-1964 than in any subsequent period.

The temporal distributions of FZRA for three climate regions (Northeast, Southeast, and South) reveal substantial variability. They all were high in 1977-1980, low in 1985-1988, and

Both snowstorms and ice storms are often accompanied or followed by extreme cold because a strong ETC (which is the meteorological cause of the snow and ice) is one of the meteorological components of the flow of extreme cold air from the Arctic. This compounds the impacts of such events in a variety of ways, including increasing the risks to human health and adversely affecting the working environment for snow removal and repair activities. While there have been no systematic studies of trends in such compound events, observed variations in these events appear to be correlated. For example, the late 1970s were characterized both by a high frequency of extreme cold (Kunkel *et al.*, 1999a) and a high frequency of high snowfall years (Kunkel *et al.*, 2007b).

2.2.3.5 CONVECTIVE STORMS
Thunderstorms in the United States are defined to be severe by the National Weather Service (NWS) if they produce hail of at least 1.9 cm

The national frequency of freezing rain days has exhibited a downward trend, being higher during 1948-1964 than in any subsequent period.

75

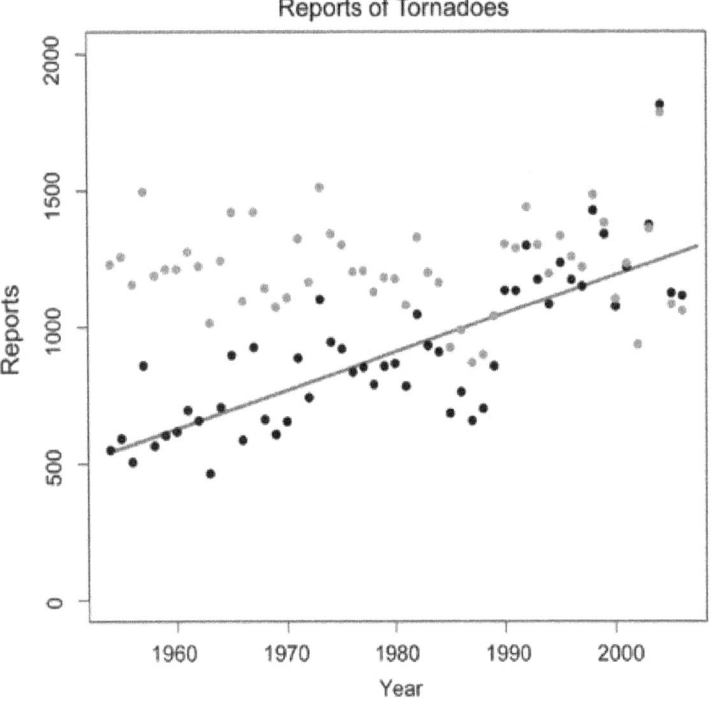

Reports of Tornadoes

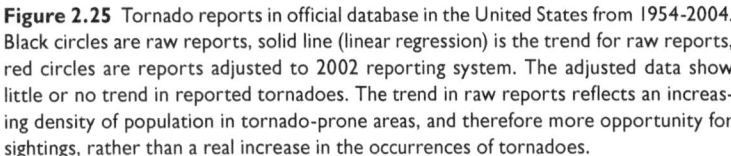

Figure 2.25 Tornado reports in official database in the United States from 1954-2004. Black circles are raw reports, solid line (linear regression) is the trend for raw reports, red circles are reports adjusted to 2002 reporting system. The adjusted data show little or no trend in reported tornadoes. The trend in raw reports reflects an increasing density of population in tornado-prone areas, and therefore more opportunity for sightings, rather than a real increase in the occurrences of tornadoes.

1955 for hail and wind. Prior to 1973, tornado reports were verified by state climatologists (Changnon, 1982). In addition, efforts to improve verification of severe thunderstorm and tornado warnings, the introduction of Doppler radars, changes in population, and increases in public awareness have led to increases in reports over the years. Changes in reporting practices have also led to inconsistencies in many aspects of the records (e. □, Brooks, 2004). Changnon and Changnon (2000) identified regional changes in hail frequency from reports made at official surface observing sites. With the change to automated surface observing sites in the 1990s, the number of hail reports at those locations dropped dramatically because of the loss of human observers at the sites. As a result, comparisons to the Changnon and Changnon work cannot be continued, although Changnon et al. (2001) have attempted to use insurance loss records as a proxy for hail occurrence.

The raw reports of annual tornado occurrences show an approximate doubling from 1954-2003 (Brooks and Dotzek, 2008), a reflection of the changes in observing and reporting. When detrended to remove this artificial trend, the data show large interannual variability, but a persistent minimum in the late 1980s (Figure 2.25). There were changes in assigning intensity estimates in the mid-1970s that resulted in tornadoes prior to 1975 being rated more strongly than those in the later part of the record (Verbout et al., 2006). More recently, there have been no tornadoes rated F5, the highest rating, since 3 May 1999, the longest gap on record. Coupled with a large decrease in the number of F4 tornadoes (McCarthy et al., 2006), it has been suggested that the strongest tornadoes are now being rated lower than practice prior to 2000.

A data set of F2 and stronger tornadoes extending back before the official record (Grazulis, 1993) provides an opportunity to examine longer trends. This examination[49] of the record

(3/4 inch) in diameter, wind gusts of at least 25.5 m per second (50 kt), or a tornado. Currently, reports come from a variety of sources to the local NWS forecast offices that produce a final listing of events for their area. Over the years, procedures and efforts to produce that listing have changed. Official data collection in near- real time began in 1953 for tornadoes and

[49] This analysis used the technique described in Brooks et al. (2003a) to estimate the spatial distribution over

from 1921-1995 indicates that the variability between periods was large, without significant long-term trends (Concannon *et al.*, 2000).

The fraction of strong tornadoes (F2 and greater) that have been rated as violent (F4 and greater) has been relatively consistent in the U.S. from the 1950s through the 1990s[50] (Brooks and Doswell, 2001)[51]. There were no significant changes in the high-intensity end of these distributions from the 1950s through the 1990s, although the distribution from 2000 and later may differ.

Reports of severe thunderstorms without tornadoes have increased even more rapidly than tornado reports (Doswell *et al.*, 2005, 2006). Over the period 1955-2004, this increase was approximately exponential, resulting in an almost 20-fold increase over the period. The increase is mostly in marginally severe thunderstorm reports (Brooks, 2007). An overall increase is seen, but the distribution by intensity is similar in the 1970s and post-2000 eras for the strongest 10% of reports of hail and wind. Thus, there is no evidence for a change in the severity of events, and the large changes in the overall number of reports make it impossible to detect if meteorological changes have occurred.

Environmental conditions that are most likely associated with severe and tornado-producing thunderstorms have been derived from reanalysis data (Brooks *et al.*, 2003b), and counts of the frequency of favorable environments for significant severe thunderstorms[52] have been determined for the area east of the Rocky Mountains in the U.S. for the period 1958-1999 (Brooks and Dotzek, 2008). The count of favorable environments decreased from the late 1950s into the early 1970s and increased after that through the 1990s, so that the frequency was approximately the same at both ends of the analyzed period. Given the high values seen at the beginning of the reanalysis era, it is

likely that the record is long enough to sample natural variability, so that it is possible that even though the 1973-1999 increase is statistically significant, it does not represent a departure from natural variability. The time series of the count of reports of very large hail (7 cm diameter and larger) shows an inflection at about the same time as the inflection in the counts of favorable environments. A comparison of the rate of increase of the two series suggested that the change in environments could account for approximately 7% of the change in reports from the mid-1970s through 1999, with the rest coming from non-meteorological sources. Changes in tornado reports do not correspond to the changes in overall favorable severe thunderstorm environment, in part because the discrimination of environments favorable for tornadoes in the reanalysis data is not as good as the discrimination of severe thunderstorm environments (Brooks *et al.*, 2003a).

There is no evidence for a change in the severity of tornadoes and severe thunderstorms, and the large changes in the overall number of reports make it impossible to detect if meteorological changes have occurred.

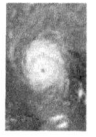

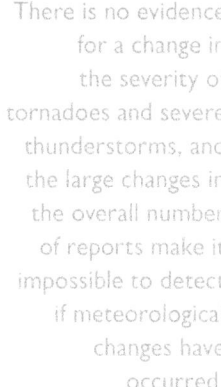

---

different periods.

[50] Note that consistent overrating will not change this ratio.

[51] Feuerstein *et al.* (2005) showed that the distribution in the U.S. and other countries could be fit to Weibull distributions with the parameters in the distribution converging as time goes along, which they associated with more complete reporting of events.

[52] Hail of at least 5 cm diameter, wind gusts of at least 33 m s[-1], and/or a tornado of F2 or greater intensity.

2.3 KEY UNCERTAINTIES RELATED TO MEASURING SPECIFIC VARIATIONS AND CHANGE

For trends beginning in the late 1800s the changing spatial coverage of the data set is a concern.

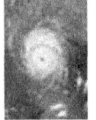

In this section, we review the statistical methods that have been used to assess uncertainties in studies of changing extremes. The focus of the discussion is on precipitation events, though similar methods have also been used for temperature.

2.3.1 Methods Based on Counting Exceedances Over a High Threshold

Most existing methods follow some variant of the following procedure, given by Kunkel *et al.* (1999b). First, daily data are collected, corrected for biases such as winter undercatchment. Only stations with nearly complete data are used (typically, "nearly complete" means no more than 5% missing values). Different event durations (for example, 1-day or 7-day) and different return periods (such as 1-year or 5-year) are considered. For each station, a threshold is determined according to the desired return value. For example, with 100 years of data and

a 5-year return value, the threshold is the 20th largest event. The number of exceedances of the threshold is computed for each year, and then averaged either regionally or nationally. The averaging is a weighted average in which, first, simple averaging is used over climate divisions (typically there are about seven climate divisions in each state), and then, an area-weighted average is computed over climate divisions, either for one of the nine U.S. climate regions or the whole contiguous United States. This averaging method ensures that parts of the country with relatively sparse data coverage are adequately represented in the final average. Sometimes (*e.*□Groisman *et al.*, 2005; Kunkel *et al.*, 2007a) the climate divisions are replaced by 1° by 1° grid cells. Two additional refinements used by Groisman *et al.* (2005) are: (i) to replace the raw exceedance counts for each year by anomalies from a 30-year reference period, computed separately for each station and (ii) to assess the standard error of the regional average using spatial statistics techniques. This calculation is based on an exponentially decreasing spatial covariance function with a range of the order 100-500 km and a nugget:sill ratio (the proportion of the variability that is not spatially correlated) between 0 and 85%, depending on the region, season, and threshold.

Once these spatially averaged annual exceedance counts or anomalies are computed, the next step is to compute trends. In most studies, the emphasis is on linear trends computed either by least squares regression or by the Kendall slope method, in which the trend is estimated as the median of all possible slopes computed from pairs of data points. The standard errors of the trends should theoretically be corrected for autocorrelation, but in the case of extreme events, the autocorrelation is usually negligible (Groisman *et al.*, 2004).

One of the concerns about this methodology is the effect of changing spatial coverage of the data set, especially for comparisons that go back to the late years of the 19th century. Kunkel *et al.* (2007a) generated simulations of the 1895-2004 data record by first randomly sampling complete years of data from a modern network of 6,351 stations for 1971-2000, projecting to a random subnetwork equivalent in size and spatial extent to the historical

data network, then using repeat simulations to calculate means and 95% confidence intervals for five 22-year periods. The confidence intervals were then superimposed on the actual 22-year means calculated from the observational data record. The results for 1-year, 5-year, and 20-year return values show clearly that the most recent period (1983-2004) has the highest return values of the five periods, but they also show the second highest return values in 1895-1916 with a sharp drop thereafter, implying a role that is still not fully explained due to natural variability.

Some issues that might justify further research include the following:

1. Further exploration of why extreme precipitation apparently decreases after the 1895-1916 period before the recent (post-1983) rise when they exceeded that level. For example, if one breaks the data down into finer resolution spatially, does one still see the same effect?

2. What about the effect of large-scale circulation effects such as ENSO events, AMO, PDO, *et*□? These could potentially be included as covariates in a time series regression analysis, thus allowing one to "correct" for circulation effects in measuring the trend.

3. The spatial analyses of Groisman *et al.* (2005) allow for spatial correlation in assessing the significance of trends, but they don't do the logical next step, which is to use the covariance function to construct optimal interpolations (also known as "kriging") and thereby produce more detailed spatial maps. This is something that might be explored in the future.

2.3.2 The GEV Approach

An alternative approach to extreme value assessment is through the Generalized Extreme Value (GEV) distribution[53] and its variants. The

GEV combines three "types" of extreme value distributions that in earlier treatments were often regarded as separate families (*e.*□, Gumbel, 1958). The distribution is most frequently applied to the annual maxima of a meteorological or hydrological variable, though it can also be applied to maxima over other time periods (*e.*□, one month or one season). With minor changes in notation, the distributions are also applicable to minima rather than maxima. The parameters may be estimated by maximum likelihood, though there are also a number of more specialized techniques such as L-moments estimation. The methods have been applied in climate researchers by a number of authors including Kharin and Zwiers (2000), Wehner (2004, 2005), Kharin *et al.* (2007).

The potential advantage of GEV methods over those based on counting threshold exceedances is that by fitting a probability distribution to the extremes, one obtains more information that is less sensitive to the choice of threshold, and can also derive other quantities such as the □-year return value □□, calculated by solving the equation □□□□)=1-1/T. Trends in the □-year return value (for typical values of □, *e.*□, 1, 10, 25, or 100 years) would be particularly valuable as indicators of changing extremes in the climate.

---

[53] The basic GEV distribution is given by the formula (see, *e.*□, Zwiers and Kharin, 1998):
$F(x) = exp\{-[1-k(x-\xi)/\alpha]^{1/k}$□ in which ξ plays the role of a centering or location constant, α determines the scale, and □ is a key parameter that determines the shape of the distribution. (Other authors have used different

notations, especially for the shape parameter.) The range of the distribution is:
□<$\xi+\alpha/k$ when □□□□>$\xi+\alpha/k$ when $k<0$, $-\infty<x<\infty$ when □□□ in which case the formula reduces to □□□ $= exp\{-exp[-(x-\xi)/\alpha]\}$ and is known as the Gumbel distribution.

Direct application of GEV methods is often inefficient because they only use very sparse summaries of the data (typically one value per year), and need reasonably long time series before they are applicable at all. Alternative methods are based on exceedances over thresholds, not just counting exceedances but also fitting a distribution to the excess over the threshold. The most common choice of distribution of excess is the Generalized Pareto distribution or GPD, which is closely related to the GEV (Pickands, 1975; Davison and Smith, 1990). Some recent overviews of extreme value distributions, threshold methods, and a variety of extensions are by Coles (2001) and Smith (2003).

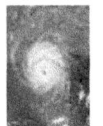

Much of the recent research (e.□, Wehner, 2005; Kharin *et al.*, 2007) has used model output data, using the GEV to estimate, for example, a 20-year return value at each grid cell, then plotting spatial maps of the resulting estimates. Corresponding maps based on observational data must take into account the irregular spatial distribution of weather stations, but this is also possible using spatial statistics (or kriging) methodology. For example, Cooley *et al.* (2007) have applied a hierarchical modeling approach to precipitation data from the Front Range of Colorado, fitting a GPD to threshold exceed-ances at each station and combining results from different stations through a spatial model to compute a map of 25-year return values. Smith *et al.* (2008) applied similar methodol-ogy to data from the whole contiguous United States, producing spatial maps of return values and also calculating changes in return values over the 1970-1999 period.

Causes of Observed Changes in Extremes and Projections of Future Changes

Convening Lead Author: William J. Gutowski, Jr., Iowa State Univ.

Lead Authors: Gabriele C. Hegerl, Univ. Edinburgh; Greg J. Holland, NCAR; Thomas R. Knutson, NOAA; Linda O. Mearns, NCAR; Ronald J. Stouffer, NOAA; Peter J. Webster, Ga. Inst. Tech.; Michael F. Wehner, Lawrence Berkeley National Laboratory; Francis W. Zwiers, Environment Canada

Contributing Authors: Harold E. Brooks, NOAA; Kerry A. Emanuel, Mass. Inst. Tech.; Paul D. Komar, Oreg. State Univ.; James P. Kossin, Univ. Wisc., Madison; Kenneth E. Kunkel, Univ. Ill. Urbana-Champaign, Ill. State Water Survey; Ruth McDonald, Met Office, United Kingdom; Gerald A. Meehl, NCAR; Robert J. Trapp, Purdue Univ.

KEY FINDINGS

Attribution of Observed Changes

Changes in some weather and climate extremes are attributable to human-induced emissions of greenhouse gases.

- Human-induced warming has likely caused much of the average temperature increase in North America over the past 50 years. This affects changes in temperature extremes.
- Heavy precipitation events averaged over North America have increased over the past 50 years, consistent with the observed increases in atmospheric water vapor, which have been associated with human-induced increases in greenhouse gases.
- It is very likely that the human-induced increase in greenhouse gases has contributed to the increase in sea surface temperatures in the hurricane formation regions. Over the past 50 years there has been a strong statistical connection between tropical Atlantic sea surface temperatures and Atlantic hurricane activity as measured by the Power Dissipation Index (which combines storm intensity, duration, and frequency). This evidence suggests a human contribution to recent hurricane activity. However, a confident assessment of human influence on hurricanes will require further studies using models and observations, with emphasis on distinguishing natural from human-induced changes in hurricane activity through their influence on factors such as historical sea surface temperatures, wind shear, and atmospheric vertical stability

Projected Changes

- Future changes in extreme temperatures will generally follow changes in average temperature:
  - Abnormally hot days and nights and heat waves are very likely to become more frequent.
  - Cold days and cold nights are very likely to become much less frequent.
  - The number of days with frost is very likely to decrease.
- Droughts are likely to become more frequent and severe in some regions as higher air temperatures increase the potential for evaporation.
- Over most regions, precipitation is likely to be less frequent but more intense, and precipitation extremes are very likely to increase.
- For North Atlantic and North Pacific hurricanes and typhoons:
  - It is likely that hurricane/typhoon wind speeds and core rainfall rates will increase in response to human-caused warming. Analyses of model simulations suggest that for each 1°C increase in tropical sea surface temperatures, hurricane surface wind speeds will increase by 1 to 8% and core rainfall rates by 6 to 18%.

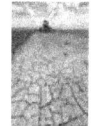

○ Frequency changes are currently too uncertain for confident projec-tions.
○ The spatial distribution of hurricanes/typhoons will likely change.
○ Storm surge levels are likely to increase due to projected sea level rise, though the degree of projected increase has not been adequately studied.

• There are likely to be more frequent deep low-pressure systems (strong storms) outside the tropics, with stronger winds and more extreme wave heights.

3.1 INTRODUCTION

Understanding physical mechanisms of extremes involves processes governing the timing and location of extreme behavior, such as El Niño-Southern Oscillation (ENSO) cycles, as well as the mechanisms of extremes themselves (e.□, processes producing heavy precipitation). This includes processes that create an environment conducive to extreme behavior, processes of the extreme behavior itself, and the factors that govern the timing and location of extreme events.

A deeper understanding of physical mechanisms is of course important for understanding why extremes have occurred in the past and for predicting their occurrence in the future. Climate models also facilitate better understanding of the physical mechanisms of climate change. However, because climate-change projections simulate conditions many decades into the future, strict verification of projections is not always possible. Other means of attaining

Climate models are important tools for understanding the causes of observed changes in extremes, as well as projecting future changes.

confidence in projections are therefore needed. Confidence in projected changes in extremes increases when the physical mechanisms producing extremes in models are consistent with observed behavior. This requires careful analysis of the observed record as well as model output. Assessment of physical mechanisms is also necessary to determine the realism of changes in extremes. Physical consistency of simulations with observed behavior is necessary though not sufficient evidence for accurate projections.

Climate model results are used throughout this report. State-of-the-art climate models are based on physical principles expressed as equations, which the model represents on a grid and integrates forward in time. There are also processes with scales too small to be resolved by the model's grid. The model represents these processes through combinations of observations and physical theory usually called parameterizations. These parameterizations influence the climate model projections and are the source of a substantial fraction of the range associated with models' projections of future climate. For more details, see SAP 3.1, SAP 3.2, and Randall *et al.* (2007).

Once developed, the climate model simulations are evaluated against observations. In general, models do a good job of simulating many of the large-scale features of climate (Randall *et al.*, 2007). An important application is the use of climate models to project changes in extremes. Evaluation of observed changes in extremes and model simulation of observed behavior is essential and an important part of this report. As noted above, such evaluation is a necessary step toward confidence in projected changes in extremes.

3.2 WHAT ARE THE PHYSICAL MECHANISMS OF OBSERVED CHANGES IN EXTREMES?

3.2.1 Detection and Attribution: Evaluating Human Influences on Climate Extremes Over North America

Climate change detection, as discussed in this chapter, is distinct from the concept that is used in Chapter 2. In that chapter, detection refers

to the identification of change in a climate record that is statistically distinguishable from the record's previous characteristics. A typical example is the detection of a statistically significant trend in a temperature record. Here, detection and attribution involves the assessment of observed changes in relation to those that are expected to have occurred in response to external forcing. Detection of climatic changes in extremes involves demonstrating statistically significant changes in properties of extremes over time. Attribution further links those changes with variations in climate forcings, such as changes in greenhouse gases (GHGs), solar radiation, or volcanic eruptions. Attribution is a necessary step toward identifying the physical causes of changes in extremes. Attribution often uses quantitative comparison between climate-model simulations and observations, comparing expected changes due to physical understanding integrated in the models with those that have been observed. By comparing observed changes with those anticipated to result from external forcing, detection and attribution studies also provide an assessment of the performance of climate models in simulating climate change. The relationships between observed and simulated climate change that are diagnosed in these studies also provide an important means of constraining projections of future change made with those models.

A challenge arises in attribution studies if the available evidence linking cause and effect in the phenomenon of interest is indirect. For example, global climate models are able to simulate the expected responses to anthropogenic and natural forcing in sea surface temperature (SST) in tropical cyclone formation regions, and thus these expected responses can be compared with observed SST changes to make an attribution assessment of the causes of the observed changes in SSTs. However, global climate models are not yet able to simulate tropical cyclone activity with reasonable fidelity, making an "end-to-end" attribution of cause and effect (Allen, 2003) more difficult. In this case, our judgment of whether an external influence has affected tropical cyclone frequency or intensity depends upon our understanding of the links between SST variability and tropical cyclone variability on the one hand, and upon the understanding of how external influence has altered

SSTs on the other. In these cases, it is difficult to quantify the magnitude and significance of the effect of the external influences on the phenomenon of interest. Therefore, attribution assessments become difficult, and estimates of the size of the anthropogenic contribution to an observed change cannot be readily derived. This problem often occurs in climate-change impacts studies and has sometimes been termed "joint attribution" (Rosenzweig et al., 2007).

3.2.1.1 DETECTION AND ATTRIBUTION: HUMAN-INDUCED CHANGES IN AVERAGE CLIMATE THAT AFFECT CLIMATE EXTREMES

This section discusses the present understanding of the causes of large-scale changes in the climatic state over North America. Simple statistical reasoning indicates that substantial changes in the frequency and intensity of extreme events can result from a relatively small shift in the average of a distribution of temperatures, precipitation, or other climate variables (Mearns et al., 1984; Katz and Brown, 1992). Expected changes in temperature extremes are largely, but not entirely, due to changes in seasonal mean temperatures. Some differences between changes in means and extremes arise because moderate changes are expected in the shape of the temperature distribution affecting climate extremes; for example, due to changes in snow cover, soil moisture, and cloudiness (e.□, Hegerl et al., 2004; Kharin et al., 2007). In contrast, increases in mean precipitation are expected to increase the precipitation variance, thus increasing precipitation extremes, but decreases in mean precipitation do not necessarily

Substantial changes in the frequency and intensity of extreme events can result from a relatively small shift in the average of a distribution of temperatures, precipitation, or other climate variables.

imply that precipitation extremes will decrease because of the different physical mechanisms that control mean and extreme precipitation (e.□, Allen and Ingram, 2002; Kharin *et al.*, 2007). Therefore, changes in the precipitation background state are also interesting for understanding changes in extremes, although more difficult to interpret (Groisman *et al.*, 1999). Relevant information about mean temperature changes appears in Chapter 2. More detailed discussion of historical mean changes appears in CCSP Synthesis and Assessment Products 1.1, 1.2, and 1.3.

Global-scale analyses using space-time detection techniques have robustly identified the influence of anthropogenic forcing on the 20th century near-surface temperature changes. This result is robust to applying a variety of statistical techniques and using many different climate simulations (Hegerl *et al.*, 2007). Detection and attribution analyses also indicate that over the past century there has likely been a cooling influence from aerosols and natural forcings counteracting some of the warming influence of the increasing concentrations of greenhouse gases. Spatial information is required in addition to temporal information to reliably detect the influence of aerosols and distinguish them from the influence of increased greenhouse gases.

A number of studies also consider sub-global scales. Studies examining North America find a detectable human influence on 20th century temperature changes, either by considering the 100-year period from 1900 (Stott, 2003) or the 50-year period from 1950 (Zwiers and Zhang, 2003; Zhang *et al.*, 2006). Based on such studies, a substantial part of the warming over North America has been attributed to human influence (Hegerl *et al.*, 2007).

Further analysis has compared simulations using changes in both anthropogenic (greenhouse gas and aerosol) and natural (solar flux and volcanic eruption) forcings with others that neglect anthropogenic changes. There is a clear separation in North American temperature changes of ensembles of simulations including just natural forcings from ensembles of simulations containing both anthropogenic and natural forcings (Karoly *et al.*, 2003; IADAG,

A human influence has been detected in trends of a U.S. climate extremes index that shows increasing extreme heat and extreme precipitation.

2005; Karoly and Wu, 2005; Wang *et al.*, 2006; Knutson *et al.*, 2006; Hegerl *et al.*, 2007), especially for the last quarter of the 20th century, indicating that the warming in recent decades is inconsistent with natural forcing alone.

Attribution of observed changes on regional (subcontinental) scales has generally not yet been accomplished. One reason is that as spatial scales considered become smaller, the uncertainty becomes larger (Stott and Tett, 1998; Zhang *et al.*, 2006) because internal climate variability is typically larger than the expected responses to forcing on these scales. Also, small-scale forcings and model uncertainty make attribution on these scales more difficult. Therefore, interpreting changes on sub-continental scales is difficult (see discussion in Hegerl *et al.*, 2007). In central North America, there is a relatively small warming over the 20th century compared to other regions around the world (Hegerl *et al.*, 2007) and the observed changes lie (just) within the envelope of changes simulated by models using natural forcing alone. In this context, analysis of a multi-model ensemble by Kunkel *et al.* (2006) for a central U.S. region suggests that the region's warming from 1901 to 1940 and cooling from 1940 to 1979 may have been a consequence of unforced internal variability.

Burkholder and Karoly (2007) detected an anthropogenic signal in multidecadal trends of a U.S. climate extremes index. The observed increase is largely due to an increase in the number of months with monthly mean daily maximum and daily minimum temperatures that are much above normal and an increase in the area of the U.S. that experienced a greater than normal proportion of their precipitation from extreme one-day events. Twentieth century simulations from coupled climate models show a similar, significant increase in the same U.S. climate extremes index for the late 20th century. There is some evidence of an anthropogenic signal in regions a few hundred kilometers across (Karoly and Wu, 2005; Knutson *et al.*, 2006; Zhang *et al.*, 2006; Burkholder and Karoly, 2007), suggesting the potential for progress in regional attribution if careful attention is given to the choice of appropriate time scales, region sizes, and fields analyzed, and if all relevant forcings are considered.

Warming from greenhouse gas increases is expected to increase the moisture content of the atmosphere. Human-induced warming has been linked to water vapor increases in surface observations (Willett *et al.*, 2007) and satellite observations over the oceans (Santer *et al.*, 2007). The greater moisture content should yield a small increase in global mean precipitation. More important, the increase in water-holding capacity of the atmosphere is expected to more strongly affect changes in heavy precipitation, for which the Clausius-Clapeyron relation provides an approximate physical constraint (e. □, Allen and Ingram, 2002). Observed changes in moisture content and mean and extreme precipitation are generally consistent with these expectations (Chapter 2 this report, Trenberth *et al.*, 2007). In addition, greenhouse gas increases are also expected to cause increased horizontal transport of water vapor that is expected to lead to a drying of the subtropics and parts of the tropics (Kumar *et al.* □2004; Neelin *et al.* □2006), and a further increase in precipitation in the equatorial region and at high latitudes (Emori and Brown □2005; Held and Soden □2006).

Several studies have demonstrated that simulated global land mean precipitation in climate model simulations including both natural and anthropogenic forcings is significantly correlated with the observations (Allen and Ingram □2002; Gillett *et al.* □2004; Lambert *et al.* □2004), thereby detecting external influence in observations of precipitation. This external influence on global land mean precipitation during the 20th century is dominated by volcanic forcing. Anthropogenic influence on the spatial distribution of global land precipitation, as represented by zonal-average precipitation changes, has also been detected (Zhang *et al.*, 2007). Both changes are significantly larger in observations than simulated in climate models, raising questions about whether models underestimate the response to external forcing in precipitation changes (see also Wentz *et al.*, 2007). Changes in North American continental-mean rainfall have not yet been formally attributed to anthropogenic influences. A large part of North America falls within the latitude band identified by Zhang *et al.* (2007) where the model simulated response to forcing is not in accord with the observed response. However,

both models and observations show a pattern of increasing precipitation north of 50°N and decreasing precipitation between 0-30°N, and this, together with agreement on decreasing precipitation south of the equator, provides support for the detection of a global anthropogenic influence

.

3.2.1.2 CHANGES IN MODES OF CLIMATE-SYSTEM BEHAVIOR AFFECTING CLIMATE EXTREMES

North American extreme climate is also substantially affected by changes in atmospheric circulation (e. □, Thompson and Wallace, 2001). Natural low frequency variability of the climate system is dominated by a small number of large-scale circulation patterns such as the ENSO, the Pacific Decadal Oscillation (PDO), and the Northern Annular Mode (NAM). The impact of these modes on terrestrial climate on annual to decadal time scales can be profound. In particular, there is considerable evidence that the state of these modes affects substantially the likelihood of extreme temperature (Thompson and Wallace, 2001; Kenyon and Hegerl, 2008), droughts (Hoerling and Kumar, 2003), and short-term precipitation extremes (e. □, Gershunov and Cayan, 2003; Eichler and Higgins, 2006) over North America.

Some evidence of anthropogenic influence on these modes appears in surface-pressure analyses. Gillett *et al.* (2003, 2005) diagnosed anthropogenic influence on Northern Hemisphere sea-level pressure change, although the model-simulated change is not as large as has been observed. Model-simulated changes in

The water-holding capacity of the atmosphere has increased in response to human-induced warming, allowing more and heavier precipitation.

extremes related to circulation changes may therefore be affected. The change in sea-level pressure largely manifests itself through an intensification of the NAM, with reduced pressure above the pole and equatorward displacement of mass, changes that are closely related to changes in another circulation pattern, the North Atlantic Oscillation (NAO) (Hurrell 1996). However, apart from the NAM, the extent to which modes of variability are excited or altered by external forcing remains uncertain. While some modes might be expected to change as a result of anthropogenic effects such as the intensified greenhouse effect, there is little *a priori* expectation about the direction or magnitude of such changes. In addition, models may not simulate well the behavior of these modes in some regions and seasons.

ENSO is the leading mode of variability in the tropical Pacific, and it has impacts on climate around the globe (Trenberth *et al.*, 2007, see also Chapter 1 this report). There have been multidecadal oscillations in the ENSO index throughout the 20th century, with more intense El Niño events since the late 1970s, which may reflect in part a mean warming of the eastern equatorial Pacific (Mendelssohn *et al.* 2005). There is presently no clear consensus on the possible impact of anthropogenic forcing on observed ENSO variability (Merryfield, 2006; Meehl *et al.*, 2007a).

Decadal variability in the North Pacific is characterized by variations in the strength of the Aleutian Low coupled to changes in North Pacific SST. The leading mode of decadal variability in the North Pacific is usually termed the PDO and has a spatial structure in the atmosphere and upper North Pacific Ocean similar to the pattern that is associated with ENSO. Pacific decadal variability can also be characterized by changes in sea-level pressure in the North Pacific, termed the North Pacific Index (Deser *et al.*, 2004). One recent study showed a consistent tendency towards the positive phase of the PDO in observations and model simulations that included anthropogenic forcing (Shiogama *et al.* 2005), though differences between the observed and simulated PDO patterns, and the lack of additional studies, limit confidence in these findings.

ENSO and Pacific decadal variability affect the mean North American climate and its extremes (*e.* , Kenyon and Hegerl, 2008), particularly when both are in phase, at which time considerable energy is propagated from tropical and northern Pacific sources towards the North American land mass (Yu *et al.*, 2007; Yu and Zwiers, 2007).

The NAM is an approximately zonally symmetric mode of variability in the Northern Hemisphere (Thompson and Wallace 1998), and the NAO (Hurrell 1996) may be viewed as its Atlantic counterpart. The NAM index exhibited a pronounced trend towards its positive phase between the 1960s and the 1990s, corresponding to a decrease in surface pressure over the Arctic and an increase over the subtropical North Atlantic (*e.* , Hurrell 1996; Thompson *et al.* 2000; Gillett *et al.* 2003). Several studies have shown this trend to be inconsistent with simulated internal variability (Osborn *et al.* 1999; Gillett *et al.* 2000; Gillett *et al.* 2002; Osborn 2004; Gillett 2005) and similar to, although larger than, simulated changes in coupled climate models in response to 20th century forcing, particularly greenhouse gas forcing and ozone depletion (Gillett *et al.* 2002; Osborn 2004; Gillett, 2005; Hegerl *et al.*, 2007). The mechanisms underlying Northern Hemisphere circulation changes also remain open to debate (see *e.* , Hoerling *et al.* 2005; Hurrell *et al.* 2005; Scaife *et al.*, 2005).

Over the period 1968–1997, the trend in the NAM was associated with approximately 50% of the winter surface warming in Eurasia, a decrease in winter precipitation over Southern Europe, and an increase over Northern Europe due to the northward displacement of the storm track (Thompson *et al.* 2000). Such a change would have substantial influence on North America, too, reducing the probability of cold extremes in winter even over large areas (for example, Thompson and Wallace, 2001; Kenyon and Hegerl, 2008), although part of the northeastern U.S. tends to show a tendency for more cold extremes with the NAO trend (Wettstein and Mearns, 2002).

3.2.2 Changes in Temperature Extremes

As discussed in Chapter 2, observed changes in temperature extremes are consistent with the observed warming of the climate (Alexander *et al.* 2006). Globally, in recent decades, there has been an increase in the number of hot extremes, particularly very warm nights, and a reduction in the number of cold extremes, such as very cold nights, and a widespread reduction in the number of frost days in mid-latitude regions.

There is now evidence that anthropogenic forcing has likely affected extreme temperatures. Christidis *et al.* (2005) analyzed a new dataset of gridded daily temperatures (Caesar *et al.* 2006) using the indices shown by Hegerl *et al.* (2004) to have potential for attribution; namely, the average temperature of the most extreme 1, 5, 10 and 30 days of the year. Christidis *et al.* (2005) detected robust anthropogenic changes in a global analysis of indices of extremely warm nights using fingerprints from the Hadley Centre Coupled Model, version 3 (HadCM3), with some indications that the model over-estimates the observed warming of warm nights. Human influence on cold days and nights was also detected, but in this case the model under-estimated the observed changes significantly, so in the case of the coldest day of the year, anthropogenic influence was not detected in observed changes in extremely warm days. Tebaldi *et al.* (2006) find that changes simulated by an ensemble of eight global models that include anthropogenic and natural forcing changes agrees well with observed global trends in heat waves, warm nights, and frost days over the last four decades.

North American observations also show a general increase in the number of warm nights, but with a decrease in the center of the continent that models generally do not reproduce (e.□, Christidis *et al.*, 2005). However, analysis for North America of models (Table 3.1) used by Tebaldi *et al.* (2006) shows reasonable agreement between observed and simulated changes in the frequency of warm nights, number of frost days, and growing season length over the latter half of the 20th century when averaged over the continent (Figure 3.1a,b,c). There is also good agreement between the observed and ensemble mean simulated spatial pattern of change in frost days (Figure 3.2a,b) over the latter half of the 20th century. Note that the observational estimate has a much greater degree of temporal (Figure 3.1) and spatial (Figure 3.2) variability than the model result. The model result is derived from an ensemble of simulations produced by many models, some of which contributed multiple realizations. Averaging over many simulations reduces much of the spatial and temporal variability that arises from internal climate variability. The variability of individual model realizations is comparable to the single set of observations, which is well bounded by the two standard deviation confidence interval about the model ensemble average. Furthermore, Meehl *et al.* (2007b) demonstrate that ensemble simulations using two coupled climate models driven by human and natural forcings approximate well the observed changes, but when driven with natural forcings only cannot reproduce the observed changes, indicating a human

Table 3.1 Models and scenarios used for computing the Frich *et al.* (2002) indices for North America that appear in this document.

Scenario	Models
SRES A1B	ccsm3.0
	cnrm
	gfdl2.0
	gfdl2.1
	inmcm3
	ipsl
	miroc3_2_medres
	miroc3_2_hires
	mri_cgcm2_3_2a
SRES A2	cnrm
	gfd2.0
	gfdl2.1
	inmcm3
	ipsl
	miroc3_2_medre
	mri_cgcm2_3_2a
SRES B1	ccsm3.0
	cnrm
	gfdl2.0
	gfdl2.1
	inmcm3
	ipsl
	miroc3_2_medres
	miroc3_2_hires

Analyses using climate models and observations provide evidence of a human influence on temperature extremes in North America.

contribution to observed changes in heat waves, frost days, and warm nights. Output from one of these ensembles, produced by the Parallel Climate Model, also shows significant trends in the Karl-Knight heat-wave index (Karl and Knight, 1997) in the eastern half of the U.S. for 1961-1990 that are similar to observed trends (Figure 3.3).

There have also been some methodological advances whereby it is now possible to estimate the impact of external forcing on the risk of a particular extreme event. For example, Stott *et al.* (2004), assuming a model-based estimate of temperature variability, estimate that past human influence may have more than doubled the risk of European mean summer temperatures as high as those recorded in 2003. Such a meth-

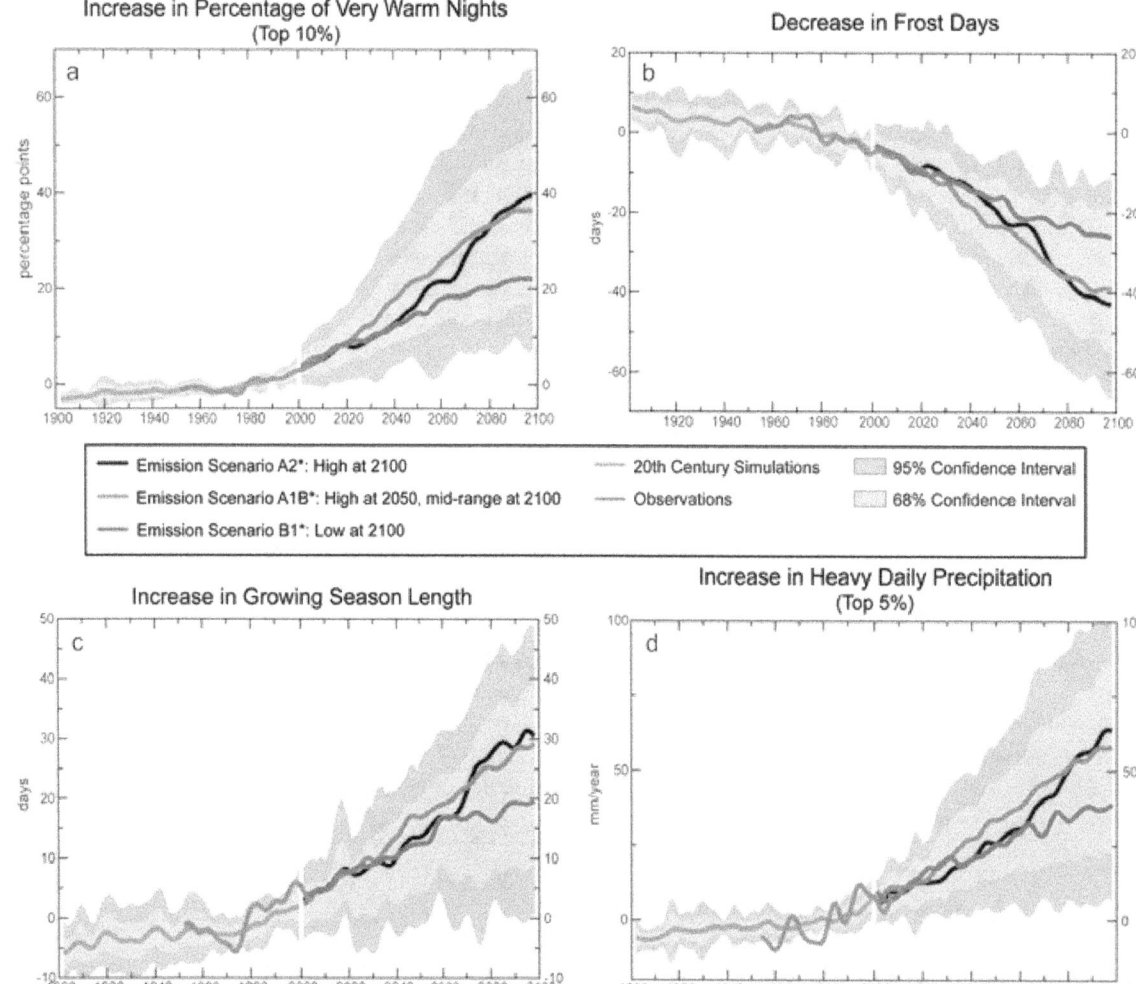

Figure 3.1 Indices (Frich *et al.*, 2002) averaged over North America for model simulations and observations for the 20th and 21st centuries showing changes relative to 1961-1990 in the a) percentage of days in a year for which daily low temperature is in the top 10% of warm nights for the period 1961-1990, b) number of frost days per year, c) growing season length (days), and d) sum of precipitation on days in the top 5% of heavy precipitation days for the period 1961-1990. In the 20th century, the confidence intervals are computed from the ensemble of 20th century simulations. In the future, the bounds are from an ensemble of simulations that used the A1B, A2, or B1 scenarios*. The bounds are the max (or min) standard deviation plus (or minus) signal over all three scenarios. The model plots are obtained from the CMIP-3 multi-model data set at PCMDI and the observations are from Peterson *et al.* (2008).

*Three future emission scenarios from the IPCC Special Report on Emissions Scenarios:
   A2 black line: emissions continue to increase rapidly and steadily throughout this century.
   A1B red line: emissions increase very rapidly until 2030, continue to increase until 2050, and then decline.
   B1 blue line: emissions increase very slowly for a few more decades, then level off and decline.
More details on the above emission scenarios can be found in the IPCC Summary for Policymakers (IPCC, 2007).

odology has not yet been applied extensively to North American extremes, though Hoerling *et al.* (2007) have used the method to conclude that the increase in human induced greenhouse gases has substantially increased the risk of a very hot year in the U.S., such as that experienced in 2006.

Decrease in Number of Frost Days Per Year

Simulated 20th Century Trend

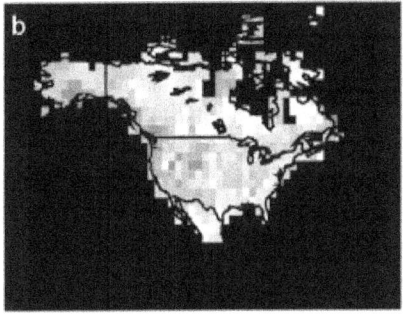

Observed 20th Century Trend

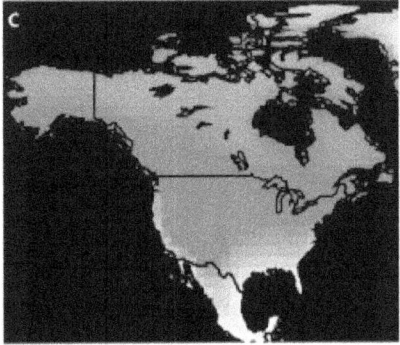

Projected 21st Century Trend

Figure 3.2 Indices (Frich *et al.*, 2002) for frost days over North America for model simulations and observations: a) 20th century trend for model ensemble, b) Observed 20th century trend, and c) 21st century trend for emission scenario A2 from model ensemble. The model plots are obtained from the CMIP-3 multi-model data set at PCMDI and the observations are from Peterson *et al.* (2008).

Changes in Heat Waves

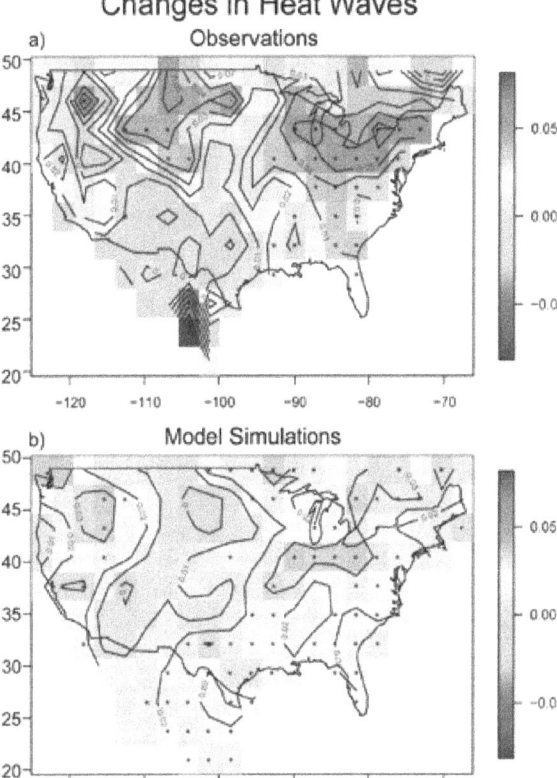

a) Observations

b) Model Simulations

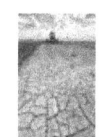

Figure 3.3 Trends in the Karl-Knight heat-wave index (Karl and Knight, 1997) for 1961-1990 in observations (top panel) and in an ensemble of climate simulations by the Parallel Climate Model (bottom panel). Dots mark trends that are significant at the 95% level.

3.2.3 Changes in Precipitation Extremes

3.2.3.1 HEAVY PRECIPITATION

Allen and Ingram (2002) suggest that while global annual mean precipitation is constrained by the energy budget of the troposphere, extreme precipitation is constrained by the atmospheric moisture content, as governed by the Clausius-Clapeyron equation, though this constraint may be most robust in extratropical regions and seasons where the circulation's fundamental dynamics are not driven by latent heat release (Pall *et al.*, 2007). For a given change in temperature, the constraint predicts a larger change in extreme precipitation than in mean precipitation, which is consistent with changes in precipitation extremes simulated by the ensemble of General Circulation Models (GCMs) available for the

IPCC Fourth Assessment Report (Kharin *et al.*, 2007). Emori and Brown (2005) discuss physical mechanisms governing changes in the dynamic and thermodynamic components of mean and extreme precipitation, and conclude that changes related to the dynamic component (*i.e.*, that due to circulation change) are secondary factors in explaining the larger increase in extreme precipitation than mean precipitation seen in models. On the other hand, Meehl *et al.* (2005) demonstrate that while tropical precipitation intensity increases are related to water vapor increases, mid-latitude intensity increases are related to circulation changes that affect the distribution of increased water vapor.

Climatological data show that the most intense precipitation occurs in warm regions (Easterling *et al.* 2000) and diagnostic analyses have shown that even without any change in total precipitation, higher temperatures lead to a greater proportion of total precipitation in heavy and very heavy precipitation events (Karl and Trenberth 2003). In addition, Groisman *et al.* (1999) have demonstrated empirically, and Katz (1999) theoretically, that as total precipitation increases, a greater proportion falls in heavy and very heavy events if the frequency of raindays remains constant. Trenberth *et al.* (2005) point out that a consequence of a global increase in precipitation intensity should be an offsetting global decrease in the duration or frequency of precipitation events, though some regions could have differing behavior, such as reduced total precipitation or increased frequency of precipitation.

Simulated changes in globally averaged annual mean and extreme precipitation appear to be quite consistent between models. The greater and spatially more uniform increases in heavy precipitation as compared to mean precipitation may allow extreme precipitation change to be more robustly detectable (Hegerl *et al.* 2004).

Evidence for changes in observations of short-duration precipitation extremes varies with the region considered (Alexander *et al.* 2006) and the analysis method that is employed (*e.* , Trenberth *et al.*, 2007). Significant increases in observed extreme precipitation have been reported over the United States, where the

increase is qualitatively similar to changes expected under greenhouse warming (*e.* , Karl and Knight 1998; Semenov and Bengtsson 2002; Groisman *et al.* 2005). However, a quantitative comparison between area-based extreme events simulated in models and station data remains difficult because of the different scales involved (Osborn and Hulme 1997; Kharin and Zwiers, 2005), and the pattern of changes does not match observed changes. Part of this difference is expected since most current GCMs do not simulate small-scale (< 100 km) variations in precipitation intensity, as occurs with convective storms. Nevertheless, when compared with a gridded reanalysis product (ERA40), the ensemble of currently available Atmosphere-Ocean General Circulation Models (AOGCMs) reproduces observed precipitation extremes reasonably well over North America (Kharin *et al.*, 2007). An attempt to detect anthropogenic influence on precipitation extremes using global data based on the Frich *et al.* (2002) indices used fingerprints from atmospheric model simulations with prescribed sea surface temperature (Kiktev *et al.* 2003). This study found little similarity between patterns of simulated and observed rainfall extremes. This is in contrast to the qualitative similarity found in other studies (Semenov and Bengtsson, 2002; Groisman *et al.*, 2005; Figure 3.4). Tebaldi *et al.* (2006) reported that an ensemble of eight global climate models simulating the 20th century showed a general tendency toward more frequent heavy-precipitation events over the past four decades, most coherently in the high latitudes of the Northern Hemisphere, broadly consistent with observed changes (Groisman *et al.* 2005). This is also seen when analyzing these models for North America (Figure 3.1d). The pattern similarity of change in precipitation extremes over this period is more difficult to assess, particularly on continental and smaller scales.

3.2.3.2 RUNOFF AND DROUGHT

Changes in runoff have been observed in many parts of the world, with increases or decreases corresponding to changes in precipitation. Climate models suggest that runoff will increase in regions where precipitation increases faster than evaporation, such as at high northern latitudes (Milly *et al.* 2005; Wu *et al.* 2005). Gedney *et al.* (2006a) attributed increased continental

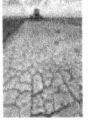

Significant increases in observed extreme precipitation have been reported over the United States, where the increase is similar to changes expected under greenhouse warming.

Projected Increases in Very Heavy Rainfall Events
(Heaviest 0.3%)

Climate Model 1

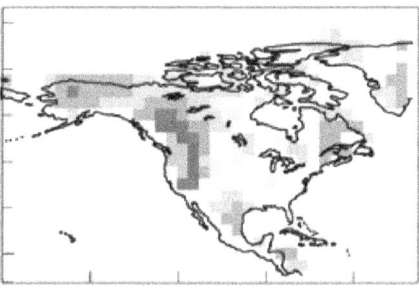

Climate Model 2

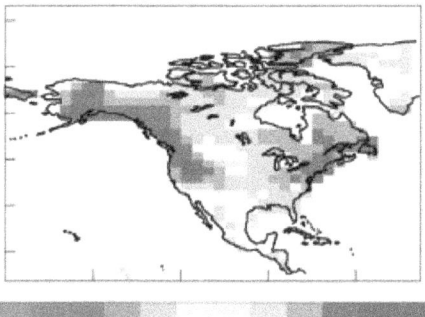

-0.8   -0.6   -0.4   -0.2   0   0.2   0.4   0.6   0.8

Change in the number of times/yr that very heavy rainfall occurs

Observed Increases in Very Heavy Rainfall Events

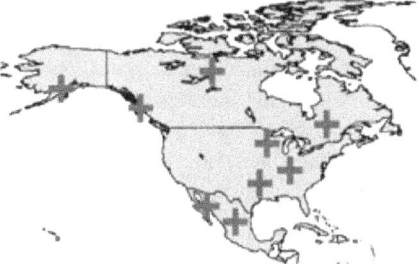

Figure 3.4 Top two panels: Projected changes at the time of carbon dioxide (CO_2) doubling in the number of occurrences per year of the heaviest daily rainfall events in two climate models. For each model, these events were defined as those that exceed the 99.7 percentile of simulated daily rainfall in a 1xCO_2 control simulation. An increase of 0.3 events per year corresponds to a 100% increase in the frequency of such events in locations where there are 100 precipitation days per year. Model 1 is the CGCM2 and Model 2 is the HadCM3. The bottom panel shows areas of North America which have observed increases in very heavy rainfall. After Groisman et al. (2005), Stone et al. (2000, updated), and Cavazos and Rivas (2004).

runoff in the latter decades of the 20th century in part to suppression of transpiration due to carbon dioxide (CO_2)-induced stomatal closure. However, their result is subject to considerable uncertainty in the runoff data (Peel and McMahon, 2006; Gedney et al., 2006b). Qian et al. (2006) simulate observed runoff changes in response to observed temperature and precipitation alone, and Milly et al. (2005) demonstrate that 20th century runoff trends simulated by several global climate models are significantly correlated with observed runoff trends. Wu et al. (2005) find that observed increases in Arctic river discharge are simulated in a global climate model with anthropogenic and natural forcing, but not in the same model with natural forcings only. Anthropogenic changes in runoff may be emerging, but attribution studies specifically on North American runoff are not available.

Mid-latitude summer drying is another anticipated response to greenhouse gas forcing (Meehl et al., 2006), and drying trends have been observed in both the Northern and Southern Hemispheres since the 1950s (Trenberth et al., 2007). Burke et al. (2006), using the HadCM3 model with all natural and anthropogenic external forcings and a global Palmer Drought Severity Index (PDSI) dataset compiled from observations by Dai et al. (2004), detect the influence of anthropogenic forcing in the observed global trend towards increased drought in the second half of the 20th century, although the model trend was weaker than observed and the relative contributions of natural external forcings and anthropogenic forcings was not assessed. Nevertheless, this supports the conclusion that anthropogenic forcing has influenced the global occurrence of drought. However, the spatial pattern of observed PDSI change over North America is dissimilar to that in the coupled model, so no anthropogenic influence has been detected for North America alone.

Nevertheless, the long term trends in the precipitation patterns over North America are well reproduced in atmospheric models driven with observed changes in sea surface temperatures (Schubert et al., 2004; Seager et al., 2005), indicating the importance of sea surface temperatures in determining North American drought (see also, for example, Hoerling and Kumar,

Mid-latitude summer drying is another anticipated response to the human-induced increase in greenhouse gases, and global drying trends have been observed since the 1950s.

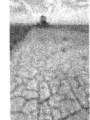

The long-term trends in precipitation patterns over North America are well reproduced in atmospheric models driven with observed changes in sea-surface temperatures.

2003). Specifically, Schubert *et al.* (2004) and Seager *et al.* (2005), using AGCMs forced with observed SSTs, show that some SST anomaly patterns, particularly in the tropical Pacific, can produce drought over North America. Using the observed SST anomalies, both studies successfully reproduce many aspects of the 1930s drought. Only the Seager *et al.* (2005) model simulates the 1950s drought over North America, indicating that more modeling studies of this kind are needed.

3.2.4 Tropical Cyclones

Long-term (multidecadal to century) scale observational records of tropical cyclone activity (frequency, intensity, power dissipation, *et□*) were described in Chapter 2. Here, discussion focuses on whether any changes can be attributed to particular causes, including anthropogenic forcings. Tropical cyclones respond to their environment in quite different manners for initial development, intensification, determination of overall size, and motion. Therefore, this section begins with a brief summary of the major physical mechanisms and understanding.

3.2.4.1 CRITERIA AND MECHANISMS FOR TROPICAL CYCLONE DEVELOPMENT

Gray (1968) drew on a global analysis of tropical cyclones and a large body of earlier work to arrive at a set of criteria for tropical cyclone development, which he called Seasonal Genesis Parameters:

- Sufficient available oceanic energy for the cyclone to develop, usually defined as a requirement for ocean temperatures > 26°C down to a depth of 60 m;
- Sufficient cyclonic (counterclockwise in Northern Hemisphere, clockwise in Southern Hemisphere) rotation to enhance the capacity for convective heating to accelerate the vertical winds;
- A small change in horizontal wind with height (weak shear) so that the upper warming can become established over the lower vortex;
- A degree of atmospheric moist instability to enable convective clouds to develop;
- A moist mid-level atmosphere to inhibit the debilitating effects of cool downdrafts; and
- Some form of pre-existing disturbance, such as an easterly wave, capable of development into a tropical cyclone.

A more recent study by Camargo *et al.* (2007) has developed a new genesis index, which is based on monthly mean values of 850 hPa relative vorticity, 700 hPa humidity, 850-250 hPa wind shear, and Potential Intensity (Bister and Emanuel, 1997). Some skill has been demonstrated in applying it to reanalysis data and global climate models to estimate the frequency and location of storms.

In the North Atlantic, the bulk of tropical cyclone developments arise from easterly waves, though such development is a relatively rare event, with only around 10-20% of waves typically developing into a tropical cyclone (Dunn, 1940; Frank and Clark, 1980; Pasch *et al.*, 1998; Thorncroft and Hodges, 2001). Thus, any large-scale mechanism that can help produce more vigorous easterly waves leaving Africa, or provide an environment to enhance their development, is of importance. ENSO

is a major influence; during El Niño years, tropical cyclone development is suppressed by a combination of associated increased vertical wind shear, general drying of the mid-levels, and oceanic cooling (e.□, Gray, 1984). The Madden-Julian Oscillation (MJO) influences cyclogenesis in the Gulf of Mexico region on 1-2 month time scales (Maloney and Hartmann, 2000). Approximately half of the North Atlantic tropical cyclone developments are associated with upper-level troughs migrating into the tropics (e.□, Pasch *et al.*, 1998; Davis and Bosart, 2001, 2006). The large-scale zonal wind flow may also modulate development of easterly wave troughs into tropical cyclones (Holland, 1995; Webster and Chang, 1988). The easterly wave development process is particularly enhanced in the wet, westerly phase of the MJO.

The eastern and central North Pacific experience very little subtropical interaction and appear to be dominated by easterly wave development (e.□, Frank and Clark, 1980). The two major environmental influences are the ENSO and MJO, associated with the same effects as described for the North Atlantic. The MJO is a particularly large influence, being associated with a more than 2:1 variation in tropical cyclone frequency between the westerly-easterly phases (Liebmann *et al.*, 1994; Molinari and Vollaro, 2000).

Suitable conditions in the western Pacific development region are present throughout the year. Developments in this region are associated with a variety of influences, including easterly waves, monsoon development, and mid-latitude troughs (e.□□Ritchie and Holland, 1999). The dominant circulation is the Asiatic monsoon, and tropical cyclones typically form towards the eastern periphery of the main monsoonal trough or further eastward (Holland, 1995), though development can occur almost anywhere (e.□, Lander, 1994a). ENSO has a major impact, but it is opposite to that in the eastern Pacific and Atlantic, with western Pacific tropical cyclone development being enhanced during the El Niño phase (Chan, 1985; Lander, 1994b; Wang and Chan, 2002).

□e□□□ale i□l□e□□e□include those that occur on scales similar to, or smaller than, the tropical

cyclone circulation and seem to be operative in some form or other to all ocean basins. These influences include interactions amongst the vorticity fields generated by Mesoscale Convective Complexes (MCCs), which may enhance cyclogenesis under suitable atmospheric conditions, but also may introduce a stochastic element in which the interactions may also inhibit short-term development (Houze, 1977; Zipser, 1977; Ritchie and Holland, 1997; Simpson *et al.*, 1997; Ritchie, 2003; Bister and Emanuel, 1997; Hendricks *et al.*, 2004; Montgomery *et al.*, 2006) and inherent barotropic instability (e.□, Schubert *et al.*, 1991; Ferreira and Schubert, 1997).

3.2.4.1.1 Factors Influencing Intensity and Duration

Once a cyclone develops, it proceeds through several stages of intensification. The maximum achievable intensity of a tropical cyclone appears to be limited by the available energy in the ocean and atmosphere. This has led to various thermodynamic assessments of the Potential Intensity (PI) that can be achieved by a cyclone for a given atmospheric/oceanic thermodynamic state (Emanuel, 1987, 1995, 2000; Holland, 1997; Tonkin *et al.*, 2000; Rotunno and Emanuel, 1987). The basis for these assessments is characteristically the sea surface temperature and the thermodynamic structure of the near-cyclone atmospheric environment, with particular emphasis on the temperature at the outflow level of air ascending in the storm core.

In most cases, tropical cyclones do not reach this thermodynamic limit, due to a number of processes that have a substantial negative influence on intensification. Major negative impacts may include: vertical shear of the horizontal wind (Frank and Ritchie, 1999; DeMaria, 1996); oceanic cooling by cyclone-induced mixing of cool water from below the mixed layer to the surface (Price, 1981; Bender and Ginis, 2000; Schade and Emanuel, 1999); potential impacts of sea spray on the surface exchange process (Wang *et al.*, 2001; Andreas and Emanuel, 2001); processes that force the cyclone into an asymmetric structure (Wang, 2002; Corbosiero and Molinari, 2003); ingestion of dry air, perhaps also with suspended dust (Neal and Holland, 1976; Dunion and Velden,

During El Niño years, tropical cyclone development in the Atlantic is suppressed by a combination of increased vertical wind shear, general drying of the mid-levels, and oceanic cooling.

2004); and internal processes. Since many of these factors tend to be transitory in nature, the longer a cyclone can spend in a region with plentiful thermodynamic energy, the better its chances of approaching the PI. This is reflected in, for example, the observation that over 80% of major hurricanes in the North Atlantic occur in systems that formed at low latitudes in the eastern region, the so-called Cape Verde storms.

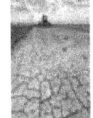

> Determining the causal influences on the observed changes in tropical cyclone characteristics is currently a subject of vigorous community debate.

A weakening tropical cyclone may merge with an extratropical system, or it may redevelop into a baroclinic system (Jones *et al.*, 2003). Since the system carries some of its tropical vorticity and moisture, it can produce extreme rains and major flooding. The transition is also often accompanied by a rapid acceleration in translation speed, which leads to an asymmetric wind field with sustained winds that may be of hurricane force on the right (left) side of the storm track in the Northern (Southern) Hemisphere, despite the overall weakening of the cyclone circulation.

3.2.4.1.2 Movement Mechanisms

Tropical cyclones are steered by the mean flow in which they are embedded, but they also propagate relative to this mean flow due to dynamical effects (Holland, 1984; Fiorino and Elsberry, 1989). This combination leads to the familiar hyperbolic (recurving) track of tropical cyclones as storms initially move westward, embedded in the low-latitude easterly flow, then more poleward and eventually eastward, as they encounter the mid-latitude westerlies.

An important result of this pattern of movement is that storms affecting the Caribbean, Mexico, Gulf States, Lower Eastern Seaboard, and Pacific Trust Territories have mostly developed in low latitudes (which also comprise the most intense systems). Eastern Pacific cyclones tend to move away from land, and those that recurve are normally suffering from combined negative effects of cold water and vertical shear. Upper Eastern U.S. Seaboard and Atlantic Canada cyclones are typically recurving and undergoing various stages of extratropical transition.

3.2.4.2 ATTRIBUTION PREAMBLE

Determining the causal influences on the observed changes in tropical cyclone characteristics is currently a subject of vigorous community debate. Chief among the more contentious topics are data deficiencies in early years, natural variability on decadal time scales, and trends associated with greenhouse warming. A summary of the published contributions to this debate at the end of 2006 is contained in a report and accompanying statement that was composed by attendees at the World Meteorological Organization's Sixth International Workshop on Tropical Cyclones (IWTC-VI) held in November 2006 (WMO, 2006) and later endorsed by the American Meteorological Society. The IPCC has also assessed understanding of the issues through roughly the same point in time and assigned probabilities to human influence on observed and future changes (IPCC, 2007). In both cases, these documents represent consensus views.

One area of interest is the North Pacific. Emanuel (2005) and Webster *et al.* (2005) have shown a clear increase in the more intense Northwest Pacific tropical cyclones (as seen in category 4 and 5 frequency or PDI) since the commencement of the satellite era. These increases have been closely related to concomitant changes in SSTs in this region. However, there are concerns about the quality of the data (WMO, 2006) and there has been little focused research on attributing the changes in this region. For these reasons, this report accepts the overall findings of WMO (2006) and IPCC (2007) as they relate to the North Pacific. These include a possible increase in intense tropical cyclone activity, consistent with Emanuel (2005) and Webster *et al.* (2005), but no clear trend in tropical cyclone numbers. The remainder of the attribution section on tropical cyclones concentrates on the North Atlantic, where the available data and published work enables more detailed attribution analysis compared to other basins.

3.2.4.3 ATTRIBUTION OF NORTH ATLANTIC CHANGES

Chapter 2 provides an overall summary of the observed variations and trends in storm frequency, section 3.3.9.6 considers future scenarios, and Holland and Webster (2007) present a detailed analysis of the changes in North Atlantic tropical storms, hurricanes, and major hurricanes over the past century, together with a critique of the potential attribution mechanisms. Here we examine these changes in terms of the potential causative mechanisms.

3.2.4.3.1 Storm Intensity

There has been no distinct trend in the mean intensity of all storms, hurricanes, or major hurricanes (Chapter 2). Holland and Webster (2007) also found that there has been a marked oscillation in major hurricane proportions, which has no observable trend. The attribution of this oscillation has not been adequately defined, but it is known that it is associated with a similar oscillation in the proportion of hurricanes that develop in low latitudes and thus experience environmental conditions that are more conducive to development into an intense system than those at more poleward locations. The lack of a mean intensity trend or a trend in major hurricane proportions is in agreement with modeling and theoretical studies that predict a relatively small increase of up to 5% for the observed 0.5 to 0.7°C trend in tropical North Atlantic SSTs (See Section 3.3.9.2 for more details on this range; see also Henderson-Sellers et al., 1998; Knutson et al., 1998, 2001; Knutson and Tuleya, 2004, 2008).

Multidecadal increases of maximum intensity due to multidecadal increases of SST may play a relatively small role in increases of overall hurricane activity and increases in frequency (discussed in the next section), for which variations in duration due to large-scale circulation changes may be the dominant factors. The relationship between SST, circulation patterns, and hurricane activity variability is not as well

understood as the thermodynamic relationships that constrain maximum intensity.

3.2.4.3.2 Storm Frequency and Integrated Activity Measures

Emanuel (2005, 2007) examined a PDI, which combines the frequency, lifetime, and intensity, and is related to the cube of the maximum winds summed over the lifetime of the storm. In Chapter 2, it was concluded that there has been a substantial increase in tropical cyclone activity, as measured by the PDI, since about 1970, strongly correlated with low-frequency variations in tropical Atlantic SSTs. It is likely that hurricane activity (PDI) has increased substantially since the 1950s and '60s in association with warmer Atlantic SSTs. It is also likely that PDI has generally tracked SST variations on multidecadal time scales in the tropical Atlantic since 1950. Holland and Webster (2007) have shown that the PDI changes have arisen from a combination of increasing frequency of tropical cyclones of all categories: tropical storms, hurricanes, and major hurricanes and a multidecadal oscillation in the proportion of major hurricanes. They found no evidence of a trend in the major hurricane proportions or in overall intensity, but a marked trend in frequency.

While there is a close statistical relationship between low frequency variations of tropical cyclone activity (e.□, the PDI and storm frequency) and SSTs (Chapter 2), this almost certainly arises from a combination of factors, including joint relationships to other atmospheric processes that effect cyclone development, such as vertical windshear (Shapiro, 1982; Kossin and Vimont, 2007; Goldenberg et al., 2001; Shapiro and Goldenberg, 1998). It is also notable that the recent SST increases have been associated with a concomitant shift towards increased developments in low latitudes and the eastern Atlantic, regions where the conditions are normally more conducive to cyclogenesis and intensification (Holland and Webster, 2007, Chapter 2). Thus, over the past 50 years there is a strong statistical connection between tropical Atlantic sea surface temperatures and Atlantic hurricane activity as measured by the PDI. The North Atlantic is the region where the relationship between hurricane power and tropical sea surface temperature is most significant, but some correlation is evident in the western

Recent increases in sea surface temperatures have been associated with a shift towards increased tropical cyclone developments in the low latitudes and eastern Atlantic.

North Pacific as well (Emanuel, 2007), while no such relationship is observed in the Eastern North Pacific. In general, tropical cyclone power dissipation appears to be modulated by potential intensity, vertical wind shear, and low-level vorticity of the large-scale flow, while sea surface temperature itself should be regarded as a co-factor rather than a causative agent (Emanuel, 2007). The observed complex relationship between power dissipation and environmental determinants is one reason why a confident attribution of human influence on hurricanes is not yet possible.

Low-frequency variations in Atlantic tropical cyclone activity have previously been attributed to a natural variability in Atlantic SSTs associated with the Atlantic Multidecadal Oscillation (Bell and Chelliah, 2006; Goldenberg et al., 2001). However, these studies either did not consider the trends over the 20th century in SST (Goldenberg et al., 2001) or did not cover a long enough period to confidently distinguish between oscillatory (internal climate variability) behavior and radiatively forced variations or trends. For example, the multidecadal AMM2

mode in Bell and Chelliah (2006) first obtains substantial amplitude around 1970. Their circulation-based indices are of insufficient length to determine whether they have a cyclical or trend-like character, or some combination thereof.

While there is undoubtedly a natural variability component to the observed tropical Atlantic SSTs, a number of studies suggest that increasing greenhouse gases have contributed to the warming that has occurred, especially over the past 30-40 years. For example, Santer et al. (2006) have shown that the observed trends in Atlantic tropical SSTs are unlikely to be caused entirely by internal climate variability, and that the pronounced Atlantic warming since around 1970 that is reproduced in their model is predominantly due to increased greenhouse gases. These conclusions are supported by several other studies that use different methodologies (e. , Knutson et al., 2006; Trenberth and Shea, 2006; Mann and Emanuel, 2006; Karoly and Wu, 2005). There is also evidence for a detectable greenhouse gas-induced SST increase in the Northwest Pacific tropical cyclogenesis region (Santer et al., 2006; see also Knutson et al., 2006 and Karoly and Wu, 2005).

We conclude that there has been an observed SST increase of 0.5-0.7°C over the past century in the main development region for tropical cyclones in the Atlantic. Based on comparison of observed SST trends and corresponding trends in climate models with and without external forcing, it is very likely that human-caused increases in greenhouse gases have contributed to the increase in SSTs in the North Atlantic and the Northwest Pacific hurricane formation regions over the 20th century.

Chapter 2 also notes that there have been fluctuations in the number of tropical storms and hurricanes from decade to decade and that data uncertainty is larger in the early part of the record compared to the satellite era beginning in 1965. Even taking these factors into account, Chapter 2 concludes that it is likely that the annual numbers of tropical storms, hurricanes and major hurricanes in the North Atlantic have increased over the past 100 years, a time in which Atlantic sea surface temperatures also increased. The evidence is less compelling for significant trends beginning in the late 1800s. The existing data for hurricane counts and one adjusted record of tropical storm counts both indicate no significant linear trends beginning from the mid- to late 1800s through 2005. In general, there is increasing uncertainty in the data as one proceeds back in time. There is no evidence for a long-term increase in North American mainland land-falling hurricanes.

Attribution of these past changes in tropical storm/hurricane activity (e.□, PDI) and frequency to various climate forcings is hampered by the lack of adequate model simulations of tropical cyclone climatologies. In the case of global scale temperature, increased formal detection-attribution studies have detected strong evidence for the presence of the space-time pattern of warming expected due to greenhouse gas increases. These studies find that other plausible explanations, such as solar and volcanic forcing together with climate variability alone, fail to explain the observed changes sufficiently. The relatively good agreement between observed and simulated trends based on climate model experiments with estimated past forcings lends substantial confidence to attribution statements for SST. However, since adequate model-based reconstructions of historical tropical cyclone variations are not currently available, we do not have estimates of expected changes in tropical cyclone variations due to a complete representation of the changes in the physical system that would have been caused by greenhouse gas increases and other forcing changes. (Relevant anthropogenic forcing includes changes in greenhouse gas concentrations, as well as aerosols [e.□, Santer *et al.*, 2006]. In addition, stratospheric ozone decreases and other factors associated with recent cooling upper atmospheric [at approximately 100 mb] temperature changes may be important [Emanuel, 2007]). Given this state of the sciences, we therefore must rely on statistical analyses, existing modeling, and theoretical information and expert judgement to make the following attribution assessment:

> It is very likely that the human-induced increase in greenhouse gases has contributed to the increase in sea surface temperatures in the hurricane formation

regions. Over the past 50 years there has been a strong statistical connection between tropical Atlantic sea surface temperatures and Atlantic hurricane activity as measured by the Power Dissipation Index (which combines storm intensity, duration, and frequency). This evidence suggests a human contribution to recent hurricane activity. However, a confident assessment of human influence on hurricanes will require further studies using models and observations, with emphasis on distinguishing natural from human-induced changes in hurricane activity through their influence on factors such as historical sea surface temperatures, wind shear, and atmospheric vertical stability.

3.2.4.3.3 Storm Duration, Track, and Extratropical Transition

There has been insufficient research on these important aspects of tropical cyclones to arrive at any firm conclusions about possible changes.

3.2.5 Extratropical Storms

Chapter 2 documents changes in strong extratropical storms during the 20th century, especially for the oceanic storm track bordering North America. Changes include altered intensity and tracks of intense storms (Wang *et al.*, 2006; Caires and Sterl, 2005). Analysis of physical mechanisms is lacking. Natural cycles of large-scale circulation affect variability, through the NAO (e.□, Lozano and Swail, 2002; Caires and Sterl, 2005) or the related NAM (Hurrell, 1995; Ostermeier and Wallace, 2003). Changes in sea-surface temperature (Graham and Diaz, 2001) and baroclinicity (Fyfe, 2003) may also play a role. Analysis of a multi-century GCM simulation by Fischer-Bruns *et al.* (2005) suggests that changes in solar activity and volcanic activity have negligible influence on strong-storm activity. However, it is likely that anthropogenic influence has contributed to extratropical circulation change during the latter half of the 20th century (Hegerl *et al.*, 2007; see also Gillett *et al.*, 2003, 2005), which would have influenced storm activity. On the other hand, the WASA Group (1998), using long records of station data, suggest that observed changes in storminess in Northern Europe

The Power Dissipation Index, which combines storm frequency, lifetime, and intensity, has increased substantially since about 1970, strongly correlated with changes in Atlantic tropical sea surface temperatures.

over the latter part of the 20th century are not inconsistent with natural internal low-frequency variability. However, analyses based on direct observations suffer from incomplete spatial and temporal coverage, especially in storm-track regions over adjacent oceans, and generally cover regions that may be too small to allow detection of externally forced signals (Hegerl *et al.*, 2007). Studies of global reanalysis products generally cover less than 50 years. While 50-year records are generally considered adequate for detection and attribution research (Hegerl *et al.*, 2007), a difficulty with reanalysis products is that they are affected by inhomogeneities resulting from changes over time in the type and quantity of data that is available for assimilation (e. □, Trenberth *et al.*, 2005).

A number of investigations have considered the climate controls on the storm intensities or on the decadal trends of wave heights generated by those storms. Most of this attention has been on the North Atlantic, and as noted above, the important role of the NAO has been recognized (e. □, Neu, 1984; WASA Group, 1998; Gulev and Grigorieva, 2004). Fewer investigations have examined the climate controls on the storms and waves in the North Pacific, and with less positive conclusions (Graham and Diaz, 2001; Gulev and Grigorieva, 2004). In particular, definite conclusions have not been reached concerning the climate factor producing the progressive increase seen in wave heights, apparently extending at least back to the 1960s.

A clear influence on the wave conditions experienced along the west coast of North America is occurrences of major El Niños such as those in 1982-83 and 1997-98. Both of these events in particular brought extreme wave conditions

> As a consequence of rising temperatures, the amount of moisture in the atmosphere is likely to rise much faster than total precipitation. This should lead to an increase in the intensity of thunderstorms.

to south central California, attributed primarily to the more southerly tracks of the storms compared with non-El Niño years. Allan and Komar (2006) found a correlation between the winter-averaged wave heights measured along the west coast and the multivariate ENSO index, showing that while the greatest increase in wave heights during El Niños takes place at the latitudes of south central California, some increase occurs along the entire west coast, evidence that the storms are stronger as well as having followed more southerly tracks. The wave climates of the west coast therefore have been determined by the decadal increase found by Allan and Komar (2000, 2006), but further enhanced during occurrences of major El Niños.

3.2.6 CONVECTIVE STORMS

Trenberth *et al.* (2005) point out that as a consequence of rising temperatures, the amount of moisture in the atmosphere is likely to rise much faster than total precipitation. This should lead to an increase in the intensity of storms, accompanied by decreases in duration or frequency of events. Environmental conditions that are most often associated with severe and tornado-producing thunderstorms have been derived from reanalysis data (Brooks *et al.*, 2003). Brooks and Dotzek (2008) applied those relationships to count the frequency of favorable environments for significant severe thunderstorms (hail of at least 5 cm diameter, wind gusts of at least 33 meters per second, and/or a tornado of F2 or greater intensity) for the area east of the Rocky Mountains in the U.S. for the period 1958-1999. The count of favorable environments decreased by slightly more than 1% per year from 1958 until the early-to-mid 1970s, and increased by approximately 0.8% per year from then until 1999, so that the frequency was approximately the same at both ends of the analyzed period. They went on to show that the time series of the count of reports of very large hail (7 cm diameter and larger) shows an inflection at about the same time as the inflection in the counts of favorable environments. A comparison of the rate of increase of the two series suggested that the change in environments could account for approximately 7% of the change in reports from the mid-1970s through 1999, with the rest coming from non-meteorological sources.

3.3 PROJECTED FUTURE CHANGES IN EXTREMES, THEIR CAUSES, MECHANISMS, AND UNCERTAINTIES

Projections of future changes of extremes are relying on an increasingly sophisticated set of models and statistical techniques. Studies assessed in this section rely on multi-member ensembles (3 to 5 members) from single models, analyses of multi-model ensembles ranging from 8 to 15 or more AOGCMs, and a perturbed physics ensemble with a single mixed layer model with over 50 members. The discussion here is intended to identify the characteristics of changes of extremes in North America and set them in the broader global context.

3.3.1 Temperature
The IPCC Third Assessment Report concluded that it is very likely that there is an increased risk of high temperature extremes (and reduced risk of low temperature extremes), with more extreme heat episodes in a future climate. This latter result has been confirmed in subsequent studies (e.g., Yonetani and Gordon, 2001), which are assessed in the 2007 IPCC report (Meehl et al, 2007a; Christensen et al. 2007). An ensemble of more recent global simulations projects marked an increase in the frequency of very warm daily minimum temperatures (Figure 3.1a). Kharin et al. (2007) show in a large ensemble of models that over North America, future increases in summertime extreme maximum temperatures follow increases in mean temperature more closely than future increases in wintertime extreme minimum temperatures. This increase in winter extreme temperatures is substantially larger than increases in mean temperature, indicating the probability of very cold extremes decreases and hot extremes increases. They also show a large reduction in the wintertime cold temperature extremes in regions where snow and sea ice decrease due to changes in the effective heat capacity and albedo of the surface. Kharin and Zwiers (2005) show in a single model that summertime warm temperature extremes increase more quickly than means in regions where the soil dries due to a smaller fraction of surface energy used for evaporation. Clark et al. (2006), using another model, show that an ensemble of doubled-CO_2 simulations produces increases in summer warm extremes for North America that are roughly the same as increases in median daily maximum temperatures. Hegerl et al. (2004) show that for both models, differences in warming rates between seasonal mean and extreme temperatures are statistically significant over most of North America in both seasons; detecting changes in seasonal mean temperature is not a substitute for detecting changes in extreme temperatures.

Events that are rare could become more commonplace. Recent studies using both individual models (Kharin and Zwiers, 2005) and an ensemble of models (Wehner 2005, Kharin, et al., 2007) show that events that currently reoccurr on average once every 20 years (i.e., have a 5% chance of occurring in a given year) will become signficantly more frequent over North America. For example, by the middle of the 21st century, in simulations of the SRES A1B scenario, the recurrence period (or expected average waiting time) for the current 20-year extreme in daily average surface-air temperature reduces to three years over most of the continental United States and five years over most of Canada (Kharin, et al., 2007). By the end of the century (Figure 3.5a), the average reoccurrence time may further reduce to every other year or less (Wehner, 2005).

Similar behavior occurs for seasonal average temperatures. For example, Weisheimer and Palmer (2005) examined changes in extreme seasonal (DJF and JJA) temperatures in 14 models for 3 scenarios. They showed that by the end of 21st century, the probability of such extreme warm seasons is projected to rise in many areas including North America. Over the North American region, an extreme seasonal temperature event that occurs 1 out of 20 years in the present climate becomes a 1 in 3-year event in the A2 scenario by the end of this century. This result is consistent with that from the perturbed physics ensemble of Clark et al. (2006) where, for nearly all land areas, extreme JJA temperatures were at least 20 times, and in some areas 100 times, more frequent compared to the control ensemble mean, making these changes greater than the ensemble spread.

Others have examined possible future cold-air outbreaks. Vavrus et al. (2006) analyzed

In a future warmer world, the probability of very cold extremes decreases, and that of hot extremes increases.

seven AOGCMs run with the A1B scenario, and defined a cold air outbreak as two or more consecutive days when the daily temperatures were at least two standard deviations below the present-day winter-time mean. For a future warmer climate, they documented a decline in frequency of 50 to 100% in Northern Hemisphere winter in most areas compared to present-day, with some of the smallest reductions occurring in western North America due to atmospheric circulation changes (blocking and ridging on the West Coast) associated with the increase of GHGs.

Several recent studies have addressed explicitly possible future changes in heat waves (very high temperatures over a sustained period of days), and found that in a future climate there is an increased likelihood of more intense, longer-lasting and more frequent heat waves (Meehl and Tebaldi 2004; Schär *et al.* 2004; Clark *et al.* 2006). Meehl and Tebaldi (2004) related summertime heat waves to circulation patterns in the models and observations. They found that the more intense and frequent summertime heat waves over the southeast and western U.S. were related in part to base state circulation changes due to the increase in GHGs. An additional factor for extreme heat is drier soils in a future warmer climate (Brabson *et al.* 2005; Clark *et al.* 2006).

The "Heat Index", a measure of the apparent temperature felt by humans that includes moisture influences, was projected in a model study from the Geophysical Fluid Dynamics Laboratory to increase substantially more than the air temperature in a warming climate in many regions (Delworth *et al.*, 1999). The regions most prone to this effect included humid regions of the tropics and summer hemisphere extratropics, including the Southeast U.S. and Caribbean. A multi-model ensemble showed

In a future climate there is an increased likelihood of more intense, longer-lasting, and more frequent heat waves.

Extreme Temperature and Precipitation Events are Projected to Become More Common

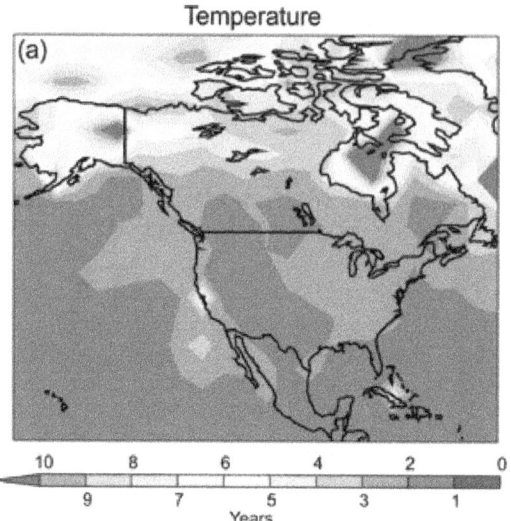

Figure 3.5 Simulations for 2090-2099 indicating how currently rare extremes (a 1-in-20-year event) are projected to become more commonplace. a) A day so hot that it is currently experienced once every 20 years would occur every other year or more by the end of the century. b) Daily total precipitation events that occur on average every 20 years in the present climate would, for example, occur once every 4-6 years for Northeast North America. These results are based on a multi-model ensemble of global climate models.

that simulated heat waves increase during the latter part of the 20th century, and are projected to increase globally and over most regions including North America (Tebaldi *et al.* 2006), though different model parameters can influence the range in the magnitude of this response (Clark *et al.* 2006).

Warm episodes in ocean temperatures can stress marine ecosystems, causing impacts such as coral bleaching (e. □, Liu *et al.*, 2006; Chapter 1, Box 1.4, this report). Key factors appear to be clear skies, low winds, and neap tides occurring near annual maximum temperatures since they promote heating with little vertical mixing of warm waters with cooler, deeper layers (Strong *et al.*, 2006). At present, widespread bleaching episodes do not appear to be related to variability such as ENSO cycles (Arzayus and Skirving, 2004) or PDO (Strong *et al.*, 2006), though the strong 1997-98 El Niño did produce sufficient warming to cause substantial coral bleaching (Hoegh-Guldberg, 1999, 2005; Wilkinson, 2000). The 2005 Caribbean coral bleaching event has been linked to warm ocean temperatures that appear to have been partially due to long-term warming associated with anthropogenic forcing and not a manifestation of unforced climate variability alone (Donner *et al.*, 2007). Warming trends in the ocean increase the potential for temperatures to exceed thresholds for mass coral bleaching, and thus may greatly increase the frequency of bleaching events in the future, depending on the ability of corals and their symbiotic algae to adapt to increasing water temperatures (see Donner *et al.*, 2007 and references therein).

A decrease in diurnal temperature range in most regions in a future warmer climate was reported in Cubasch *et al.* (2001) and is substantiated by more recent studies (e. □, Stone and Weaver□2002), which are assessed in the 2007 IPCC report (Meehl *et al.*, 2007a; Christensen *et al.*, 2007). However, noteworthy departures from this tendency have been found in the western portion of the United States, particularly the Southwest, where increased diurnal temperature ranges occur in several regional (e. □, Bell *et al.*, 2004; Leung *et al.*, 2004) and global (Christensen *et al.*, 2007) climate-change simulations. Increased diurnal temperature range often occurs in areas that experience drying in the summer.

3.3.2 Frost
As the mean climate warms, the number of frost days is expected to decrease (Cubasch *et al.*, 2001). Meehl *et al.* (2004) have shown that there would indeed be decreases in frost days in a future warmer climate in the extratropics,

particularly along the northwest coast of North America, with the pattern of the decreases dictated by the changes in atmospheric circulation from the increase in GHGs. Results from a multi-model ensemble show simulated and observed decreases in frost days for the 20th century continuing into the 21st century over North America and most other regions (Meehl *et al.*, 2007a, Figure 3.1b). By the end of the 21st century, the number of frost days averaged over North America is projected to decrease by about one month in the three future scenarios considered here.

In both the models and the observations, the number of frost days decreased over the 20th century (Figure 3.1b). This decrease is generally related to warming climate, although the pattern of the warming and pattern of the frost-days changes (Figure 3.2) are not well correlated. The decrease in the number of frost days per year is biggest in the Rockies and along the west coast of North America. The projected pattern of change in 21st century frost days is similar to the 20th century pattern, but much larger in magnitude. In some places, by 2100, the number of frost days is projected to decrease by more than two months.

These changes would have a large impact on biological activity (See Chapter 1 for more discussion). An example of a positive impact is that there would be an increase in growing season length, directly related to the decrease in frost days per year. A negative example is that fruit trees would suffer because they need a certain number of frost periods per winter season to set their buds, and in some places, this threshold would no longer be met. Note also that changes in wetness and CO_2 content of the air would also affect biological responses.

3.3.3 Growing Season Length
A quantity related to frost days in many mid- and high latitude areas, particularly in the Northern Hemisphere, is growing season length as defined by Frich *et al.* (2002), and this has been projected to increase in a future climate in most areas (Tebaldi *et al.*□2006). This result is also shown in a multi-model ensemble where the simulated increase in growing season length in the 20th century continues into the 21st century over North America and most other

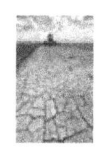

By 2100, the number of frost days averaged across North America is projected to decrease by one month, with decreases of more than two months in some places. These changes would have a large impact on biological activity.

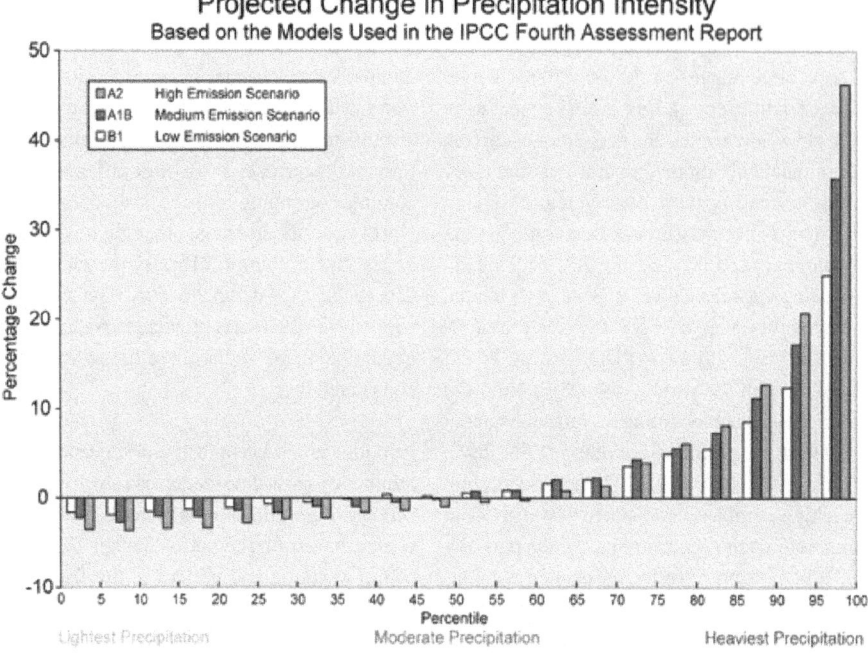

Projected Change in Precipitation Intensity
Based on the Models Used in the IPCC Fourth Assessment Report

Figure 3.6 Projected changes in the intensity of precipitation, displayed in 5% increments, based on a suite of models and three emission scenarios. As shown here, the lightest precipitation is projected to decrease, while the heaviest will increase, continuing the observed trend. The higher emission scenarios yield larger changes. Figure courtesy of Michael Wehner.

The water available from spring snowmelt runoff will likely decrease, affecting water-resource management and exacerbating the impacts of droughts.

regions (Meehl *et al.*, 2007a, Figure 3.1c). The growing season length has increased by about one week over the 20th century when averaged over all of North America in the models and observations. By the end of the 21st century, the growing season is on average more than two weeks longer than present day. (For more discussion on the reasons these changes are important, see Chapter 1 this report.)

3.3.4 Snow Cover and Sea Ice
Warming generally leads to reduced snow and ice cover (Meehl *et al.*, 2007a). Reduction in perennial sea ice may be large enough to yield a summertime, ice-free Arctic Ocean in this century (Arzel *et al.*, 2006; Zhang and Walsh, 2006). Summer Arctic Ocean ice also may undergo substantial, decadal-scale abrupt changes rather than smooth retreat (Holland *et al.*, 2006). The warming may also produce substantial reduction in the duration of seasonal ice in lakes across Canada and the U.S. (Hodgkins *et al.*, 2002; Gao and Stefan, 2004; Williams *et al.*, 2004; Morris *et al.*, 2005) and in rivers (Hodgkins *et al.*, 2003; Huntington *et al.*, 2003). Reduced sea ice in particular may produce more strong storms over the ocean (Section 3.3.10).

Reduced lake ice may alter the occurrence of heavy lake-effect snowfall (Section 3.3.8).

The annual cycle of snow cover and river runoff may be substantially altered in western U.S. basins (Miller *et al.*, 2003; Leung *et al.*, 2004). The water available from spring snowmelt runoff will likely decrease, affecting water-resource management and exacerbating the impacts of droughts.

3.3.5 Precipitation
Climate-model projections continue to confirm the conclusion that in a future climate warmed by increasing GHGs, precipitation intensity (*i.e.*, precipitation amount per event) is projected to increase over most regions (Wilby and Wigley 2002; Kharin and Zwiers 2005; Meehl *et al.* 2005; Barnett *et al.* 2006) and the increase of precipitation extremes is greater than changes in mean precipitation (Figure 3.6; Kharin and Zwiers 2005). Currently, rare precipitation events could become more commonplace in North America (Wehner, 2005; Kharin *et al.*, 2007). For example, by the middle of the 21st century, in simulations of the SRES A1B scenario, the recurrence period for the current 20-year

extreme in daily total precipitation reduces to between 12 and 15 years over much of North America (Kharin, *et al.*, 2007). By the end of the century (Figure 3.5b), the expected average reoccurrence time may further reduce to every six to eight years (Wehner, 2005; Kharin, *et al.* 2007). Note the area of little change in expected average reoccurrence time in the central United States in Figure 3.5b.

As discussed in section 3.2.3 of this chapter and in Hegerl *et al.* (2007), the substantial increase in precipitation extremes is related to the fact that the energy budget of the atmosphere constrains increases of large-scale mean precipitation, but extreme precipitation responds to increases in moisture content and thus the nonlinearities involved with the Clausius-Clapeyron relationship and the zero lower bound on precipitation rate. This behavior means that for a given increase in temperature, increases in extreme precipitation are relatively larger than the mean precipitation increase (*e.*, Allen and Ingram 2002), so long as the character of the regional circulation does not change substantially (Pall *et al.*, 2007). Additionally, timescale can play a role whereby increases in the frequency of seasonal mean rainfall extremes can be greater than the increases in the frequency of daily extremes (Barnett *et al.* 2006).

The increase in mean and extreme precipitation in various regions has been attributed to contributions from both dynamic (circulation) and thermodynamic (moisture content of the air) processes associated with global warming (Emori and Brown, 2005), although the precipitation mean and variability changes are largely due to the thermodynamic changes over most of North America. Changes in circulation also contribute to the pattern of precipitation intensity changes over Northwest and Northeast North America (Meehl *et al.* 2005). Kharin and Zwiers (2005) showed that changes to both the location and scale of the extreme value distribution produced increases of precipitation extremes substantially greater than increases of annual mean precipitation. An increase in the scale parameter from the gamma distribution represents an increase in precipitation intensity, and various regions such as the Northern Hemisphere land areas in winter showed particularly high values of increased

scale parameter (Semenov and Bengtsson 2002; Watterson and Dix 2003). Time slice simulations with a higher resolution model (at approximately 1°) show similar results using changes in the gamma distribution, namely increased extremes of the hydrological cycle (Voss *et al.* 2002).

3.3.6 Flooding and Dry Days
Changes in precipitation extremes have a large impact on both flooding and the number of precipitation-free days. The discussion of both is combined because their changes are related, despite the fact that this seems counterintuitive.

A number of studies have noted that increased rainfall intensity may imply increased flooding. McCabe *et al.* (2001) and Watterson (2005) showed there was an increase in extreme rainfall intensity in extratropical surface lows, particularly over Northern Hemisphere land. However, analyses of climate changes due to increased GHGs gives mixed results, with increased or decreased risk of flooding depending on the model analyzed (Arora and Boer, 2001; Milly *et al.*, 2002; Voss *et al.*, 2002).

Global and North American averaged time series of the Frich *et al.* (2002) indices in the multi-model analysis of Tebaldi *et al.* (2006) show simulated increases in heavy precipitation during the 20th century continuing through the 21st century (Meehl *et al.*, 2007a; Figure 3.1d), along with a somewhat weaker and less consistent trend for increasing dry periods between rainfall events for all scenarios (Meehl *et al.*, 2007a). Part of the reason for these results is that precipitation intensity increases almost everywhere, but particularly at mid and high latitudes, where mean precipitation increases (Meehl *et al.* 2005).

There are regions of increased runs of dry days between precipitation events in the subtropics and lower midlatitudes, but a decreased number of consecutive dry days at higher midlatitudes and high latitudes where mean precipitation increases. Since there are areas of both increases and decreases of consecutive dry days between precipitation events in the multi-model average, the global mean trends are smaller and less consistent across models. Consistency of

Intense and heavy rainfall events with high runoff amounts are projected to be interspersed with longer relatively dry periods, increasing the risk of both very dry conditions and flooding.

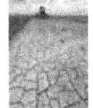

response in a perturbed physics ensemble with one model shows only limited areas of increased frequency of wet days in July, and a larger range of changes of precipitation extremes relative to the control ensemble mean in contrast to the more consistent response of temperature extremes (discussed above), indicating a less consistent response for preciptitation extremes in general compared to temperature extremes (Barnett *et al.* 2006).

Associated with the risk of drying is a projected increase in the chance of intense precipitation and flooding. Though somewhat counter-intuitive, this is because precipitation is projected to be concentrated into more intense events, with longer periods of little precipitation in between. Therefore, intense and heavy episodic rainfall events with high runoff amounts are interspersed with longer relatively dry periods with increased evapotranspiration, particularly in the subtropics (Frei *et al.* 1998; Allen and Ingram 2002; Palmer and Räisänen 2002; Christensen and Christensen 2003; Beniston 2004; Christensen and Christensen 2004; Pal *et al.* 2004; Meehl *et al.* 2005). However, increases in the frequency of dry days do not necessarily mean a decrease in the frequency of extreme high rainfall events depending on the threshold used to define such events (Barnett *et al.* 2006). Another aspect of these changes has been related to the mean changes of precipitation,

with wet extremes becoming more severe in many areas where mean precipitation increases, and dry extremes becoming more severe where the mean precipitation decreases (Kharin and Zwiers 2005; Meehl *et al.* 2005; Räisänen 2005; Barnett *et al.* 2006). However, analysis of a 53-member perturbed-physics ensemble indicates that the change in the frequency of extreme precipitation at an individual location can be difficult to estimate definitively due to model parameterization uncertainty (Barnett *et al.* 2006).

3.3.7 Drought
A long-standing result (e. ., Manabe *et al.*, 1981) from global coupled models noted in Cubasch *et al.* (2001) has been a projected increase of summer drying in the midlatitudes in a future warmer climate, with an associated increased likelihood of drought. The more recent generation of models continues to show this behavior (Burke *et al.* 2006; Meehl *et al.* 2006, 2007a; Rowell and Jones 2006). For example, Wang (2005) analyzed 15 recent AOGCMs to show that in a future warmer climate, the models simulate summer dryness in most parts of the northern subtropics and midlatitudes, but there is a large range in the amplitude of summer dryness across models. Hayhoe *et al.* (2007) found, in an ensemble of AOGCMs, an increased frequency of droughts lasting a month or longer in the northeastern U.S. Droughts associated with summer drying could result in regional vegetation die-offs (Breshears *et al.* 2005) and contribute to an increase in the percentage of land area experiencing drought at any one time. For example, extreme drought increases from 1% of present day land area (by definition) to 30% by the end of the century in the Hadley Centre AOGCM's A2 scenario (Burke *et al.* 2006). Drier soil conditions can also contribute to more severe heat waves as discussed above (Brabson *et al.* 2005).

A recent analysis of Milly *et al.* (2005) shows that several AOGCMs project greatly reduced annual water availability over the Southwest United States, the Caribbean, and in parts of Mexico in the future (Figure 3.7). In the historical context, this area is subject to very severe and long lasting droughts (Cook *et al.*, 2004). The tree-ring record indicates that the late 20th century was a time of greater-than-average

Percentage Change in Annual Runoff (2090-2099)

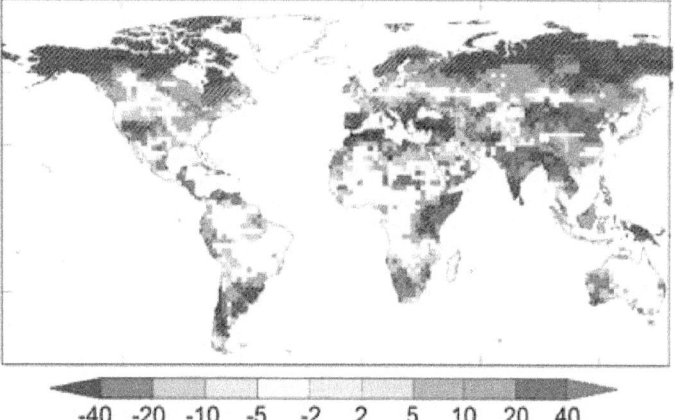

-40 -20 -10 -5   -2   2   5   10   20   40

Figure 3.7 Change in annual runoff (%) for the period 2090-2099, relative to 1980-1999. Values are obtained from the median in a multi-model dataset that used the A1B emission scenario. White areas indicate where less than 66% of the models agree in the sign of change, and stippled areas indicate where more than 90% of the models agree in the sign of change. Derived from the analysis of Milly *et al.* (2005).

water availability. However, the consensus of most climate-model projections is for a reduction of cool season precipitation across the U.S. Southwest and northwest Mexico (Christensen *et al.*, 2007). This is consistent with a recent 10-year shift to shorter and weaker winter rainy seasons and an observed northward shift in northwest Pacific winter storm tracks (Yin, 2005). Reduced cool season precipitation promotes drier summer conditions by reducing the amount of soil water available for evapotranspiration in summer.

The model projections of reduced water availability over the Southwest United States and Mexico needs further study. The uncertainty associated with these projections is related to the ability of models to simulate the precipitation distribution and variability in the present climate and to correctly project the response to future changes. For example, the uncertainty associated with the ENSO response to climate change (Zelle *et al.*, 2005; Meehl *et al.*, 2007a) also impacts the projections of future water availability in the Southwest United States and northern Mexico (*e.*□, Meehl *et al.*, 2007c). Changes in water availability accompanied by changes in seasonal wind patterns, such as Santa Ana winds, could also affect the occurrence of wildfires in the western United States (*e.*□, Miller and Schlegel, 2006). See Chapter 1 for more discussion on the importance of drought.

3.3.8 Snowfall

Extreme snowfall events could change as a result of both precipitation and temperature change. Although reductions in North American snow depth and areal coverage have been projected (Frei and Gong, 2005; Bell and Sloan, 2006; Déry and Wood, 2006), there appears to be little analysis of changes in extreme snowfall. An assessment of possible future changes in heavy lake-effect snowstorms (Kunkel *et al.*, 2002) from the Great Lakes found that surface air temperature increases are likely to be the dominant factor. They examined simulations from two different climate models and found that changes in the other factors favorable for heavy snow events were relatively small. In the snowbelts south of Lakes Ontario, Erie, and Michigan, warming decreases the frequency of temperatures in the range of -10°C to 0°C

that is favorable for heavy lake-effect snowfall. Thus, decreases in event frequency are likely in these areas. However, in the northern, colder snowbelts of the Great Lakes, such as the Upper Peninsula of Michigan, moderate increases in temperature have minor impacts on the frequency of favorable temperatures because in the present climate temperatures are often too cold for very heavy snow; warming makes these days more favorable, balancing the loss of other days that become too warm. Thus, the future frequency of heavy events may change little in the northern snowbelts of lake-effects regions.

Increased temperature suggests that heavy snow events downwind of the Great Lakes will begin later in the season. Also, increased temperature with concomitant increased atmospheric moisture implies that in central and northern Canada, Alaska, and other places cold enough to snow (*e.g.*, high mountains), the intensity of heavy snow events may increase.

3.3.9 Tropical Cyclones (Tropical Storms and Hurricanes)
3.3.9.1 INTRODUCTION

In response to future anthropogenic climate warming, tropical cyclones could potentially change in a number of important ways, including frequency, intensity, size, duration, tracks, area of genesis or occurrence, precipitation, and storm surge characteristics.

Overarching sources of uncertainty in future projections of hurricanes include uncertainties in future emission scenarios for climatically important radiative forcings, global-scale climate sensitivity to these forcings and the limited capacity of climate models to adequately simulate intense tropical cyclones. The vulnerability to

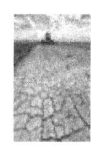

Extreme snowfall events could change as a result of both precipitation and temperature change.

The high sensitivity of tropical storm and hurricane activity in the Atlantic basin to modest environmental variations suggests the possibility of strong sensitivity of hurricane activity to human-caused climate change.

storm surge flooding from future hurricanes will very likely increase, in part due to continuing global sea level rise associated with anthropogenic warming, modulated by local sea level changes due to other factors such as local land elevation changes and regionally varying sea level rise patterns. These related topics are covered in more detail in CCSP Synthesis and Assessment Products 2-1, 3-2, and 4-1, and the IPCC Fourth Assessment Report chapters on climate sensitivity, future emission scenarios, and sea level rise. An additional assessment of the state of understanding of tropical cyclones and climate change as of 2006 was prepared by the tropical cyclone community (WMO, 2006; section 3.2.4 of this CCSP document). This CCSP document summarizes some of the earlier findings but also highlights new publications since these reports.

Future projections of hurricanes will depend upon not only global mean climate considerations, but also on regional-scale projections of a number of aspects of climate that can potentially affect tropical cyclone behavior. These include:

- The local potential intensity (Emanuel 2005; 2006, Holland 1997), which depends on SSTs, atmospheric temperature and moisture profiles, and near-surface ocean temperature stratification;
- Influences of vertical wind shear, large-scale vorticity, and other circulation features (Gray, 1968, 1984; Goldenberg *et al.*, 2001; Bell and Chelliah, 2006); and,
- The characteristics of precursor disturbances such as easterly waves and their interaction with the environment (Dunn, 1940, Frank and Clark, 1980, Pasch *et al.*, 1998, Thorncroft and Hodges, 2001).

Details of future projections in regions remote from the tropical cyclone basin in question may also be important. For example, El Niño fluctuations in the Pacific influence Atlantic basin hurricane activity (Chapter 2 this report, Section 3.2 of this chapter). West African monsoon activity has been correlated with Atlantic hurricane activity (Gray, 1990), as have African dust outbreaks (Evan *et al.*, 2006). Zhang and

Delworth (2006) show how a warming of the northern tropical Atlantic SST relative to the southern tropical Atlantic produces atmospheric circulation features, such as reduced vertical wind shear of the mean wind field, that are correlated with low-frequency variations in major hurricane activity (Goldenberg *et al.*, 2001).

The high sensitivity of tropical storm and hurricane activity in the Atlantic basin to modest environmental variations suggests the possibility of strong sensitivity of hurricane activity to anthropogenic climate change, though the nature of such changes remains to be determined. Confidence in any future projections of anthropogenic influence on Atlantic hurricanes will depend on the reliability of future projections of the local thermodynamic state (*e.g.*, potential intensity) as well as circulation changes driven by both local and remote influences, as described above. Projected effects of global warming on El Niño remain uncertain (Timmermann, 1999; Zelle *et al.*, 2005; Meehl *et al.*, 2007a). There is climate model-based evidence that the time-mean climate late in the 21st century will be characterized by higher tropical-cyclone potential intensity in most tropical-cyclone regions, and also tend toward having a decreased east-west overturning circulation in the Pacific sector in the 21st century, with likely consequences for vertical wind shear and other characteristics in the tropical Atlantic (Vecchi and Soden, 2007).

Even assuming that the climate factors discussed above can be projected accurately, additional uncertainties in hurricane future projections arise from uncertainties in understanding and modeling the response of hurricanes to changing environmental conditions. This is exacerbated by projections that the large-scale conditions for some factors, such as decadal means and seasonal extremes of SSTs, will be well outside the range of historically experienced values. This raises questions of the validity of statistical models trained in the present day climate (Ryan *et al.*, 1992; Royer *et al.*, 1998); thus, the emphasis here is placed on physical models and inferences as opposed to statistical methods and extrapolation. Thus, we consider projections based on global and regional nested modeling frameworks as well as more idealized modeling or theoretical frameworks developed specifically for hurricanes. The idealized approaches include potential intensity theories as well as empirical indices which attempt to relate tropical cyclone frequency to large-scale environmental conditions. Global and regional nested models simulate the development and life cycle of tropical cyclone-like phenomena that are typically much weaker and with a larger spatial scale than observed tropical cyclones. These model storms are identified and tracked using automated storm tracking algorithms, which typically differ in detail between studies, but include both intensity and "warm-core" criteria which must be satisfied. Models used for existing studies vary in horizontal resolution, with the low-resolution models having a grid spacing of about 300 km, medium resolution with grid spacing of about 120 km, and high resolution with grid spacing of 20-50 km.

3.3.9.2 Tropical Cyclone Intensity

Henderson-Sellers *et al.* (1998), in an assessment of tropical cyclones and climate change, concluded that the warming resulting from a doubling of CO_2 would cause the potential intensity of tropical cyclones to remain the same or increase by 10 to 20%. (Their estimate was given in terms of central pressure fall; all other references to intensity in this section will refer to maximum surface winds, except where specifically noted otherwise.) They also noted limitations of the potential intensity theories, such as sea spray influences and ocean interactions.

Further studies using a high resolution hurricane prediction model for case studies or idealized experiments under boundary conditions provided from high CO_2 conditions (Knutson *et al.*, 1998; Knutson and Tuleya, 1999, 2004, 2008) have provided additional model-based evidence to support these theoretical assessments. For a CO_2-induced tropical SST warming of 1.75°C, they found a 14% increase in central pressure fall (Figure 3.8) and a 6% increase in maximum surface wind or a maximum wind speed sensitivity of about 4% per degree Celsius (Knutson and Tuleya, 2008). For the pressure fall sensitivity, they reported that their model result (+14%) was intermediate between that of two potential intensity theories (+8% for Emanuel's MPI and +16% for Holland's MPI) applied to the same large scale environments. In a related study,

Knutson *et al.* (2001) demonstrated that inclusion of an interactive ocean in their idealized hurricane model did not significantly affect the percentage increase in hurricane intensity associated with CO_2-induced large-scale SST warming. Caveats to these idealized studies are the simplified climate forcing (CO_2 only versus a mixture of forcings in the real world) and neglect of potentially important factors such as vertical wind shear and changes in tropical cyclone distribution.

Global climate model experiments have historically been performed at resolutions which precluded the simulation of realistic hurricane intensities (*e.g.*, major hurricanes). To date, the highest resolution tropical cyclone/climate change experiment published is that of Oouchi *et al.* (2006). Under present climate conditions, they simulated tropical cyclones with central pressures as low as about 935 hPa and surface wind speeds as high as about 53 m per second Oouchi *et al.* report a 14% increase in the annual maximum tropical cyclone intensity globally, and a 20% increase in the Atlantic, both in response to a greenhouse-warming experiment with global SSTs increasing by about 2.5°C. A

There is climate model-based evidence that the average climate late in the 21st century will be characterized by higher tropical-cyclone potential intensity in most tropical-cyclone regions.

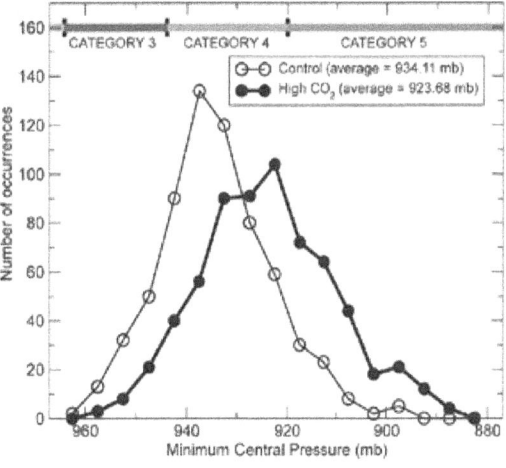

Figure 3.8 Frequency histograms of hurricane intensities in terms of central pressure (mb) aggregated across all idealized hurricane experiments in the Knutson and Tuleya (2004) study. The light curve shows the histogram from the experiments with present-day conditions, while the dark curve is for high CO_2 conditions (after an 80-year warming trend in a +1% per year CO_2 experiment). The results indicate that hurricanes in a CO_2-warmed climate will have significantly higher intensities (lower central pressures) than hurricanes in the present climate.

Changes in Aspects of Climate That Regulate Hurricane Development

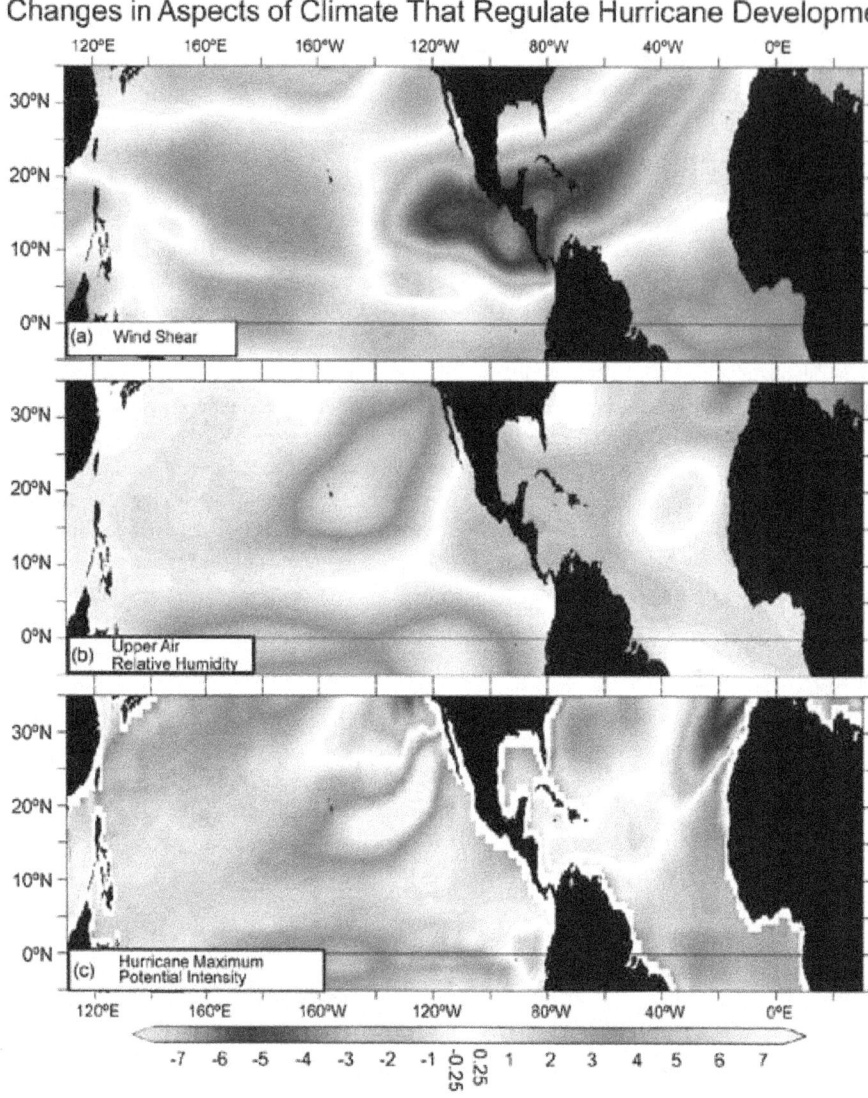

Figure 3.9 Percent changes in June-November ensemble mean a) vertical wind shear (multiplied by -1), b) mid-tropospheric relative humidity, and c) maximum potential intensity of tropical cyclones for the period 2081-2100 minus the period 2001-2021 for an ensemble of 18 GCMs, available in the IPCC AR4 archive, using the A1B scenario. The percentage changes are normalized by the global surface air temperature increase projected by the models. For the wind shear (a), blue areas denote regions with projected increases in vertical wind shear, a factor that is detrimental for hurricane development. Adapted from Vecchi and Soden (2007).

notable aspect of their results is the finding that the occurrence rate of the most intense storms increased despite a large reduction in the global frequency of tropical cyclones. Statistically significant intensity increases in their study were limited to two of six basins (North Atlantic and South Indian Ocean). As will be discussed in the next section on frequency projections, a caveat for the Oouchi *et al.* (2006) results for the Atlantic basin is the relatively short

sample periods they used for their downscaling experiments. Bengtsson *et al.* (2007) also find a slightly reduced tropical cyclone frequency in the Atlantic coupled with an increase in the intensities (measured in terms of relative vorticity) of the most intense storms. The latter finding only became apparent at relatively high model resolution (at approximately 30-40 km grid).

Other studies using comparatively lower resolution models have reported tropical-cyclone intensity results. However, the simulated response of intensity to changes in climate in lower resolution models may not be reliable as they have not been able to simulate the marked difference in achievable tropical-cyclone intensities for different SST levels (*e.g.*, Yoshimura *et al.*, 2006) as documented for observed tropical cyclones (DeMaria and Kaplan, 1994; Whitney and Hobgood, 1997; Baik and Paek, 1998). Given this important caveat, the lower resolution model results for intensity are mixed: Tsutsui (2002) and McDonald *et al.* (2005) report intensity increases under warmer climate conditions, while Sugi *et al.* (2002), Bengtsson *et al.* (2006), and Hasegawa and Emori (2005; western North Pacific only), and Chauvin *et al.* (2006; North Atlantic only) found either no increase or a decrease of intensity.

Vecchi and Soden (2007) present maps of projected late 21st century changes in Emanuel's potential intensity, vertical wind shear, vorticity, and mid-tropospheric relative humidity as obtained from the latest (IPCC, 2007) climate models (Figure 3.9). While their results indicate an increase in potential intensity in most tropical cyclone regions, the Atlantic basin in particular displays a mixture with about two-thirds of the area showing increases and about one-third slight decreases. In some regions, they also found a clear tendency for increased vertical wind shear and reduced mid-tropospheric relative humidity – factors that are detrimental for tropical cyclone development. In the Gulf of Mexico and closer to the U.S. and Mexican coasts, the potential intensity generally increases. The net effect of these composite changes remains to be modeled in detail, although existing global modeling studies (Oouchi *et al.*, 2006; Bengtsson *et al.*, 2007) suggest increases in the intensities and frequencies of the strongest storms. An area average of the Vecchi and Soden (2007) 18 model composite MPI changes (for the region 90°W-60°W, 18°N-35°N) yields a negligible wind speed sensitivity of about 1%/°C (G. Vecchi, personal communication, 2007). This region is north of the Caribbean and includes regions of the Gulf of Mexico and along the U.S. East Coast that are most relevant to U.S. and Mexican landfalls. It is noted that the potential intensity technique used by Vecchi and Soden (Emanuel MPI) was found to be the least sensitive of three estimation techniques in the multi-model study of Knutson and Tuleya (2004). In the Eastern Pacific, the potential intensity is projected to increase across the entire basin, although the vertical wind shear increases may counteract this to some extent.

Influence of Climatic Factors that Contribute to Hurricane Development

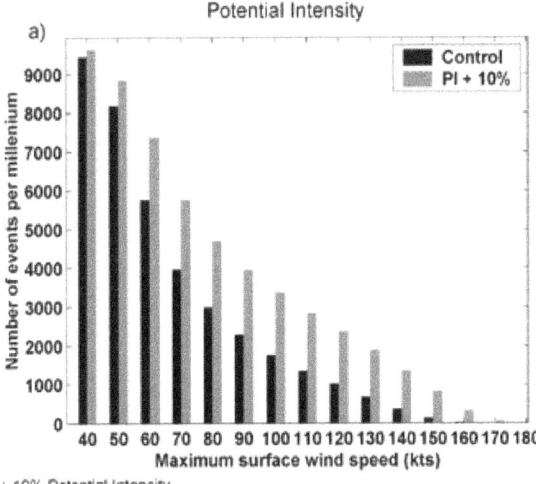

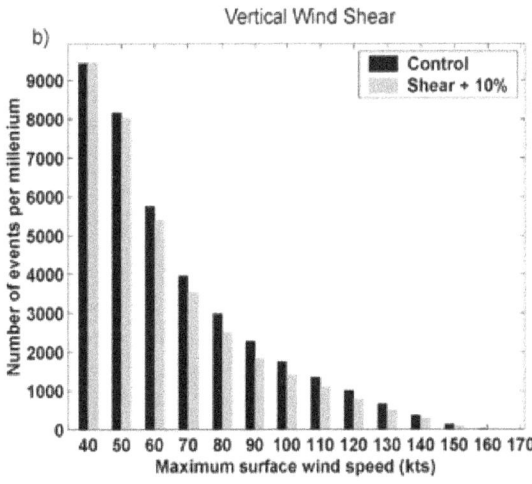

+ 10% Potential Intensity
+ 65% Simulated Power Dissipation Index

Figure 3.10 Number of events per 1000 years with peak wind speeds exceeding the value on the x-axis. Results obtained by running a simple coupled hurricane intensity prediction model over a set of 3000 synthetic storm tracks for the North Atlantic. The grey bars depict storms for present day climate conditions. The black bars depict storms for similar conditions except that the potential intensity (a) or vertical wind shear (b) of the environment is increased everywhere by 10%. From Emanuel (2006).

A more recent idealized calculation by Emanuel *et al.* (2006) finds that artificially increasing the modeled potential intensity by 10% leads to a marked increase in the occurrence rate of relatively intense hurricanes (Figure 3.10a), and to a 65% increase in the PDI. Increasing vertical wind shear by 10% leads to a much smaller decrease in the occurrence rate of relatively intense hurricanes (Figure 3.10b) and a 12% reduction in the PDI. This suggests that increased potential intensity in a CO_2-warmed climate implies a much larger percentage change in potential destructiveness of storms from wind damage than the percentage change in wind speed itself. Emanuel (1987; 2005) estimated a wind speed sensitivity of his MPI to greenhouse gas-induced warming of about 5% per °C; based on one climate model.

The statistical analysis of Jagger and Elsner (2006) provides some support for the notion of more intense storms occurring with higher global temperatures, based on observational analysis. However, it is not yet clear if the empirical relationship they identified is specifically related to anthropogenic influences on global temperature.

In summary, theory and high-resolution idealized models indicate increasing intensity and frequency of the strongest hurricanes and typhoons in a CO_2-warmed climate. Parts of the Atlantic basin may have small decreases in the upper limit intensity, according to one multi-model study of theoretical potential intensity. Expected changes in tropical cyclone intensity and their confidence are therefore assessed as follows: in the Atlantic and North Pacific basins, some increase of maximum surface wind speeds of the strongest hurricanes and typhoons is likely. We estimate the likely range for the intensity increase (in terms of maximum surface winds) to be about 1 to 8% per °C tropical sea surface warming over most tropical cyclone regions. This range encompasses the broad range of available credible estimates, from the relatively low 1.3% per °C area average estimate by Vecchi (personal communication, 2007) of Vecchi and Soden (2007) to the higher estimate (5% per °C) of Emanuel (1987, 2005), and includes some additional subjective margin of error in this range. The ensemble sensitivity estimate from the dynamical hurricane model-

ing study of Knutson and Tuleya (2004) of 3.7% per °C is near the middle of the above range. Furthermore, the available evidence suggests that maximum intensities may decrease in some regions, particularly in parts of the Atlantic basin, even though sea surfaces are expected to warm in all regions.

This assessment assumes that there is no change in geographical distribution of the storms (i.e., the storms move over the same locations, but with a generally warmer climate). On the other hand, there is evidence (Holland and Webster, 2007) that changes in distribution (*e.g.*, tropical-cyclone development occurring more equatorward, or poleward of present day) have historically been associated with large changes in the proportion of major hurricanes. It is uncertain how such distributions will change in the future (see below), but such changes potentially could strongly modify the projections reported here.

3.3.9.3 TROPICAL CYCLONE FREQUENCY AND AREA OF GENESIS

In contrast to the case for tropical-cyclone intensity, the existing theoretical frameworks for relating tropical-cyclone frequency to global climate change are relatively less well-developed. Gray (1979) developed empirical relationships that model the geographical variation of tropical-cyclone genesis in the present climate relatively well, but several investigators have cautioned against the use of these relationships in a climate change context (Ryan *et al.*, 1992; Royer *et al.*, 1998). Royer *et al.* proposed a modified form of the Gray relationships based on a measure of convective rainfall as opposed to SST or oceanic heat content, but this alternative has not been widely tested. They showed that tropical-cyclone frequency results for a future climate scenario depended strongly on whether the modified or unmodified genesis parameter approach was used. More recently, Emanuel and Nolan (2004) and Nolan *et al.* (2006) have developed a new empirical scheme designed to be more appropriate for climate change application (see also Camargo *et al.*, 2006), but tropical-cyclone frequency/climate change scenarios with this framework have not been published to date.

Models indicate increasing intensity and frequency of the strongest hurricanes and typhoons in a CO_2-warmed climate.

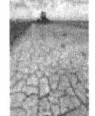

Vecchi and Soden (2007) have assessed the different components of the Emanuel and Nolan (2004) scheme using outputs from the IPCC AR4 models. Their results suggest that a decrease in tropical cyclone frequency may occur over some parts of the Atlantic basin associated with a SW-NE oriented band of less favorable conditions for tropical cyclogenesis and intensification, including enhanced vertical wind shear, reduced mid-tropospheric relative humidity, and slight decrease in potential intensity. The enhanced vertical shear feature (present in about 14 of 18 models in the Caribbean region) also extends into the main cyclogenesis region of the Eastern Pacific basin. Physically, this projection is related to the weakening of the east-west oriented Walker Circulation in the Pacific region, similar to that occurring during El Niño events. During El Niño conditions in the present-day climate, hurricane activity is reduced, as occurred, for example, in the latter part of the 2006 season. While this projection may appear at odds with observational evidence for an increase in Atlantic tropical cyclone counts during the past century (Holland and Webster, 2007; Vecchi and Knutson, 2008), there is evidence that this has occurred in conjunction with a regional decreasing trend in storm occurrence and formation rates in the western part of the Caribbean and Gulf of Mexico (Vecchi and Knutson, 2008; Holland, 2007). Earlier, Knutson and Tuleya (2004) had examined the vertical wind shear of the zonal wind component for a key region of the tropical Atlantic basin using nine different coupled models from the CMIP2+ project. Their analysis showed a slight preference for increased vertical shear under high CO_2 conditions if all of the models are considered, and a somewhat greater preference for increased shear if only the six models with the most realistic present-day simulation of shear in the basin are considered. Note that these studies are based on different sets of models, and that a more idealized future forcing scenario was used in the earlier Knutson and Tuleya study.

Alternative approaches to the empirical analysis of large-scale fields are the global and regional climate simulations, in which the occurrence of model tropical cyclones can be tracked. Beginning with the early studies of Broccoli and Manabe (1990), Haarsma et al. (1993), and Bengtsson et al. (1996), a number of investigators have shown that global models can generate tropical cyclone-like disturbances in roughly the correct geographical locations with roughly the correct seasonal timing. The annual occurrence rate of these systems can be quite model dependent (Camargo et al., 2005), and is apparently sensitive to various aspects of model physics (e.g., Vitart et al., 2001).

The notion of using global models to simulate the climate change response of tropical cyclone counts is given some support by several studies showing that such models can successfully simulate certain aspects of interannual to interdecadal variability of tropical-cyclone occurrence seen in the real world (Vitart et al., 1997; Camargo et al., 2005; Vitart and Anderson, 2001). A recent regional model dynamical downscaling study (Knutson et al., 2007) with an 18 km grid model and a more idealized modeling approach (Emanuel et al., 2008) both indicate that the increase in hurricane activity in the Atlantic from 1980-2005 can be reproduced in a model using specified SSTs and large-scale historical atmospheric information from reanalyses.

Since tropical cyclones are relatively rare events and can exhibit large interannual to interdecadal variability, large sample sizes (i.e., many seasons) are typically required to test the significance of any changes in a model simulation against the model's "natural variability."

The most recent future projection results obtained from medium and high resolution (120 km-20 km) GCMs are summarized in Table 3.2. Among these models, those with higher resolution indicate a consistent signal of fewer tropical cyclones globally in a warmer climate, while two lower resolution models find essentially no change. There are, however, regional variations in the sign of the changes, and these vary substantially between models (Table 3.2). For the North Atlantic in particular, more tropical cyclones are projected in some models, despite a large reduction globally (Sugi et al., 2002; Oouchi et al., 2006), while fewer Atlantic tropical cyclones are projected by other models (e.g., McDonald et al., 2005; Bengtsson et al., 2007). It is not clear at present how the Sugi et al. (2002) and Oouchi et al. (2006) results for

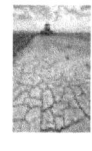

While there is recent observational evidence for an increase in the number of tropical storms and hurricanes in the Atlantic, a confident assessment of future storm frequency cannot be made at this time.

Table ... Summary ... tropical cyclone frequency ... expressed as a Percent of Present day levels as simulated by several climate ...C...s under global warming conditions...

Reference	Model	Resolution	Experiment	Global	North Atlantic	NW Pacific	NE Pacific
Sugi et al. 2002	JMA time slice	T106 L21 (~120km)	10y 1xCO2, 2xCO2	_66_	**1□**	_34_	33
Tsutsui 2002	NCAR CCM2	T42 L18	10y 1xCO2 2xCO2 from 115y CO2 1% pa	102	86	111	91
McDonald et al. 2005	HadAM3 time slice	N144 L30 (~100km)	15y IS95a 1979-1994 2082-2097	_94_	_75_	_70_	**1□□**
Hasegawa and Emori 2005	CCSR/NIES/FRCGC time slice	T106 L56 (~120km)	5x20y at 1xCO2 7x20y at 2xCO2			96	
Yoshimura et al. 2006	JMA time slice	T106 L21 (~120km)	10y 1xCO2, 2xCO2	_85_			
Bengtsson et al. 2006	ECHAM5-OM	T63 L31 1.5° L40	AIB 3 members 30y 20C and 21C	94			
Oouchi et al. 2006	MRI/JMA time slice	TL959 L60 (~20km)	10y AIB 1982-1993 2080-2099	_70_	**1□□**	_62_	_66_
Chauvin et al. 2006	ARPEGE-Climate time slice	Stretched non-uniform grid (~50 km)	10y CNRM SRES-B2: Hadley SRES-A2:		**11□** _75_		
Bengtsson et al. 2007	ECHAM5 time slice	up to T319 (down to ~30-40 km grid)	20yr, AIB scenario	---	_87_	_72_	_107_

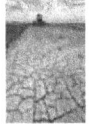

Bold = significantly **more** tropical cyclones in the future simulation
Italic = significantly _fewer_ tropical cyclones in the future simulation
Plain text = not significant or significance level not tested

the Atlantic reconciles with the tendency for increased vertical wind shear projected for parts of that basin by most recent models (Vecchi and Soden, 2007). For example, Oouchi et al. (2006) do not analyze how Atlantic vertical wind shear changed in their warming experiment. However, their results suggest that a future increase in tropical cyclone frequency in the Atlantic is at least plausible based on current models. Chauvin et al. (2006) and Emanuel et al. (2008) find, in multi-model experiments, that the sign of the changes in tropical cyclone frequency in the North Atlantic basin depends on the climate model used. All of the results cited here should be treated with some caution, as it is not always clear that these changes are greater than the model's natural variability, or that the natural variability or the tropical-cyclone genesis process are being properly simulated in the models, particularly for the Atlantic basin. For example, Oouchi et al. (2006) sample relatively short periods (20 years) from a single pair of experiments to examine greenhouse gas-induced changes, yet internal multidecadal variability in the Atlantic in their model conceivably could produce changes in tropical-cyclone-relevant fields (such as wind shear) between two 20-year periods that are larger than those for the radiative perturbation they are focusing on in their study.

From the above summarized results, it is not clear that current models provide a confident assessment of even the sign of change of tropical cyclone frequency in the Atlantic, East Pacific, or Northwest Pacific basins. From an observational perspective, recent studies (Chapter 2 this report) report that there has been a long term increase in Atlantic tropical-cyclone counts since the late 1800s, although the magnitude, and in some cases, statistical significance of the trend depends on adjustments for missing storms early in the record.

Based on the above available information, we assess that it is unknown how late 21st century tropical cyclone frequency in the Atlantic and North Pacific basins will change compared to the historical period (approximately 1950-2006).

3.3.9.4 TROPICAL CYCLONE PRECIPITATION

The notion that tropical cyclone precipitation rates could increase in a warmer climate is based on the hypothesis that moisture convergence into tropical cyclones will be enhanced by the increased column integrated water vapor – with the increased water vapor being extremely likely to accompany a warming of tropical SSTs. The increased moisture convergence would then be expected to lead to enhanced precipitation rates. This mechanism has been discussed in the context of extreme precipitation in general by Trenberth (1999), Allen and Ingram (2002), and Emori and Brown (2005). In contrast to the near-storm or storm core precipitation rate, accumulated rainfall at a locality along the storm's path is strongly dependent upon the speed of the storm, and there is little guidance at present on whether any change in this factor is likely in a future warmed climate.

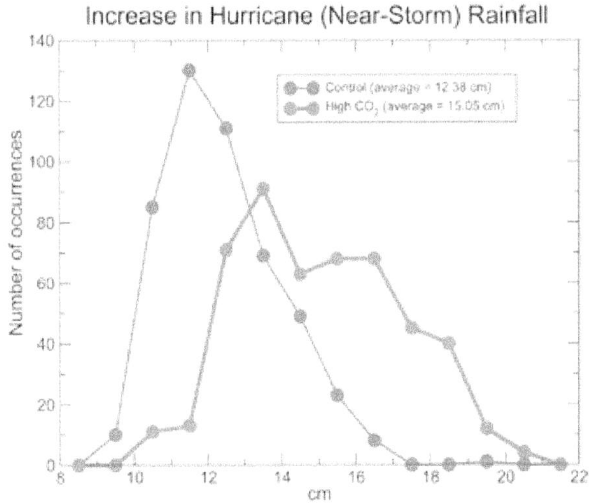

An enhanced near-storm tropical rainfall rate for high CO_2 conditions has been simulated, for example, by Knutson and Tuleya (2004, 2008) based on an idealized version of the GFDL hurricane model. The latter study reported an increase of 21.6% for a 1.75°C tropical SST warming (Figure 3.11), or about 12% per degree Celsius SST increase. Using a global model, Hasegawa and Emori (2005) found an increase in tropical-cyclone-related precipitation in a warmer climate in the western North Pacific basin, despite a decrease in tropical-cyclone intensity there in their model. Chauvin *et al.* (2006) found a similar result in the North Atlantic in their model, and Yoshimura *et al.* (2006) found a similar result on a global domain. There are issues with all of these modeling studies as they are, of course, low resolution, and thus generally depend on parameterization of much of the rainfall within the grid box. Further, there is a tendency towards tropical cyclone rainfall simulations that have a high bias in core rainfall rates (*e.g.* Marchok *et al.*, 2007). Nevertheless, the consistent result of an increased rainfall with greenhouse warming over a number of models, together with the theoretical expectations that this will occur, lends credibility to there being a real trend.

Based on the modeling studies to date, the relatively straightforward proposed physical mechanism, and the observed increases in extremely heavy rainfall in the United States (although not established observationally for hurricane-related rainfall [Groismann *et al.*, 2004]), it is likely that hurricane related rainfall (per storm) will increase in this century. Note that if the frequency of tropical cyclones decreases, the total rainfall from tropical cyclones may decrease. The expected general magnitude of the change for storm core rainfall rates is about +6% to +18% per degree Celsius increase in tropical SST.

It is likely that hurricane related rainfall (per storm) will increase in this century.

Increase in Hurricane (Near-Storm) Rainfall

Control (average = 12.38 cm)
High CO₂ (average = 15.05 cm)

Figure 3.11 As in Figure 3.8, but for near-hurricane precipitation, estimated as the average precipitation rate for the 102 model grid points (32,700 km² area) with highest accumulated rainfall over the last 6 hours of the 5-day idealized hurricane experiments in Knutson and Tuleya (2004). The results indicate that hurricanes in a CO_2-warmed climate will have substantially higher core rainfall rates than those in the present climate. From Knutson and Tuleya (2008).

3.3.9.5 TROPICAL CYCLONE SIZE, DURATION, TRACK, STORM SURGE, AND REGIONS OF OCCURRENCE

In this section, other possible impacts of greenhouse gas induced climate warming on tropical cyclones are briefly assessed. The assessment is highly preliminary, and the discussion for these relatively brief, owing to the lack of detailed studies on these possible impacts at this time.

Wu and Wang (2004) explored the issue of tropical cyclone track changes in a climate change context. Based on experiments derived from one climate model, they found some evidence for inferred track changes in the Northwest Pacific, although the pattern of changes was fairly complex.

Concerning storm duration, using an idealized hurricane simulation approach in which the potential intensity of a large sample of Atlantic basin storms with synthetically generated storm tracks was artificially increased by 10%, Emanuel (2006) found that the average storm lifetime of all storms increased by only 3%, whereas the average duration at hurricane intensity for those storms that attained hurricane intensity increased by 15%. However, in the Atlantic and Northeast Pacific, future changes in duration

Storm surge levels are likely to increase due to projected sea level rise.

are quite uncertain, owing to the uncertainties in formation locations and potential circulation changes mentioned previously.

Few studies have attempted to assess possible future changes in hurricane size. Knutson and Tuleya (1999) noted that the radius of hurricane-force winds increased a few percent in their experiments in which the intensities also increased a few percent.

Changes in tropical cyclone activity may be particularly apparent near the wings of the present climatological distributions. For example, locations near the periphery of current genesis regions may experience relatively large fractional changes in activity.

Storm surge depends on many factors, including storm intensity, size and track, local bathymetry, and the structure of coastal features such as wetlands and river inlets. Unknowns in storm frequency, tracks, size, and future changes to coastal features lead to considerable uncertainty in assessing storm surge changes. However, the high confidence of there being future sea level rise, as well as the likely increase of intensity of the strongest hurricanes, leads to an assessment that storm surge levels are likely to increase, though the degree of projected increase has not been adequately studied.

3.3.9.6 RECONCILIATION OF FUTURE PROJECTIONS AND PAST VARIATIONS

In this section, we comment on reconciling the future projections discussed above with the past-observed variations in tropical cyclone activity. A confident assessment of human influence on hurricanes will require further studies with models and observations, with emphasis on how human activity has contributed to the observed changes in hurricane activity through its influence on factors such as historical SSTs, wind shear, and vertical stability.

No published model study has directly simulated a substantial century-scale rise in Atlantic tropical cyclone counts similar to those reported for the observations (*e.g.*, Chapter 2). In fact, the 20th century behavior in tropical cyclone frequency has not yet been documented for existing models. One exception is Bengtsson *et al.* (2007) who simulate little change in

tropical cyclone frequencies comparing the late 1800s and late 1900s. A recent modeling study (Knutson *et al.*, 2007) indicates that the increase in hurricane activity in the Atlantic from 1980-2005 can be reproduced using a high-resolution nested regional model downscaling approach. However, the various changes in the large-scale atmospheric and SST forcings used to drive their regional model were prescribed from observations. Concerning future projections, the multi-model consensus of increased vertical wind shear in the IPCC AR4 models (Vecchi and Soden, 2007) further implies that it would be difficult to reconcile significant long-term increasing trends in tropical cyclone counts with existing models. If in further studies a significant anthropogenic signal were detected in observed tropical cyclone activity and confidently attributed to increasing GHGs, then this would imply that a future increase in tropical cyclone frequency in the Atlantic would be much more likely than assessed here.

☐☐☐☐ Extratr☐☐☐al ☐t☐rms

Scientists have used a variety of methods for diagnosing extratropical storms in GCM projections of future climate. These include sea-level pressure (Lambert and Fyfe, 2006), strong surface winds (Fischer-Bruns *et al.*, 2005), lower atmosphere vorticity (Bengtsson *et al.*, 2006), and significant wave heights (Wang *et al.*, 2004; Caires *et al.*, 2006). Consequently, there are no consistent definitions used to diagnose extreme extratropical storms. Some analyses do not, for example, determine events in extreme percentiles, but rather consider storms that deepen below a threshold sea-level pressure (*e.g.*, Lambert and Fyfe, 2006), though such thresholds may effectively select the most extreme percentiles.

Wave heights of course indicate strong storms

only over oceans, but the strongest extratropical storms typically occur in ocean storm tracks, so all three methods focus on similar regions. Ocean storms in the North Atlantic and North Pacific are relevant for this study because they affect coastal areas and shipping to and from North America. GCMs projecting climate change can supply sea-level pressure and surface winds, but they typically do not compute significant wave heights. Rather, empirical relationships (Wang *et al.*, 2004; Caires *et al.*, 2006) using sea-level pressure anomalies and gradients provide estimates of significant wave heights.

Despite the variety of diagnoses, some consistent changes emerge in analyses of extratropical storms under anthropogenic greenhouse warming. Projections of future climate indicate strong storms will be more frequent (Figure 3.12; Wang *et al.*, 2004; Fischer-Bruns *et al.*, 2005; Bengtsson *et al.*, 2006; Caires *et al.*, 2006; Lambert and Fyfe, 2006; Pinto *et al.*, 2007), though the overall number of storms may decrease. These changes are consistent with observed trends over the last half of the 20th century (Paciorek *et al.*, 2002). More frequent strong storms may reduce the frequency of all

> Projections of future climate indicate strong non-tropical storms will be more frequent, though the overall number of storms may decrease, consistent with observed trends over the last 50 years.

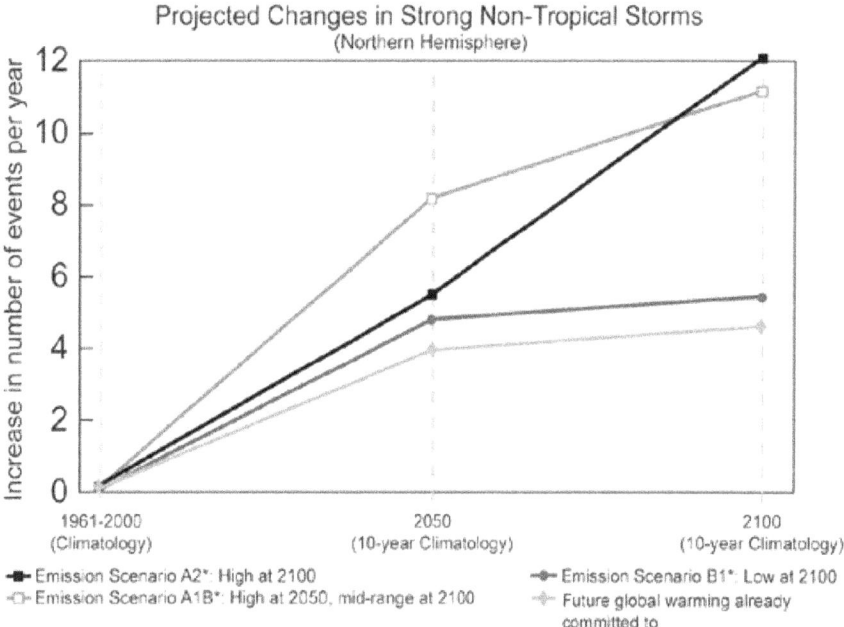

☐ig☐re ☐☐☐ The projected change in intense low pressure systems (strong storms) during the cold season for the Northern Hemisphere for various emission scenarios* (adapted from Lambert and Fyfe; 2006). Storms counted have central pressures less than 970 mb and occur poleward of 30°N during the 120-day season starting November 15. Adapted from Lambert and Fyfe (2006).

extratropical storms by increasing the stability of the atmosphere (Lambert and Fyfe 2006). Analyses of strong winds (Fischer-Bruns *et al.*, 2005, Pinto *et al.*, 2007), lower atmosphere vorticity (Bengtsson *et al.*, 2006), and significant wave heights (Wang *et al.*, 2004; Caires *et al.*, 2006) from single models suggest increased storm strength in the northeast Atlantic, but this increase is not apparent in an analysis using output from multiple GCMs (Lambert and Fyfe, 2006). Differences may be due to the focus on cold season behavior in the wind and wave analyses, whereas Lambert and Fyfe's (2006) analysis includes the entire year.

The warming projected for the 21st century is largest in the high latitudes due to a poleward retreat of snow and ice resulting in enhanced warming (Manabe and Stouffer 1980; Meehl *et al.*, 2007a). Projected seasonal changes in sea ice extent show summertime ice area declining much more rapidly than wintertime ice area, and that sea ice thins largest where it is initially the thickest, which is consistent with observed sea ice thinning in the late 20th century (Meehl *et al.*, 2007a). Increased storm strength in the northeast Atlantic found by some may be linked to the poleward retreat of arctic ice (Fischer-Bruns *et al.*, 2005) and a tendency toward less frequent blocking and more frequent positive phase of the NAM (Pinto *et al.*, 2007), though further analysis is needed to diagnose physical associations with ice line, atmospheric temperature, and pressure structures and storm behavior. Whether or not storm strength increases, the retreat of sea ice together with changing sea levels will likely increase the exposure of arctic coastlines to damaging waves and extreme erosion produced by strong storms (Lynch *et al.*, 2004; Brunner *et al.*, 2004; Cassano *et al.*, 2006), continuing an observed trend of increasing coastal erosion in arctic Alaska (Mars and Houseknecht, 2007). Rising sea levels, of course, may expose all coastlines to more extreme wave heights (*e.g.*, Cayan *et al.*, 2008).

Convective Storms

Conclusions about possible changes in convective precipitating storms and associated severe-weather hazards under elevated greenhouse gas concentrations have remained elusive. Perhaps the most important reason for this

is the mesoscale (10s of km) and smaller dynamics that control behavior of these storms, particularly the initiation of storms. Del Genio *et al.* (2007), Marsh *et al.* (2007) and Trapp *et al.* (2007) have evaluated changes in the frequency of environments that are favorable for severe thunderstorms in GCM simulations of greenhouse-enhanced climates. In all three cases, increases in the frequency of environments favorable to severe thunderstorms are seen, but the absence of the mesoscale details in the models means that the results are preliminary. Nevertheless, the approach and the use of nested models within the GCMs show promise for yielding estimates of changes in extreme convective storms.

The retreat of sea ice together with changing sea levels will likely increase the exposure of arctic coastlines to damaging waves and extreme erosion produced by strong storms, continuing an observed trend of increasing coastal erosion in arctic Alaska.

CHAPTER 4

☐easures To Improve Our Understanding of Weather and Climate Extremes

Convening Lead Author: David R. Easterling, NOAA

Lead Authors: David M. Anderson, NOAA; Stewart J.Cohen, Environment Canada and University of British Columbia; William J. Gutowski, Jr., Iowa State Univ.; Greg J. Holland, NCAR; Kenneth E. Kunkel, Univ. Ill. Urbana-Champaign, Ill. State Water Survey; Thomas C. Peterson, NOAA; Roger S. Pulwarty, NOAA; Ronald J. Stouffer, NOAA; Michael F. Wehner, Lawrence Berkeley National Laboratory

☐☐C☐☐☐☐☐☐☐

In this chapter we identify areas of research and activities that can improve our understanding of weather and climate extremes. Many of these research areas and activities are consistent with previous reports, especially the CCSP SAP 1.1 report, *Temperature Trends in the Lower Atmosphere: Steps for Understanding and Reconciling Differences,* on reconciling temperature trends between the surface and free atmosphere.

Many types of extremes, such as excessively hot and cold days, drought, and heavy precipitation show changes over North America consistent with observed warming of the climate. Regarding future changes, model projections show large changes in warm and cold days consistent with projected warming of the climate by the end of the 21st century. However, there remains uncertainty in both observed changes, due to the quality and homogeneity of the observations, and in model projection, due to constraints in model formulation, in a number of other types of climate extremes, including tropical cyclones, extratropical cyclones, tornadoes, and thunderstorms.

☐Ⅲ

> ☐he ☐☐htin☐ed de☐el☐☐ment and maintenan☐e ☐☐high ☐☐alit☐ ☐imate ☐☐ser☐ing s☐stems ☐ill im☐r☐☐e ☐☐r a☐ilit☐t☐m☐nit☐r and dete☐t ☐☐t☐re ☐hanges in ☐imate extremes☐

Recently, more emphasis has been placed on the development of true climate observing networks that adhere to the Global Climate Observing System (GCOS) Climate Monitoring Principles. This is exemplified by the establishment in the United States of the Climate Reference Network, in Canada of the Reference Climate Network, and recent efforts in Mexico to establish a climate observing network. Stations in these networks are carefully sited and instrumented and are designed to be benchmark observing systems adequate to detect the true climate signal for the region being monitored.

Similar efforts to establish a high-quality, global upper-air reference network have been undertaken under the auspices of GCOS. However, this GCOS Reference Upper-Air Network (GRUAN) is dependent on the use of current and proposed new observing

The intent of these data adjustments is to approximate homogeneous time series where the variations are only due to variations in climate and not due to the non-climatic changes discussed above. However, the use of these adjustment schemes introduces another layer of uncertainty into the results of analyses of climate variability and change. Thus, research into both the methods and quantifying uncertainties introduced through use of these methods would improve understanding of observed changes in climate.

Even with the recent efforts to develop true climate observing networks, an understanding of natural and human effects on historical weather and climate extremes is best achieved through study of very long (century-scale) records because of the presence of multidecadal modes of variability in the climate system. For many of the extremes discussed here, including temperature and precipitation extremes, storms, and drought, there are significant challenges in this regard because long-term, high-quality, homogeneous records are not available. For example, recent efforts have been made in the United States to digitize surface climate data for the 19th century; however, using these data poses several problems. The density of stations was also considerably less than in the 20th century. Equipment and observational procedures were quite variable and different than the standards established within the U.S. Cooperative Network (COOP) in the 1890s. Thus, the raw data are not directly comparable to COOP data. However, initial efforts to homogenize these data have been completed and analysis shows interesting features, including high frequencies of extreme precipitation and low frequencies of heat waves for the 1850-1905 period over the conterminous United States.

stations, whose locations will be determined through observing system simulation experiments (OSSEs) that use both climate model simulations and observations to determine where best to locate new observing stations.

However, at the present these efforts generally are restricted to a few countries and large areas of the world, even large parts of North America remain under observed. Developing climate observing networks, especially in areas that traditionally have not had long-term climate observations, would improve our ability to monitor and detect future changes in climate, including extremes.

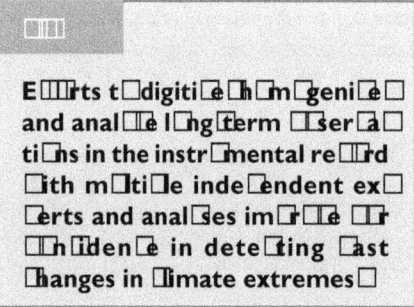

E⬚⬚rts t⬚digiti⬚e ⬚h⬚m⬚geni⬚e⬚ and anal⬚⬚e l⬚ng⬚term ⬚⬚ser⬚a⬚ ti⬚hs in the instr⬚mental re⬚⬚rd ⬚ith m⬚lti⬚le inde⬚endent ex⬚ ⬚erts and anal⬚ses im⬚r⬚⬚e ⬚⬚r ⬚⬚h⬚iden⬚e in dete⬚ting ⬚ast ⬚hanges in ⬚limate extremes⬚

Research using homogeneity-adjusted observations provides a better understanding of climate system variability in extremes. Observations of past climate have, by necessity, relied on observations from weather observing networks established for producing and verifying weather forecasts. In order to make use of these datasets in climate analyses, non-climatic changes in the data, such as changes due to station relocations, land-use change, instrument changes, and observing practices must be accounted for through data adjustment schemes.

In some cases, heterogeneous records of great length are available and useful information has been extracted. However, there are many opportunities where additional research may result in longer and higher quality records to better characterize the historical variations. For example, the ongoing uncertainty and debate about tropical cyclone trends is rooted in the heterogeneous nature of the observations and different approaches toward approximating homogeneous time series. Efforts to resolve the existing uncertainties in tropical cyclone frequency and intensity should continue by the re-examination of the heterogeneous records by a variety of experts to insure that multiple perspectives on tropical cyclone frequency are included in critical data sets and analyses. However, this notion of multiple independent experts and analyses should not be restricted to the question of tropical cyclone frequency, but should be applied to all aspects of climate research.

> Weather □ser□ng s□stems ad□hering t□standards □□□□ser□a□ti□n □□nsistent □ith the needs □□□□th the □limate and the □eather resear□h □□mm□nities im□r□□e □□r a□ilit□t□dete□t □□ser□ed □hanges in □limate extremes□

Smaller-scale storms, such as thunderstorms and tornadoes are particularly difficult to observe since historical observations have been highly dependent on population density. For example, the U.S. record of tornadoes shows a questionable upward trend that appears to be due mainly to increases in population density in tornado-prone regions. With more people in these regions, tornadoes that may have gone unobserved in earlier parts of the record are now being recorded, thus hampering any analysis of true climate trends of these storms. Since many of the observations of extreme events are collected in support of operational weather forecasting, changes in policies and procedures regarding those observations need to take climate change questions into account in order to collect high-quality, consistently collected data over time and space. Therefore, consistent standards of collection of data about tornadoes and severe thunderstorms would be beneficial. Included in this process is a need for the collection of information about reports that allows users to know the confidence levels that can be applied to reports.

However, in the absence of homogeneous observations of extremes, such as thunderstorms

and tornadoes, one promising method to infer changes is through the use of surrogate measures. For example, since the data available to study past trends in these kinds of storms suffer from the problems outlined above, an innovative way to study past changes lies in techniques that relate environmental conditions to the occurrence of thunderstorms and tornadoes. Studies along these lines could then produce better relationships than presently exist between favorable environments and storms. Those relationships could then be applied to past historical environmental observations and reanalysis data to make improved estimates of long-term trends.

> Extended re□□ntr□□ti□ns □□□ast □limate □sing □eather m□dels initiali□ed □ith h□m□gen□□s s□r□a□e □□ser□ati□ns □□□ld hel□ im□r□□e □□r □nderstanding □□ str□ng extratr□□i□al □□□□hes and □ther as□e□ts □□□limate □aria□ilit□□□

Studies of the temporal variations in the frequency of strong extratropical cyclones have typically examined the past 50 years and had to rely on reanalysis fields due to inconsistencies with the historical record. But a much longer period would enable a better understanding of possible multidecadal variability in strong storms. There are surface pressure observations extending back to the 19th century and, although the spatial density of stations decreases backwards in time, it may be possible to identify strong extratropical cyclones and make some deductions about long-term variations. Additionally, efforts to extend reanalysis products back to the early 20th century using only surface observations have recently begun. These efforts are desirable since they provide physically-consistent depictions of climate behavior and contribute to an understanding of causes of observed changes in climate extremes.

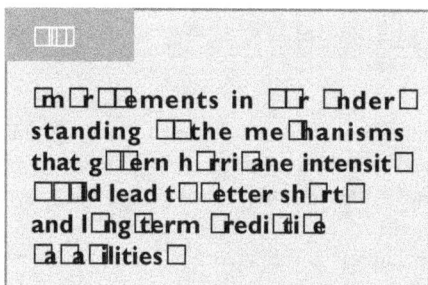

☐he ☐reati☐n ☐☐ann☐all☐re☐ s☐☐ed☐regi☐nal☐s☐ale re☐☐n☐ str☐☐ti☐ns ☐☐the ☐limate ☐☐r the ☐ast ☐☐☐☐☐ears ☐☐☐d hel☐ im☐r☐☐e ☐☐r ☐nderstanding ☐☐ ☐er☐☐ng☐term regi☐nal ☐limate ☐aria☐ilit☐☐

The development of a wide-array of climate reconstructions for the last two millennia, such as temperature, precipitation, and drought would provide a longer baseline to analyze infrequent extreme events, such as those occurring once a century or less. This and other paleoclimatic research can also answer the question of how extremes change when the global climate was warmer and colder than today.

The instrumental record of climate is generally limited to the past 150 years or so. Although there are observations of temperature and precipitation as recorded by thermometers and rain gauges for some locations prior to the early to mid-1800s, they are few and contain problems due to inconsistent observing practices, thus their utility is limited. However, the paleoclimate record covering the past 2,000 years

and beyond reveals extremes of greater amplitude and longer duration compared to events observed in the instrumental record of the past 100 years (e.g.,Woodhouse and Overpeck, 1998). The paleoclimate record also reveals that some events occur so infrequently that they may be observed only once, or even not at all, during the instrumental period. An improved array of paleo time series would improve understanding of the repeat frequency of rare events, for example, events occurring only once a century.

The frequency of some extremes appears tied to the background climate state, according to some paleoclimate records. For example, century-scale changes in the position of the subtropical high may have affected hurricane tracks and the frequency of hurricanes in the Gulf of Mexico (Elsner et al., 2000). Throughout the western United States, the area exposed to drought may have been elevated for four centuries from 900-1300 A.D., according to the Palmer Drought Severity Index reconstructed from tree rings (Cook et al., 2004). The period from 900-1300 A.D. was a period when the global mean temperature was above average (Mann et al., 1999), consistent with the possibility that changes in the background climate state can affect some extremes. The paleoclimatic record can be used to further understand the possible changes in extremes during warmer and colder climates of the past.

☐m☐r☐☐ements in ☐☐r ☐nder☐ standing ☐☐the me☐hanisms that g☐☐ern h☐☐ri☐ane intensit☐ ☐☐☐d lead t☐☐etter sh☐rt☐ and l☐ng☐term ☐redi☐☐i☐e ☐a☐a☐ilities☐

A major limitation of our current knowledge lies in the understanding of hurricane intensity together with surface wind structure and rainfall, and particularly how these relate to a combination of external forcing from the ocean and surrounding atmosphere, and potentially chaotic internal processes. This lack of understanding and related low predictive capacity has been recognized by several expert committees set up in the wake of the disastrous 2005 Atlantic hurricane season:

The National Science Board recommended that the relevant federal agencies commit to a major hurricane research program to reduce the impacts of hurricanes and encompassing all aspects of the problem: physical sciences, engineering, social, behavioral, economic, and ecological (NSB, 2006);

- The NOAA Science Advisory Board established an expert Hurricane Intensity Research Working Group that recommended specific action on hurricane intensity and rainfall prediction (NOAA SAB, 2006);

- The American Geophysical Union convened a meeting of scientific experts to produce a white paper recommending action across all science-engineering and community levels (AGU, 2006); and

- A group of leading hurricane experts convened several workshops to develop priorities and strategies for addressing the most critical hurricane issues (HiFi, 2006).

While much of the focus for these groups was on the short-range forecasting and impacts reduction aspects of hurricanes, the research recommendations also apply to longer term projections. Understanding the manner in which hurricanes respond to their immediate atmospheric and oceanic environment would improve prediction on all scales.

An issue common to all of these expert findings is the need for understanding and parameterization of the complex interactions occurring at the high-wind oceanic interface and for very high model resolution in order for forecast models to be able to capture the peak intensity and fluctuations in intensity of major hurricanes. Climate models are arriving at the capacity to resolve regional structures but not relevant details of the hurricane core region. As such, some form of statistical inference will be required to fully assess future intensity projections.

> **Esta▢ishing a gl▢▢all▢▢▢▢hsistent ▢ind de▢initi▢n ▢▢r determining h▢rri▢ane intensit▢▢▢▢▢d all▢▢ ▢▢r m▢▢e ▢▢hsistent ▢▢m▢ari▢ s▢hs a▢▢▢s the gl▢▢e▢**

A major issue with determining the intensity of hurricanes lies with the definition of wind speed. The United States uses a one-minute average and the rest of the world uses ten minutes, which causes public confusion. For example, a hurricane could be reported using one-minute

winds while it was still a tropical storm by a ten-minute standard. This is becoming more serious with the spread of global news in which a U.S. television channel can be reporting a hurricane while the local news service is not. A related issue lies with comparing aircraft and numerical model "winds" with those recorded by anemometers: an aircraft is moving through the atmosphere and reports an inherently different wind; the wind speed in a model is dependent on the grid spacing and time step used.

> **▢m▢▢▢ements in the a▢lit▢ ▢▢▢limate m▢dels t▢re▢reate the re▢ent ▢ast as ▢ell as ma▢e ▢r▢e▢ti▢hs ▢nder a ▢ariet▢▢▢▢ ▢▢r▢ing s▢enari▢s are de▢endent ▢n a▢▢ess t▢▢▢th ▢▢m▢▢tati▢nal and h▢man res▢▢r▢es▢**

The continued development and improvement of numerical climate models, and observational networks for that matter, is highly related to funding levels of these activities. A key factor is the recruitment and retention of people necessary to perform the analysis of models and observations. For the development and analysis of models, scientists are drawn to institutions with supercomputing resources. For example, the high resolution global simulations of Oouchi *et al.* (2006) to predict future hurricane activity are currently beyond the reach of U.S. tropical cyclone research scientists. This limitation is also true for other smaller-scale storm systems, such as severe thunderstorms and tornadoes. Yet, to understand how these extreme events might change in the future it is critical that climate models are developed that can realistically resolve these types of weather systems. Given sufficient computing resources, current U.S. climate models can achieve very high horizontal resolution. Current generation high performance computing (HPC) platforms are also sufficient, provided that enough access to computational cycles is made available. Furthermore, many other aspects of the climate system relevant to extreme events, such as extra-tropical cyclones, would be much better simulated in such integrations than they

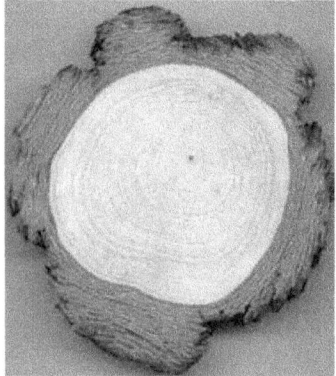

are at current typical global model resolutions.

Even atmospheric models at approximately 20 kilometer horizontal resolution are still not finely resolved enough to simulate the high wind speeds and low pressure centers of the most intense hurricanes (Category 5 on the Saffir-Simpson scale). Realistically capturing details of such intense hurricanes, such as the inner eyewall structure, will require models up to one kilometer horizontal resolution. Such ultra-high resolution global models will require very high computational rates to be viable (Wehner *et al.*, 2008). This is not beyond the reach of next generation HPC platforms but will need significant investments in both model development (human resources) as well as in dedicated computational infrastructure (Randall, 2005).

☐☐re extensi☐e a☐☐ess t☐high tem☐☐ral res☐☐ti☐n data ☐dail☐☐ h☐☐rl☐☐☐☐☐m ☐limate m☐del sim☐lati☐ns ☐th ☐☐the ☐ast and ☐☐r the ☐☐t☐re ☐☐☐☐d all☐☐ ☐☐r im☐r☐☐ed ☐nderstanding ☐☐☐☐☐ tential ☐hanges in ☐eather and ☐limate extremes☐

In order to achieve high levels of statistical confidence in analyses of climate extremes using methods such as those based on generalized extreme value theory, lengthy stationary datasets are required. Although climate model output is well suited to such analysis, the datasets are often unavailable to the research community at large. Many of the models utilized for the Intergovernmental Panel on Climate Change Fourth Assessment Report (IPCC AR4) were integrated as ensembles, permitting more robust statistical analysis. The simulations were made available at the Program for Climate Model Diagnostics and Intercomparison (PCMDI) at Law-

rence Livermore National Laboratory. However, the higher temporal resolution data necessary to analyze extreme events is quite incomplete in the PCMDI database, with only four models represented in the daily averaged output sections with ensemble sizes greater than three realizations and many models not represented at all. Lastly, a critical component of this work is the development of enhanced data management and delivery capabilities such as those in the NOAA Operational Model Archive and Distribution System (NOMADS), not only for archive and delivery of model simulations, but for reanalysis and observational data sets as well (NRC, 2006).

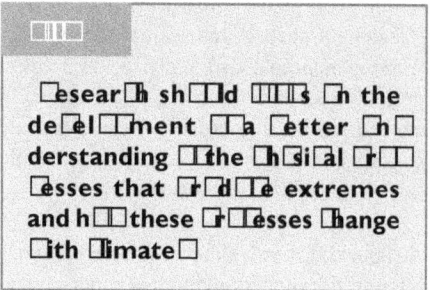

☐esear☐h sh☐☐d ☐☐☐☐s ☐n the de☐el☐☐ment ☐☐a ☐etter ☐h☐ derstanding ☐☐the ☐h☐si☐al ☐r☐☐ ☐esses that ☐r☐d☐☐e extremes and h☐☐these ☐r☐☐esses ☐hange ☐ith ☐limate☐

Analyses should include attribution of probability distribution changes to natural or anthropogenic influences, comparison of individual events in contemporary and projected climates, and the synoptic climatology of extremes and its change in projected climates. The ultimate goal should be a deeper understanding of the physical basis for changes in extremes that improves modeling and thus lends confidence in projected changes.

Literature is lacking that analyzes the physical processes producing extremes and their changes as climate changes. One area that is particularly sparse is analysis of so-called "compound extremes," events that contain more than one type of extreme, such as drought and extremely high temperatures occurring simultaneously.

A substantial body of work has emerged on attribution of changes, with a growing subset dealing with attribution of changes in extremes. Such work shows associations between climate forcing mechanisms and changes in extremes, which is an important first step toward understanding what changes in extremes are attributable to climate change. However, such work typically does not examine the coordinated physical processes linking the extreme behavior to the climate in which it occurs.

More effort should be dedicated to showing how the physical processes producing extremes are changing. Good examples are studies by Meehl and Tebaldi (2004) on severe heat waves, Meehl *et al.* (2004) on changes in frost days, and Meehl and Hu (2006) on megadroughts. Each of these examples involves diagnosing a coherent set of climate-system processes that yield the extreme behavior. An important aspect of the work is demonstrating correspondence between observed and simulated physical processes that yield extremes and, in some of these cases, evaluation of changes in the physical processes in projected climates.

More broadly, the need is for greater analysis of the physical climatology of the climate system leading to extremes. Included in this are further studies of the relationship in projected future climates between slow oscillation modes, such as PDO and AO, and variation in extremes (*e.g.*, Thompson and Wallace, 2001). Methods of synoptic climatology (*e.g.*, Cassano *et al.* 2006; Lynch *et al.* 2006) could also provide deeper physical insight into the processes producing extremes and their projected changes. Also, the development and use of environmental proxies for smaller storm systems, such as severe thunderstorms and tornadoes from regional and nested climate models, is encouraged. Finally, more probability analysis of the type applied by Stott *et al.* (2004) to the 2003 European heat wave is needed to determine how much the likelihood of individual extreme events has been altered by human influences on climate.

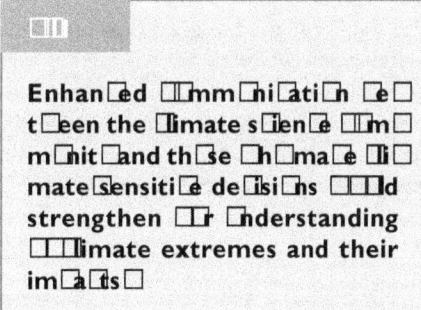

Enhanced communication between the climate science community and those climate sensitive decisions could strengthen our understanding of climate extremes and their impacts.

Because extremes can have major impacts on socioeconomic and natural systems, changes in climate extremes (frequency, timing, magnitude) will affect the ability of states, provinces and local communities to cope with rare weather events. The process of adaptation to climate change begins with addressing existing vulnerabilities to current and near-term climatic extremes and is directly linked to disaster risk management. Research and experience have shown that reducing the impacts of extremes and associated complex multiple-stress risks, require improvements in i) early warning systems, ii) information for supporting better land-use planning and resource management, iii) building codes, and iv) coordination of contingency planning for pre- and post-event mitigation and response.

Because the links between impacts and changes in extremes can be complex, unexpected, and highly nonlinear, especially when modified by human interventions over time, research into these linkages should be strengthened to better understand system vulnerabilities and capacity, to develop a portfolio of best practices, and to implement better response options. But best practices guidelines do not do any good unless they are adequately communicated to the relevant people. Therefore, mechanisms for collaboration and exchange of information among climate scientists, impacts researchers, decision makers (including resources managers, insurers, emergency officials, and planners), and the public should be developed and supported. Such mechanisms would involve multi-way information exchange systems and pathways. Better communication between these groups would help communities and individuals make the most appropriate responses to changing extremes. As climate changes, making the complexities of climate risk management explicit can transform event to event response into a learning process for informed proactive management. In such learning-by-doing approaches, the base of knowledge is enhanced through the accumulation of practical experience for risk scenario development and disaster mitigation and preparedness.

☐☐☐

☐ relia☐le data☐ase that lin☐s ☐eather and ☐limate extremes ☐ith their im☐a☐ts ☐n☐l☐ding damages and ☐☐sts ☐nder ☐hang☐ ing s☐☐i☐e☐☐h☐mi☐☐☐☐nditi☐ns☐ ☐☐☐ld hel☐☐☐r ☐nderstanding ☐☐these e☐ents☐

Many adaptations can be implemented at low cost, but comprehensive estimates of adaptation costs and benefits are currently limited, partly because detailed information is not adequately archived and made available to researchers. To address this problem, guidelines should be developed to improve the methods to collect, archive, and quality control detailed information on impacts of extreme events and sequences of extremes, including costs (of emergency responses), insured and uninsured loss estimates (including property and commercial/business), loss of life, and ecological damage, as well as the effectiveness of post event responses. Additionally, networks of systematic observations of key elements of physical, biological, and socioeconomic systems affected by climate extremes should be developed, particularly in regions where such networks are already known to be deficient.

There are increasing calls for more and improved structured processes to assess climatic risks and communicate such information for economic and environmental benefit (Pulwarty *et al.*, 2007). Fortunately, there are prototypes of such processes and programs from which lessons may be drawn. The NOAA Regional Integrated Sciences and Assessments (RISA) program represents a widely acknowledged, successful effort to increase awareness of climate risks and to foster multi-way risk communication between research and practitioners. RISAs conduct research that addresses critical complex climate sensitive issues of concern to decision makers and policy planners at a regional level. The RISAs are primarily based at universities with some team members based at government research facilities, non-profit organizations, or private sector entities. More recently, state governments, Watershed Commissions and federal agencies are requesting that the information products of programs, such as RISAs and Regional Climate Centers, be coordinated across agencies, states, and tribal nations to inform proactive risk-reduction measures and adaptation practices (Pulwarty *et al.*, 2007).

In the case of drought, the creation of the National Integrated Drought Information System (NIDIS) represents the culmination of many years of experience from scientific inquiry, monitoring insight and socioeconomic impacts. The NIDIS is based on interagency teams and working groups at federal, state, and local levels. This broad experience base and increasingly cross-sectoral vulnerability to extremes points to the need for a unified federal policy to help states and local communities prepare for and mitigate the damaging effects of drought across temporal and spatial scales (NIDIS Act 2006 Public Law 109-430). The NIDIS is a dynamic and accessible drought risk information system. It provides com-

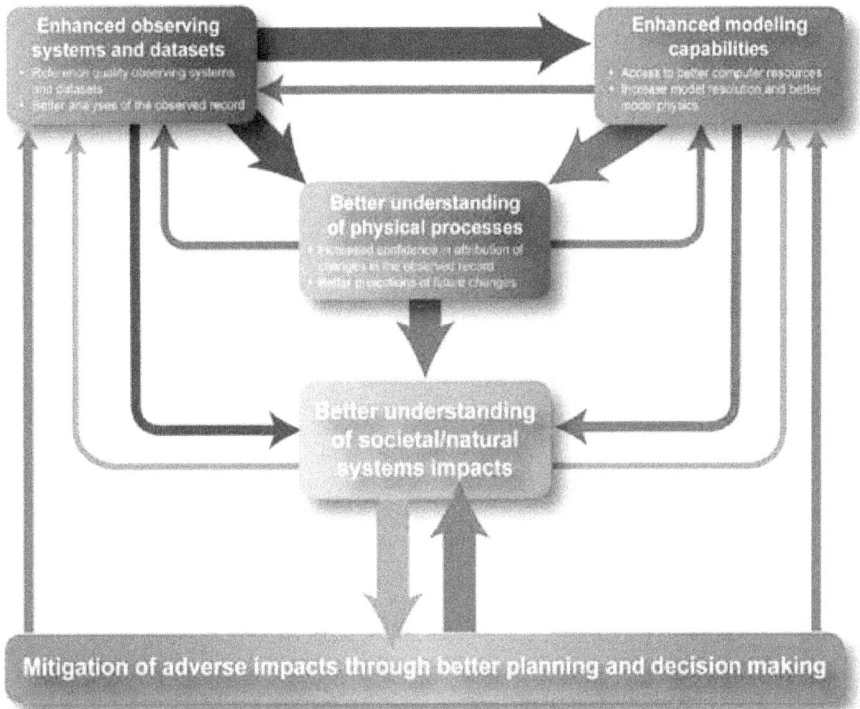

Figure 4.1 Interrelationships between inputs and components leading to better understanding. Thick arrows indicate major linkages included in this assessment. Better observing systems result in improved analyses which helps improve modeling, physical understanding, and impacts through clearer documentation of observed patterns in climate. Similarly, improved modeling helps improve physical understanding and, together, can point to deficiencies in observing systems as well as helping to understand future impacts. Lastly, a better understanding of the relationships between climate extremes and impacts can help improve observations by identifying deficiencies in observations (e.g., under-observed areas), and improve modeling efforts by identifying specific needs from model simulations for use in impacts studies.

munities engaged in drought preparedness with the capacity to determine the potential impacts of drought in their locale, and develop the decision support tools, such as risk management triggers, needed to better prepare for and mitigate the effects of drought. At present, NIDIS focuses on coordinating disparate federal, state, and local drought early warning systems and plans, and on acting as an integrated drought information clearinghouse, scaling up from county to watersheds and across timescales of climate variability and change.

Figure 4.1 shows the complex interrelationships

4.6 Summary

between the different sections and recommendations in this chapter. Enhanced observing systems and data sets allow better analyses of the observed climate record for patterns of observed variability and change. This provides information for the climate modeling community to verify that their models produce realistic simulations of the observed record, providing increased confidence in simulations of future climate. Both of these activities help improve our physical understanding of the climate which, linked with model simulations through observing system simulation experiments, helps understand where we need better observations, and leads to better formulation of model physics through process studies of observations. This link between observed and modeling patterns of climate change also provides the basis for establishing the cause and effect relationships critical for attribution of climate change to human activities. Since the ultimate goal of this assessment is to provide better information to policy and decision makers, a better understanding of the relationships between climate extremes and their impacts is critical information for reducing the vulnerability of societal and natural systems to climate extremes.

APPENDIX A

Statistical Trend Analysis

Coordinating Lead Author: Richard L. Smith, Univ. N.C., Chapel Hill

In many places in this report, but especially in Chapter 2, trends have been calculated, either based directly on some climatic variable of interest (*e.g.*, hurricane or cyclone counts) or from some index of extreme climate events. Statistical methods are used in determining the form of a trend, estimating the trend itself along with some measure of uncertainty (*e.g.*, a standard error), and in determining the statistical significance of a trend. A broad-based introduction to these concepts has been given by Wigley (2006). The present review extends Wigley's by introducing some of the more advanced statistical methods that involve time series analysis.

Some initial comments are appropriate about the purpose, and also the limitations, of statistical trend estimation. Real data rarely conform exactly to any statistical model, such as a normal distribution. Where there are trends, they may take many forms. For example, a trend may appear to follow a quadratic or exponential curve rather than a straight line, or it may appear to be superimposed on some cyclic behavior, or there may be sudden jumps (also called changepoints) as well or instead of a steadily increasing or decreasing trend. In these cases, assuming a simple linear trend (equation (1) below) may be misleading. However, the slope of a linear trend can still represent the most compact and convenient method of describing the overall change in some data over a given period of time.

In this appendix, we first outline some of the modern methods of trend estimation that involve estimating a linear or nonlinear trend in a correlated time series. Then, the methods are illustrated on a number of examples related to climate and weather extremes.

The basic statistical model for a linear trend can be represented by the equation

(1) $y_t = b_0 + b_1 t + u_t$

where t represents the year, y_t is the data value of interest (*e.g.*, temperature or some climate index in year t), b_0 and b_1 are the intercept and slope of the linear regression, and u_t represents a random error component. The simplest case is when u_t are uncorrelated error terms with mean 0 and a common variance, in which case we typically apply the standard ordinary least squares (OLS) formulas to estimate the intercept and slope, together with their standard errors. Usually the slope (b_1) is interpreted as a trend, so this is the primary quantity of interest.

The principal complication with this analysis in the case of climate data is usually that the data are autocorrelated; in other words, the terms cannot be taken as independent. This brings us within the field of statistics known as time series analysis, see *e.g.*, the book by Brockwell and Davis (2002). One common way to deal with this is to assume the values form an autoregressive, moving average process (ARMA for short). The standard ARMA(p,q) process is of the form

(2) $u_t - \varphi_1 u_{t-1} - \ldots - \varphi_p u_{t-p} = \varepsilon_t + \theta_1 \varepsilon_{t-1} + \ldots + \theta_q \varepsilon_{t-q}$

where $\varphi_1 \ldots \varphi_p$ are the autoregressive coefficients, $\theta_1 \ldots \theta_q$ are the moving average coefficients, and the ε_t terms are independent with mean 0 and common variance. The orders p and q are sometimes determined empirically, or sometimes through more formal model-determination techniques such as the Akaike Information Criterion (AIC) or the Bias-Corrected Akaike Information Criterion (AICC). The autoregressive and moving average coefficients may be determined by one of several estimation algorithms (including maximum likelihood), and the regression coefficients b_0 and b_1 by the algorithm of generalized least squares (GLS). Typically, the GLS estimates are not very different from the OLS estimates that arise when autocorrelation is ignored, but the standard errors can be very different. It is quite common that a trend that appears to be statistically significant when estimated under OLS regression is not statistically significant under GLS regression, because of the larger standard error that is usually though not invariably associated with

GLS. This is the main reason why it is important to take autocorrelation into account.

An alternative model which is an extension of (1) is

(3) $y_t = b_0 + b_1 x_{t1} + \ldots + b_k x_{tk} + u_t$

where $x_{t1} \ldots x_{tk}$ are k regression variables (covariates) and $b_1 \ldots b_k$ are the associated coefficients. A simple example is polynomial regression, where $x_{tj} = t^j$ for $j=1,\ldots,k$. However, a polynomial trend, when used to represent a nonlinear trend in a climatic dataset, often has the disadvantage that it behaves unstably at the endpoints, so alternative representations such as cubic splines are usually preferred. These can also be represented in the form of (3) with suitable $x_{t1} \ldots x_{tk}$. As with (1), the u_t terms can be taken as uncorrelated with mean 0 and common variance, in which case OLS regression is again appropriate, but it is also common to consider the u_t as autocorrelated.

There are, by now, several algorithms available that fit these models in a semiautomatic fashion. The book by Davis and Brockwell (2002) includes a CD containing a time series program, ITSM, that among many other features, will fit a model of the form (1) or (3) in which the u_t terms follow an ARMA model as in (2). The orders p and q may be specified by the user or selected automatically via AICC. Alternatively, the statistical language R (R Development Core Team, 2007) contains a function "arima" which allows for fitting these models by exact maximum likelihood. The inputs to the arima function include the time series, the covariates, and the orders p and q. The program calculates maximum likelihood/GLS estimates of the ARMA and regression parameters, together with their standard errors, and various other statistics including AIC. Although R does not contain an automated model selection procedure, it is straightforward to write a short subroutine that fits the time series model for various values of p and q (for example, all values of p and q between 0 and 10), and then identifies the model with minimum AIC. This method has been routinely used for several of the following analyses.

However, it is not always necessary to search through a large set of ARMA models. In very many cases, the AR(1) model in which p=1, q=0, captures almost all of the autocorrelation, in which case this would be the preferred approach.

In other cases, it may be found that there is cyclic behavior in the data corresponding to large-scale circulation indexes such as the Southern Oscillation Index (SOI – often taken as an indicator of El Niño) or the Atlantic Multidecadal Oscillation (AMO) or the Pacific Decadal Oscillation (PDO). In such cases, an alternative to searching for a high-order ARMA model may be to include SOI, AMO or PDO directly as one of the covariates in (2).

Two other practical features should be noted before we discuss specific examples. First, the methodology we have discussed assumes the observations are normally distributed with constant variances (homoscedastic). Sometimes it is necessary to make some transformation to improve the fit of these assumptions. Common transformations include taking logarithms or square roots. With data in the form of counts (such as hurricanes), a square root transformation is often made because count data are frequently represented by a Poisson distribution, and for that distribution, a square root transformation is a so-called variance-stabilizing transformation, making the data approximately homoscedastic.

The other practical feature that occurs quite frequently is that the same linear trend may not be apparent through all parts of the data. In that case, it is tempting to select the start and finish points of the time series and recalculate the trend just for that portion of the series. There is a danger in doing this, because in formally testing for the presence of a trend, the calculation of significance levels typically does not allow for the selection of a start and finish point. Thus, the procedure may end up selecting a spurious trend. On the other hand, it is sometimes possible to correct for this effect, for example, by using a Bonferroni correction procedure. An example of this is given in our analysis of the heatwave index dataset below.

E□□□□E I□C□□□ □□□E□ □□□□ □□EC□□□□ □□□□□

The data consist of the "cold index," 1895-2005. A density plot of the data shows that the original data are highly right-skewed, but a cube-root transformation leads to a much more symmetric distribution (Figure A.1).

We therefore proceed to look for trends in the cube root data.

A simple OLS linear regression yields a trend of -.00125 per year, standard error .00068, for which the 2-sided p-value is .067. Recomputing using the minimum-AIC ARMA model yields the optimal values p=q=3, trend -.00118, standard error .00064, p-value .066. In this case, fitting an ARMA model makes very little difference to the result compared with OLS. By the usual criterion of a .05 significance level, this is not a statistically significant result, but it is close enough that we are justified in concluding there is still some evidence of a downward linear trend. Figure A.2 illustrates the fitted linear trend on the cube root data.

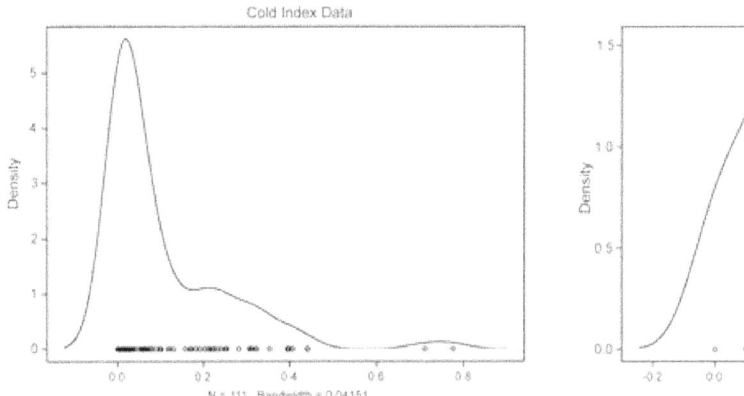

Figure Density plot for the cold index data (left), and for the cube roots of the same data (right).

EXAMPLE TWO: A COLD WAVE INDEX FOR THE CONTINENTAL USA

This example is more complicated to analyze because of the presence of several outlying values in the 1930s which frustrate any attempt to fit a linear trend to the whole series. However, a density plot of the raw data show that they are very right-skewed, whereas taking natural logarithms makes the data look much more normal (Figure A.3). Therefore, for the rest of this analysis we work with the natural logarithms of the heat wave index.

In this case, there is no obvious evidence of a linear trend either upwards or downwards. However, nonlinear trend fits suggest an oscillating pattern up to about 1960, followed by a steadier upward drift in the last four decades. For example, the solid curve in Figure A.4, which is based on a cubic spline fit with 8 degrees of freedom, fitted by ordinary linear regression, is of this form.

Motivated by this, a linear trend has been fitted by time series regression to the data from 1960-2005 (dashed straight line, Figure A.4). In this case, searching for the best ARMA model by the AIC criterion led to the ARMA(1,1) model being selected. Under this model, the fitted linear trend has a slope of 0.031 per year and a standard error of .0035. This is very highly statistically significant. Assuming normally distributed errors, the probability that such a result could have been reached by chance, if there were no trend, is of the order 10^{-18}.

We should comment a little about the justification for choosing the endpoints of the linear trend (in this case, 1960 and

2005) in order to give the best fit to a straight line. The potential objection to this is that it creates a bias associated with multiple testing. Suppose, as an artificial example, we were to conduct 100 hypothesis tests based on some sample of data, with significance level .05. This means that if there were in fact no trend present at all, each of the tests would have a .05 probability of incorrectly concluding that there was a trend. In 100 such tests, we would typically expect about 5 of the tests to lead to the conclusion that there was a trend.

A standard way to deal with this issue is the Bonferroni correction. Suppose we still conducted 100 tests, but adjusted the significance level of each test to .05/100=.0005.
Then even if no trend were present, the probability that at least one of the tests led to rejecting the null hypothesis would be no more than 100 times .0005, or .05. In other words, with the Bonferroni correction, .05 is still an upper bound on the overall probability that one of the tests falsely rejects the null hypothesis.

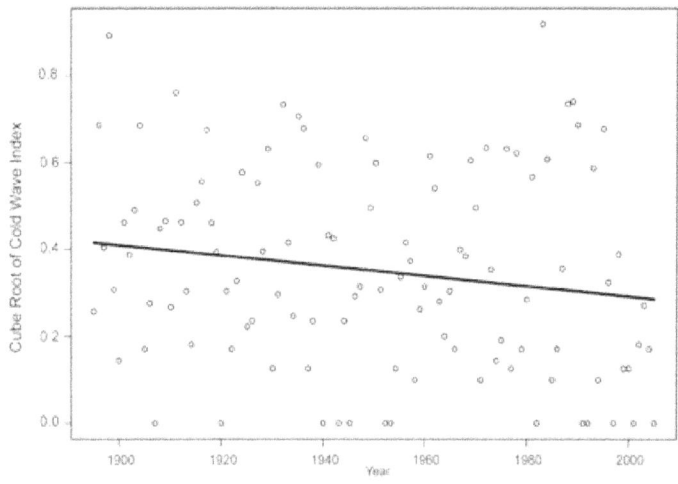

Figure Cube root of cold wave index with fitted linear trend.

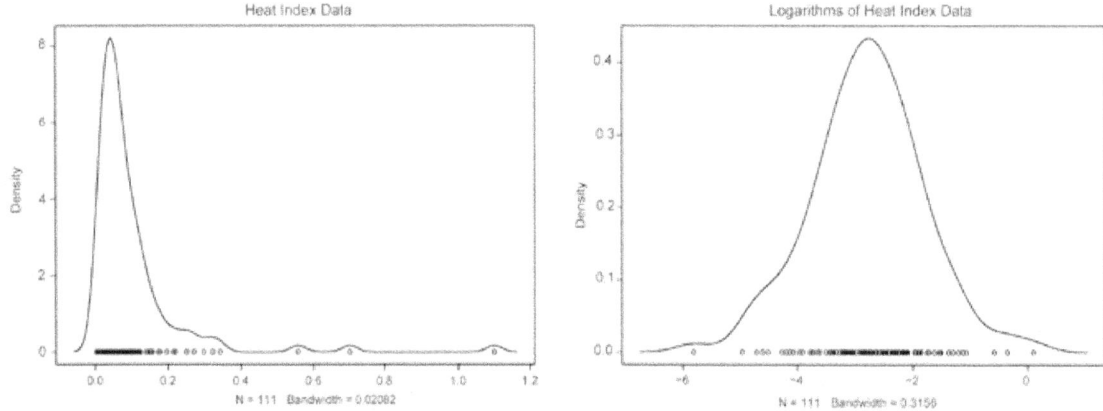

Fig□re □□□ Density plot for the heat index data (left), and for the natural logarithms of the same data (right).

In the case under discussion, if we allow for all possible combinations of start and finish dates, given a 111-year series, that makes for 111x110/2=6105 tests. To apply the Bonferroni correction in this case, we should therefore adjust the significance level of the individual tests to .05/6105=.0000082. However, this is still very much larger than 10^{-18}. The conclusion is that the statistically significant result cannot be explained away as merely the result of selecting the endpoints of the trend.

This application of the Bonferroni correction is somewhat unusual. It is rare for a trend to be so highly significant that selection effects can be explained away completely, as has been shown here. Usually, we have to make a somewhat more subjective judgment about what are suitable starting and finishing points of the analysis.

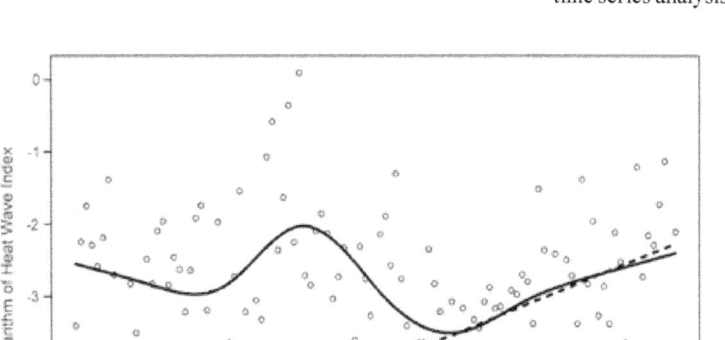

Fig□re □□□ Trends fitted to natural logarithms of heat index. Solid curve: nonlinear spline with 8 degrees of freedom fitted to the whole series. Dashed line: linear trend fitted to data from 1960-2005.

In this example, we considered the time series of 1-day heavy precipitation frequencies for a 20-year return value. In this case, the density plot for the raw data is not as badly skewed as in the earlier examples (Figure A.5, left plot), but is still improved by taking square roots (Figure A.5, right plot). Therefore, we take square roots in the subsequent analysis.

Looking for linear trends in the whole series from 1895-2005, the overall trend is positive but not statistically significant (Figure A.6). Based on simple linear regression, the estimated slope is .00023 with a standard error of .00012, which just fails to be significant at the 5% level. However, time series analysis identifies an ARMA (5, 3) model, when the estimated slope is still .00023, the standard error rises to .00014, which is again not statistically significant.

However, a similar exploratory analysis to that in Example 2 suggested that a better linear trend could be obtained starting around 1935. To be specific, we have considered the data from 1934-2005. Over this period, time series analysis identifies an ARMA(1,2) model, for which the estimated slope is .00067, standard error .00007, under which a formal test rejects the null hypothesis of no slope with a significance level of the order of 10^{-20} under normal theory assumptions. As with Example 2, an argument based on the Bonferroni correction shows that this is a clearly significant result

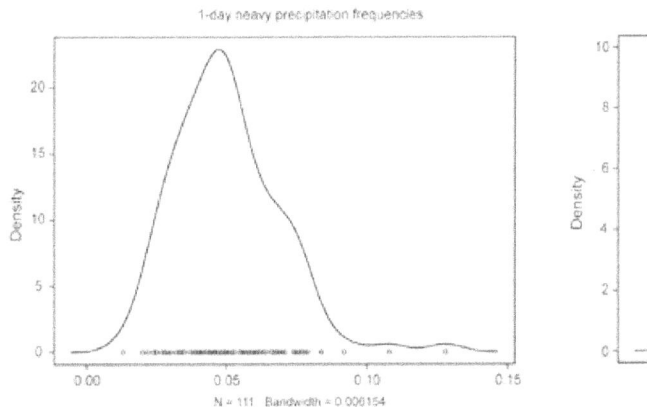 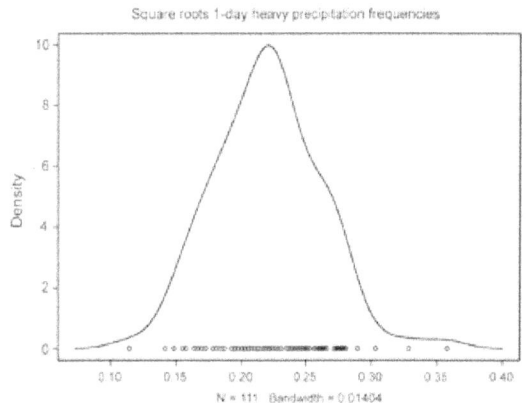

Figure A.5 Density plot for 1-day heavy precipitation frequencies for a 20-year return value (left), and for square roots of the same data (right).

even allowing for the subjective selection of start and finish points of the trend.

Therefore, our conclusion in this case is that there is an overall positive but not statistically significant trend over the whole series, but the trend post-1934 is much steeper and clearly significant.

This is a similar example based on the time series of 90-day heavy precipitation frequencies for a 20-year return value. Once again, density plots suggest a square root transformation (the plots look rather similar to Figure A.5 and are not shown here).

After taking square roots, simple linear regression leads to an estimated slope of .00044, standard error .00019, based on the whole data set. Fitting ARMA models with linear trend leads us to identify the ARMA(3,1) as the best model under AIC: in that case, the estimated slope becomes .00046 and the standard error actually goes down, to .00009. Therefore, we conclude that the linear trend is highly significant in this case (Figure A.7).

This analysis is based on historical reconstructions of tropical cyclone counts described in the recent paper of Vecchi and Knutson (2008). We consider two slightly different reconstructions of the data: the "one-encounter" reconstruction in which only one intersection of a ship and storm is required for a storm to be counted as seen, and the "two-encounter" reconstruction that requires two intersections before a storm is counted. We focus particularly on the contrast between trends over the 1878-2005 and 1900-2005 time periods, since before the start of the present analysis, Vecchi and Knutson had identified these two periods as of particular interest.

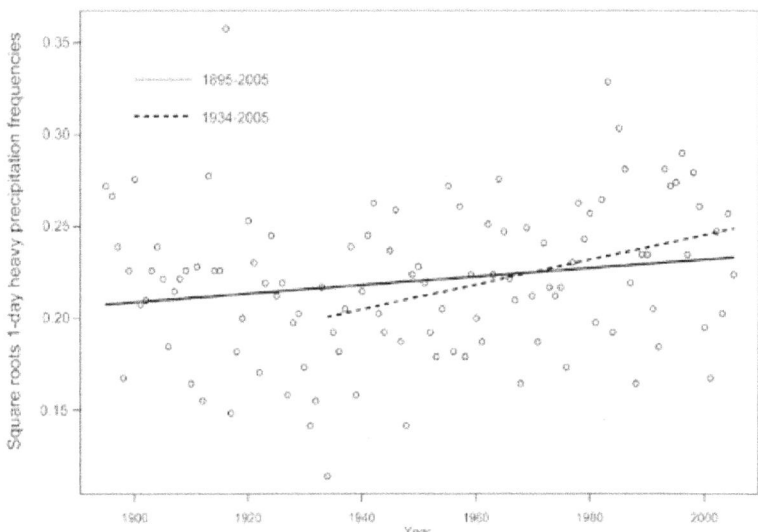

Figure A.7 Trend analysis for the square roots of 1-day heavy precipitation frequencies for a 20-year return value, showing estimated linear trends over 1895-2005 and 1934-2005.

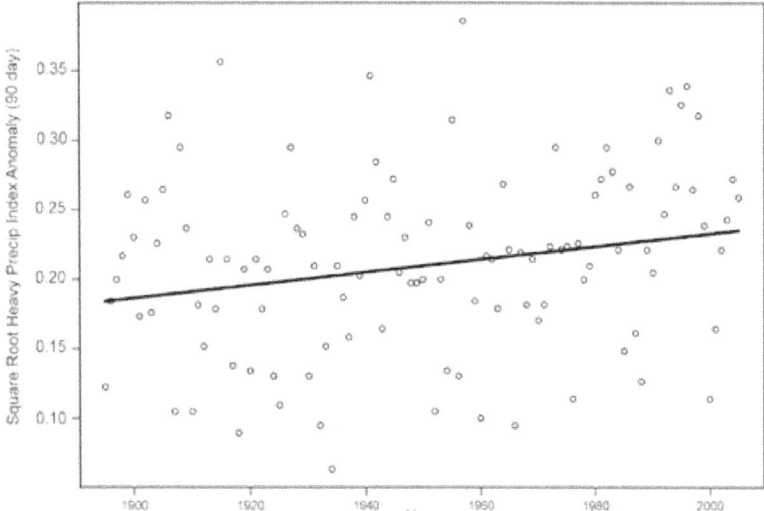

When repeated for 1900-2005, ordinary least-squares regression leads to a slope of .042, standard error .012. The same analysis based on a time series model (ARMA(9,2)) leads to a slope of .045 and a standard error of .021. Although the standard error is much bigger under the time series model, this is still significant with a p-value of about .03.

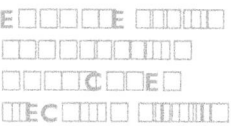

□ig□re □□□ Trend analysis for the square roots of 90-day heavy precipitation frequencies.

For 1878-2005, using the one-encounter dataset, we find by ordinary least squares a linear trend of .017 (storms per year), standard error .009, which is not statistically significant. Selecting a time series model by AIC, we identify an ARMA(9,2) model as best (an unusually large order of a time series model in this kind of analysis), which leads to a linear trend estimate of .022, standard error .022, which is clearly not significant.

When the same analysis is repeated from 1900-2005, we find by linear regression a slope of .047, standard error .012, which is significant. Time series analysis now identifies the ARMA(5,3) model as optimal, with a slope of .048, standard error .015–very clearly significant. Thus, the evidence is that there is a statistically significant trend over 1900-2005, though not over 1878-2005.

A comment here is that if the density of the data is plotted as in several earlier examples, this suggests a square root transformation to remove skewness. Of course the numerical values of the slopes are quite different if a linear regression is fitted to square root cyclones counts instead of the raw values, but qualitatively, the results are quite similar to those just cited–significant for 1900-2005, not significant for 1878-2005–after fitting a time series model. We omit the details of this.

The second part of the analysis uses the "two-encounter" data set. In this case, fitting an ordinary least-squares linear trend to the data 1878-2005 yields an estimated slope .014 storms per year, standard error .009, not significant. The time series model (again ARMA(9,2)) leads to estimated slope .018, standard error .021, not significant.

The final example is a time series of U.S. landfalling hurricanes for 1851-2006 taken from the website http://www.aoml.noaa.gov/hrd/ hurdat/ushurrlist18512005-gt.txt. The data consist of annual counts and are all between 0 and 7. In such cases a square root transformation is often performed because this is a variance stabilizing transformation for the Poisson distribution. Therefore, square roots have been taken here.

A linear trend was fitted to the full series and also for the following subseries: 1861-2006, 1871-2006, and so on up to 1921-2006. As in preceding examples, the model fitted was ARMA (p,q) with linear trend, with p and q identified by AIC.

For 1871-2006, the optimal model was AR(4), for which the slope was -.00229, standard error .00089, significant at p=.01.

For 1881-2006, the optimal model was AR(4), for which the slope was -.00212, standard error .00100, significant at p=.03.

For all other cases, the estimated trend was negative, but not statistically significant.

GLOSSARY AND ACRONYMS

☐☐☐☐☐☐☐

afforestation
the process of establishing trees on land that has lacked forest cover for a very long period of time or has never been forested

anthropogenic
human-induced

apparent consumption
the amount or quantity expressed by the following formula: production + imports – exports +/– changes in stocks

biomass
the mass of living organic matter (plant and animal) in an ecosystem; biomass also refers to organic matter (living and dead) available on a renewable basis for use as a fuel; biomass includes trees and plants (both terrestrial and aquatic), agricultural crops and wastes, wood and wood wastes, forest and mill residues, animal wastes, livestock operation residues, and some municipal and industrial wastes

carbon sequestration
the process of increasing the carbon content of a carbon reservoir other than the atmosphere; often used narrowly to refer to increasing the carbon content of carbon pools in the biosphere and distinguished from physical or chemical collection of carbon followed by injection into geologic reservoirs, which is generally referred to as "carbon capture and storage"

carbon cycle
the term used to describe the flow of carbon (in various forms such as carbon dioxide [CO_2], organic matter, and carbonates) through the atmosphere, ocean, terrestrial biosphere, and lithosphere

carbon equivalent
the amount of carbon in the form of CO_2 that would produce the same effect on the radiative balance of the Earth's climate system; applicable in this report to greenhouse gases such as methane (CH_4)

carbon intensity
the relative amount of carbon emitted per unit of energy or fuels consumed

climate change projection
This term is commonly used rather than "climate pre-diction" for longer-range predictions that are based on various scenarios of human or natural changes in the agents that drive climate

climate prediction
the prediction of various aspects of the climate of a region during some future period of time. Climate predictions are generally in the form of probabilities of anomalies of climate variables (*e.g.*, temperature, precipitation), with lead times up to several seasons

coastal waters
the region within 100 km from shore in which processes unique to coastal marine environments influence the partial pressure of CO_2 in surface sea waters

CO_2 equivalent
the amount of CO_2 that would produce the same effect on the radiative balance of the Earth's climate system as another greenhouse gas, such as CH_4

CO_2 fertilization
the phenomenon in which plant growth increases (and agricultural crop yields increase) due to the increased rates of photosynthesis of plant species in response to elevated concentrations of CO_2 in the atmosphere

decarbonization
reduction in the use of carbon-based energy sources as a proportion of total energy supplies or increased use of carbon-based fuels with lower values of carbon content per unit of energy content

deforestation
the process of removing or clearing trees from forested land

dry climates
climates where the ratio of mean annual precipitation to potential evapotranspiration is less than 1.0

ecosystem
a community (☐*e.*☐an assemblage of populations of plants, animals, fungi, and microorganisms that live in an environment and interact with one another, forming, together, a distinctive living system with its own composition, structure, environmental relations, development, and function)

and its environment treated together as a functional system of complementary relationships and transfer and circulation of energy and matter

energy intensity
the relative amount or ratio of the consumption of energy to the resulting amount of output, service, or activity (*i.e.* expressed as energy per unit of output)

feebates
systems of progressive vehicle taxes on purchases of less efficient new vehicles and subsidies for more efficient new vehicles

fossil fuels
fuels such as coal, petroleum, and natural gas derived from the chemical and physical transformation (fossilization) of the remains of plants and animals that lived during the Carboniferous Period 360–286 million years ago

global warming potential (GWP)
a factor describing the radiative forcing impact (*e.g.* warming of the atmosphere) of one unit mass of a given greenhouse gas relative to the warming caused by a similar mass of CO_2; CH_4, for example, has a GWP of 23

greenhouse gases
gases including water vapor, CO_2, CH_4, nitrous oxide, and halocarbons that trap infrared heat, warming the air near the surface and in the lower levels of the atmosphere

leakage
The part of emissions reductions in Annex B countries that may be offset by an increase of the emission in the non-constrained countries above their baseline levels. This can occur through (1) relocation of energy-intensive production in non-constrained regions; (2) increased consumption of fossil fuels in these regions through decline in the international price of oil and gas triggered by lower demand for these energies; and (3) changes in incomes (and thus in energy demand) because of better terms of trade. "Leakage" also refers to the situation in which a carbon sequestration activity (*e.g.* tree planting) on one piece of land inadvertently, directly or indirectly, triggers an activity, which in whole or part counteracts the carbon effects of the initial activity

mitigation
a human intervention to reduce the sources of, or to enhance the sinks of, greenhouse gases

net ecosystem exchange
the net flux of carbon between the land and the atmosphere, typically measured using eddy covariance techniques; note: NEE and NEP are equivalent terms but are not always iden-

tical because of measurement and scaling issues, and the sign conventions are reversed; positive values of NEE (net ecosystem exchange with the atmosphere) usually refer to carbon released to the atmosphere (*i.e.* a source), and negative values refer to carbon uptake (*i.e.* a sink)

net ecosystem production
the net carbon accumulation within the ecosystem after all gains and losses are accounted for, typically measured using ground-based techniques; by convention, positive values of NEP represent accumulations of carbon by the ecosystem, and negative values represent carbon loss

net primary production
the net uptake of carbon by plants in excess of respiratory loss

North America
the combined land area of Canada, the United States of America, and Mexico and their coastal waters

North American Carbon Program
a multidisciplinary research program, supported by a number of different U.S. federal agencies through a variety of intramural and extramural funding mechanisms and award instruments, to obtain scientific understanding of North America's carbon sources and sinks and of changes in carbon stocks needed to meet societal concerns and to provide tools for decision makers

ocean acidification
the phenomenon in which the pH of the oceans becomes more acidic due to increased levels of CO_2 in the atmosphere which, in turn, increase the amount of dissolved CO_2 in sea water

option
a choice among a set of possible measures or alternatives

peatlands
areas characterized as having an organic layer thickness of at least 30 cm (note, the current United States' and Canadian soil taxonomies specify a minimum thickness of 40 cm)

permafrost
soils or rocks that remain below 0°C for at least two consecutive years

pool/reservoir
any natural region or zone, or any artificial holding area, containing an accumulation of carbon or carbon-bearing compounds or having the potential to accumulate such substances

reforestation

the process of establishing a new forest by planting or seeding trees in an area where trees have previously been removed

sink

in general, any process, activity, or mechanism which removes a greenhouse gas or a precursor of a greenhouse gas or aerosol from the atmosphere; in this report, a sink is any regime or pool in which the amount of carbon is increasing (*i.e.* is being accumulated or stored)

source

in general, any process, activity, or mechanism which releases a greenhouse gas or a precursor of a greenhouse gas or aerosol into the atmosphere; in this report, a source is any regime or pool in which the amount of carbon is decreasing (*i.e.* is being released or emitted)

stocks

the amount or quantity contained in the inventory of a pool or reservoir

temperate zones

regions of the earth's surface located above 30° latitude and below 66.5° latitude

trend

a systematic change over time

tropical zones

regions located between the earth's equator and 30° latitude (this area includes subtropical regions)

uncertainty

a term used to describe the range of possible values around a best estimate, sometimes expressed in terms of probability or likelihood (see Preface, this report)

wet climates

climates where the ratio of mean annual precipitation to potential evapotranspiration is greater than 1.0

wetlands

areas that are inundated or saturated by surface water or groundwater at a frequency and duration sufficient to support—and that, under normal circumstances, do support—a prevalence of vegetation typically adapted for life in saturated soil conditions, including swamps, marshes, bogs, and similar areas

ACRONYMS AND ABBREVIATIONS

μatm	microatmosphere (a measure of pressure)
ACEEE	American Council for an Energy-Efficient Economy
CAFE	Corporate Average Fuel Economy
CAIT	Climate Analysis Indicators Tool
CAST	Council for Agricultural Science and Technology
CBO	U.S. Congressional Budget Office
CCSP	U.S. Climate Change Science Program
CCTP	Climate Change Technology Program
CDIAC	Carbon Dioxide Information Analysis Center
CEC	California Energy Commission
CH_4	methane
CIEEDAC	Canadian Industrial Energy End-Use Data and Analysis Centre
CO	carbon monoxide
CO_2	carbon dioxide
CO_3	carbonate
COP	Conference of Parties
DOC	dissolved organic carbon
DOE	U.S. Department of Energy
DOT	U.S. Department of Transportation
EIA	Energy Information Administration
EPA	U.S. Environmental Protection Agency
ESCOs	energy services companies
FAO	Food and Agriculture Organization
FWMS	freshwater mineral-soil
g	gram
GAO	U.S. Government Accountability Office
GDP	gross domestic product
GHG	greenhouse gas
Gt C	gigatons of carbon (billions of metric tons; *i.e.*, petagrams)
GWP	global warming potential
ha	hectare
HCO_3	bicarbonate
ICLEI	International Council for Local Environmental Initiatives (now known as International Governments for Local Sustainability)
IOOS	Integrated Ocean Observing System
IPCC	Intergovernmental Panel on Climate Change

IWG	Interlaboratory Working Group	**UNFCCC**	United Nations Framework Convention on Climate Change
kg	kilogram		
km	kilometer	**USDA**	U.S. Department of Agriculture
L	liter	**VOCs**	volatile organic compounds
LEED	Leadership in Energy and Environment Design	**WBCSD**	World Business Council for Sustainable Development
m	meter		
MAP	mean annual precipitation		
Mt C	megatons of carbon (millions of metric tons; □*e.*, teragrams)		
N₂O	nitrous oxide (also, dinitrogen oxide)		
NACP	North American Carbon Program		
NAO	North Atlantic oscillation		
NAS	U.S. National Academy of Sciences		
NASA	National Aeronautics and Space Administration		
NATS	North American Transportation Statistics		
NCAR	National Center for Atmospheric Research		
NCEP	National Centers for Environmental Prediction; National Commission on Energy Policy		
NEE	net ecosystem exchange		
NEP	net ecosystem productivity		
NGO	non-governmental organization		
NO₂	nitrogen dioxide		
NOAA	National Oceanic and Atmospheric Administration		
NOₓ	oxides of nitrogen		
NPP	net primary productivity		
NRC	National Research Council		
NRCS	Natural Resources Conservation Service		
NSF	National Science Foundation		
NWI	National Wetland Inventory		
OCCC	Ocean Carbon and Climate Change		
pCO₂	partial pressure of carbon dioxide in units of microatmospheres or ppm		
PDO	Pacific decadal oscillation		
PET	potential evapotranspiration		
PJ	petajoules		
ppm	parts per million by volume		
PPP	purchasing power parity		
RGGI	Regional Greenhouse Gas Initiative		
SAP	Synthesis and Assessment Product		
SBSTA	Subsidiary Body for Scientific and Technological Advice		
SOCCR	State of the Carbon Cycle Report		
μatm	microatmospheres or 10^-6 atmospheres		

REFERENCES

CITED REFERENCES

Acuna-Soto, R., D.W. Stahle, M.K. Cleaveland, and M.D. Therrell, 2002: Megadrought and megadeath in 16th century Mexico. *Emerging Infectious Diseases*, **8(4)**, 360-362.

Allen, C.D. and D.D. Breshears, 1998: Drought-induced shift of a forest–woodland ecotone: rapid landscape response to climate variation. *Proceedings of the National Academy of Sciences*, **95(25)**, 14839-14842.

Andreadis, K.M. and D.P. Lettenmaier, 2006: Trends in 20th century drought over the continental United States. *Geophysical Research Letters*, **33**, L10403, doi:10.1029/2006GL025711.

Arctic Climate Impact Assessment, 2004: *Impacts of a Warming Arctic*. Cambridge University Press, Cambridge, UK, and New York, 139 pp.

Arguez, A. (ed.), 2007: State of the Climate in 2006. *Bulletin of the American Meteorological Society*, **88(6)**, s1-s135.

Atherholt, T.B., M.W. LeChevallier, W.D. Norton, and J.S. Rosen, 1998: Effect of rainfall on giardia and crypto. *Journal of the American Water Works Association*, **90(9)**, 66-80.

Ausubel, J.H., 1991: Does climate still matter? *Nature*, **350(6320)**, 649-652.

Baker, A.C., 2001: Reef corals bleach to survive change. *Nature*, **411(6839)**, 765-66.

Baker, A.C., C.J. Starger, T.R. McClanahan, and P.W. Glynn, 2004: Coral reefs: corals' adaptive response to climate change. *Nature*, **430(7001)**, 741.

Balanya, J., J.M. Oller, R.B. Huey, G.W. Gilchrist, and L. Serra, 2006: Global genetic change tracks global climate warming in *Drosophila subobscura*. *Science*, **313(5794)**, 1773-1775.

Barry, R.G., 2006: The status of research on glaciers and global glacier recession: a review. *Progress in Physical Geography*, **30(3)**, 285-306.

Berg, E.E., J.D. Henry, C.L. Fastie, A.D. De Volder, and S.M. Matsuoka, 2006: Spruce beetle outbreaks on the Kenai Peninsula, Alaska, and Kluane National Park and Reserve, Yukon Territory: relationship to summer temperatures and regional differences in disturbance regimes. *Forest Ecology and Management*, **227(3)**, 219-232.

Berube, A. and B. Katz, 2005: *Katrina's window: confronting concentrated poverty across America*. The Brookings Institution, Washington DC, 13 pp.

Board on Natural Disasters, 1999: Mitigation emerges as a major strategy for reducing losses caused by natural disasters. *Science*, **284(5422)**, 1943-1947.

Bonnin, G.M., B. Lin, and T. Parzybok, 2003: Updating NOAA/NWS rainfall frequency atlases. In: *Proceedings reengineering and reestablishing the capability to estimate extreme weather and climate*. 17th Conference on Hydrology, 9-13 February 2003, Long Beach, CA. American Meteorological Society, Boston, Paper J3.20. Extended abstract available at http://ams.confex.com/ams/pdfpapers/54731. pdf

Bowden, M., R. Kates, P. Kay, W. Riebsame, H. Gould, D. Johnson, R. Warrick, and D. Weiner, 1981: The effects of climate fluctuations on human populations: two hypotheses. In: *Climate and history: studies in past climates and their impact on man* [Wigley, T.M.L., M.J. Ingram, and G. Farmer (eds.)]. Cambridge University Press, Cambridge UK, and New York, pp. 479-513.

British Columbia Ministry of Forests and Range, 2006a: *Managing mountain pine beetle attacked stands*. FOR0011-000152, British Columbia Ministry of Forestry and Range, Victoria. 1 p.

British Columbia Ministry of Forests and Range, 2006b: *Beetle blasted forest feeds deficit to the hilt*, News Release. http://www2.news.gov.bc.ca/news_releases_2005-2009/2006FOR0112-001109.htm

Brooks, H.E. and C.A. Doswell III, 2001: Normalized damage from major tornadoes in the United States: 1890–1999. *Weather and Forecasting*, **16**, 168–176.

Brunkard, J.M., J.L. Robles López, J. Ramirez, E. Cifuentes, S.J. Rothenberg, E.A. Hunsperger, C.G. Moore, R.M. Brussolo, N.A. Villarreal, B.M. Haddad, 2007: Dengue Fever seroprevalence and risk factors, Texas-Mexico border, 2004. *Emerging Infectious Diseases*, **13**,1477-1483.

Bull, J.J., 1980: Sex determination in reptiles. *Quarterly Review of Biology*, **55(1)**, 3-21.

Bull, J.J. and R.C. Vogt, 1979: Temperature-dependent sex determination in turtles. *Science*, **206(4423)**, 1186-1188.

Burton, I., 1962: *Types of agricultural occupance of flood plains in the United States*. University of Chicago Press, Chicago, 167 pp.

Burton, I., R.W. Kates, and G.F. White, 1993: *The environment as hazard*. The Guilford Press, New York, 2nd ed., 290 pp.

Campbell-Lendrum, D., A. Pruss-Ustun, and C. Corvalan, 2003: How much disease could climate change cause? In: *Climate change and human health: risks and responses* [McMichael, A.J., D.H Campbell-Lendrum, C.F. Corvalán, K.L. Ebi, A.K. Githeko, J.D. Scheraga, and A. Woodward (eds.)]. World Health Organization, Geneva, Switzerland, pp. 133-158.

Canadian Standards Association, 2001: *Canadian standards association overhead systems. Antennas and systems for loading (design tools)*. Ice and wind loads contributed by R. Morris, T. Yip, and H. Auld. Canadian Standards Association, Toronto, 118 pp.

Cathey, H.M., 1990: *USDA plant hardiness zone map*, USDA Miscellaneous Publication No. 1475. United States Department of Agriculture, Washington, DC, 1 map.

Cayan, D.R., S.A. Kammerdiener, M.D. Dettinger, J.M. Capiro, and D.H. Peterson, 2001: Changes in the onset of spring in the western United States. *Bulletin of the American Meteorological Society*, **82(2)**, 399-415.

Changnon, S.D., 2003: Measures of economic impacts of weather extremes: getting better but far from what is needed - a call for action. *Bulletin of the American Meteorological Society*, **84(9)**, 1231-1235. doi:10.1175/BAMS-84-9-1231

Checkley, W., L.D. Epstein, R.H. Gliman, D. Figueroa, R.I. Cama, J.A. Patz, and R.E. Black, 2000: Effects of El Niño and ambient temperature on hospital admissions for diarrhoeal diseases in Peruvian children. *La⬜⬜et* **355(9202)**, 442-450.

Chowdhury, A.G. and S.P. Leatherman, 2007: Innovative testing facility to mitigate hurricane-induced losses. *⬜⬜⬜a⬜a⬜ t⬜⬜⬜⬜⬜r⬜e ⬜⬜e⬜⬜a⬜⬜e⬜⬜⬜⬜al ⬜⬜⬜*, **88(25)**, 262.

Christensen, N.S., A.W. Wood, N. Voisin, D.P. Lettenmaier, and R.N. Palmer, 2004: The effects of climate change on the hydrology and water resources of the Colorado River basin. *⬜/⬜⬜ ⬜at⬜⬜⬜⬜a⬜ge*, **62(1-3)**, 337-363.

Christoplos, I., 2006: The elusive 'window of opportunity' for risk reduction in post-disaster recovery. Discussion paper at the ProVention Consortium Forum 2006 *⬜re⬜gt⬜e⬜⬜g gl⬜⬜al ⬜⬜la⬜⬜at⬜⬜ ⬜⬜ ⬜⬜a⬜⬜e⬜ ⬜⬜⬜ ⬜e⬜⬜⬜⬜*, 2-3 February 2006, Bangkok. Available at http://www.proventionconsortium.org/themes/default/pdfs/Forum06/Forum06_Session3_Recovery.pdf

Cohen, S.J. and M.W. Waddell, 2008: *⬜l⬜ate ⬜⬜a⬜ge ⬜⬜t⬜e ⬜⬜⬜ ⬜e⬜⬜⬜⬜* McGill-Queen's University Press, Montreal, 392 pp. (in press, expected July 2008).

Collins, D.J. and S.P. Lowe, 2001: *⬜ ⬜a⬜⬜ ⬜al⬜⬜at⬜⬜⬜ ⬜ata⬜et ⬜⬜⬜.⬜ ⬜⬜⬜⬜a⬜e ⬜ ⬜e⬜⬜* Casualty Actuarial Society Forum, Casualty Actuarial Society, Arlington, Va., pp. 217-251. Available at http://www.casact.org/pubs/forum/01wforum/01wf217.pdf

Colwell, R.R., 1996: Global climate and infectious disease: the cholera paradigm. *⬜⬜e⬜⬜e*, **274(5295)**, 2025-2031.

Corso, P.S., M.H. Kramer, K.A. Blair, D.G. Addiss, J.P. Davis and A.C. Haddix, 2003: Cost of illness in the 1993 waterborne *⬜⬜⬜⬜r⬜⬜⬜⬜⬜⬜⬜⬜⬜* outbreak, Milwaukee, Wisconsin. *⬜⬜e⬜g⬜⬜g ⬜⬜e⬜t⬜⬜⬜⬜⬜⬜ea⬜e* **9(4)**, 426-431.

Crossett, K.M., T.J. Culliton, P.C. Wiley, and T.R. Goodspeed, 2004: *⬜⬜⬜⬜at⬜⬜ ⬜e⬜⬜⬜ ⬜⬜⬜g t⬜e ⬜⬜a⬜al ⬜⬜⬜te⬜ ⬜tate⬜ ⬜⬜⬜⬜⬜⬜⬜⬜*National Oceanic and Atmospheric Administration, National Ocean Service, Washington, DC, 54 pp.

Curriero, F.C., J.A. Patz, J.B. Rose, and S. Lele, 2001: The association between extreme precipitations and waterborne disease outbreaks in the United States, 1948-1994. *⬜⬜e⬜⬜a⬜ ⬜⬜⬜⬜al ⬜⬜⬜⬜⬜/⬜⬜⬜ealt⬜*, **91(8)**, 1194-1199.

Cutter, S.L. and C. Emrich, 2005: Are natural hazards and disaster losses in the U.S. increasing? *⬜⬜⬜⬜⬜⬜a⬜a⬜t⬜⬜⬜⬜⬜⬜e ⬜⬜e⬜⬜a⬜⬜e⬜⬜⬜⬜⬜al ⬜⬜⬜⬜*, **86(41)**, 381.

Cutter, S.L., M. Gall and C.T. Emrich, 2008: Toward a comprehensive loss inventory of weather and climate hazards. In: *⬜l⬜⬜ate ⬜⬜⬜e⬜⬜e⬜a⬜⬜ ⬜⬜⬜⬜et⬜* [Diaz H.L. and R. J. Murnane (eds.)]. Cambridge University Press, Cambridge, UK, and New York, pp.279-295.

Donner, S.D., W.J. Skirving, C.M. Little, M. Oppenheimer, and O. Hoegh-Guldberg, 2005: Global assessment of coral bleaching and required rates of adaptation under climate change. *⬜l⬜⬜al ⬜⬜a⬜ge⬜⬜⬜l⬜g⬜*11(12), 2251-2265.

Donner, S.D., T.R. Knutson, and M. Oppenheimer, 2007: Model-based assessment of the role of human-induced climate change in the 2005 Caribbean coral bleaching event. *⬜⬜⬜ee⬜⬜g⬜ ⬜⬜ t⬜e ⬜at⬜⬜al ⬜⬜a⬜e⬜⬜⬜⬜⬜⬜e⬜*, **104(13)**, 5483-5488.

Downton, M.W., J.Z.B. Miller, and R.A. Pielke Jr., 2005: Reanalysis of U.S. National Weather Service Flood Loss Database. *⬜at⬜⬜al ⬜a⬜a⬜⬜⬜e⬜e⬜*, **6**, 13-22.

Dunn P.O. and D.W. Winkler, 1999: Climate change has affected the breeding date of tree swallows throughout North America. *⬜⬜⬜ee⬜⬜g⬜⬜⬜r⬜e ⬜⬜⬜al ⬜⬜⬜⬜et⬜⬜⬜⬜⬜⬜⬜⬜⬜⬜⬜e⬜e⬜⬜⬜* **266(1437)**, 2487-2490.

Easterling, D.R., J.L. Evans, P.Y. Groisman, T.R. Karl, K.E. Kunkel, and P. Ambenje, 2000a: Observed variability and trends in extreme climate events: A brief review. *⬜⬜llet⬜⬜⬜ ⬜⬜r⬜e ⬜⬜e⬜⬜⬜⬜ ⬜a⬜⬜ete⬜⬜⬜l⬜g⬜al ⬜⬜⬜et⬜*, **81(3)**, 417-425.

Easterling, D.R., S. Chagnon, T.R. Karl, J. Meehl, and C. Parmesan, 2000b: Climate extremes: observations, modeling, and impacts. *⬜⬜e⬜⬜e*, **289(5487)**, 2068-2074.

Ebi, K.L., D.M. Mills, J.B. Smith, and A. Grambsch, 2006: Climate change and human health impacts in the United States: an update on the results of the U.S. national assessment. *⬜⬜⬜⬜⬜⬜⬜⬜ ⬜e⬜⬜al ⬜ealt⬜⬜⬜e⬜⬜⬜e⬜t⬜⬜e⬜*114(9), 1318-1324.

Ehrlich, P.R., D.D. Murphy, M.C. Singer, C.B. Sherwood, R.R. White, and I.L. Brown, 1980: Extinction, reduction, stability and increase: the responses of checkerspot butterfly (*⬜⬜⬜⬜⬜⬜ ⬜⬜a⬜e⬜⬜⬜a*) populations to the California drought. *⬜e⬜⬜⬜g⬜a*, **46(1)**, 101-105.

Engelthaler, D.M., D.G. Mosley, J.E. Cheek, C.E. Levy, K.K. Komatsu, P. Ettestad, T. Davis, D.T. Tanda, L. Miller, J.W. Frampton, R. Porter, and R.T. Bryan⬜1999: Climatic and environmental patterns associated with hantavirus pulmonary syndrome, Four Corners region, United States. *⬜⬜e⬜g⬜⬜g ⬜⬜⬜ ⬜e⬜t⬜⬜⬜⬜⬜⬜ea⬜e⬜***5(1)**, 87-94.

FEMA (Federal Emergency Management Agency), 1995: *⬜at⬜⬜⬜⬜ al ⬜ ⬜t⬜gat⬜⬜⬜ ⬜t⬜ateg⬜⬜a⬜te⬜⬜⬜⬜⬜⬜⬜⬜⬜⬜g ⬜a⬜e⬜⬜⬜⬜⬜ ⬜⬜⬜⬜te⬜* Federal Emergency Management Agency, Washington, DC, 40 pp.

Gillett, N.P., A.J. Weaver, F.W. Zwiers, and M.D. Flannigan, 2004: Detecting the effect of climate change on Canadian forest fires. Ge⬜⬜⬜⬜al ⬜e⬜ea⬜⬜⬜ette⬜, **31**, L18211, doi:10.1029/2004GL020876.

Glantz, M.A., 1996: *⬜⬜⬜e⬜t⬜⬜⬜a⬜ge⬜⬜l ⬜⬜⬜⬜⬜⬜ ⬜⬜a⬜ ⬜⬜ ⬜⬜⬜ate a⬜⬜⬜⬜⬜et⬜* Cambridge University Press, Cambridge, UK, and New York, 194 pp.

Glantz, M.A., 2003: *⬜l⬜⬜ate ⬜⬜a⬜⬜⬜ ⬜ ⬜⬜⬜⬜e⬜* Island Press, Washington, DC, 291 pp.

Glass, G.E., J.E. Cheek, J.A. Patz, T.M. Shields, T.J. Doyle, D.A. Thoroughman, D.K. Hunt, R.E. Enscore, K.L. Gage, C. Ireland, C.J. Peters, and R. Bryan, 2000: Using remotely sensed data to identify areas of risk for hantavirus pulmonary syndrome. *⬜⬜e⬜g⬜⬜g ⬜⬜e⬜t⬜⬜⬜⬜⬜ea⬜e⬜***6(3)**, 238-247.

Goklany, I.M. and S.R. Straja, 2000: U.S. trends in crude death rates due to extreme heat and cold ascribed to weather, 1979-97. *⬜e⬜⬜⬜⬜l⬜g⬜* **7(S1)**, 165-173.

Gubler, D.J., P. Reiter, K.I. Ebi, W. Yap, R. Nasci, and J.A. Patz, 2001: Climate variability and change in the United States: potential impacts on vector- and rodent-borne diseases. *⬜⬜⬜⬜⬜⬜⬜ ⬜e⬜⬜al ⬜ealt⬜⬜⬜e⬜⬜⬜e⬜t⬜⬜e⬜*, **109, (supplement 2)**, 223-233.

Guha-Sapir, D., D. Hargitt, and P. Hoyois, 2004: *⬜⬜⬜⬜t⬜ ⬜ea⬜⬜ ⬜⬜at⬜⬜al ⬜⬜a⬜te⬜⬜⬜⬜⬜⬜⬜⬜e ⬜⬜⬜⬜⬜e⬜* UCL Presses, Universitaires de Louvain, Louvain-la-Neuve, Belgium, 188 pp.

Hawkins, B.A. and M. Holyoak, 1998: Transcontinental crashes of insect populations? *⬜⬜e⬜⬜a⬜ ⬜at⬜⬜al⬜⬜***152(3)**, 480-484.

Hazards and Vulnerability Research Institute, 2007: *⬜e ⬜a⬜ t⬜al ⬜a⬜a⬜⬜⬜e⬜⬜a⬜⬜⬜⬜e⬜⬜ata⬜a⬜e ⬜⬜⬜t⬜e ⬜⬜⬜te⬜⬜tate⬜*

Version 5.1 [Online Database]. University of South Carolina, Columbia, SC. Available at http://www.sheldus.org

Heinz Center, The H. John Heinz III Center for Science, Economics and the Environment, 2002: *The State of the Nation's Ecosystems: Measuring the Lands, Waters, and Living Resources of the United States.* Cambridge University Press, Cambridge, UK, and New York, 288 pp.

Herring, D., 1999: Evolving in the presence of fire. *Earth Observatory.* Available at http:earthobservatory.nasa.gov/Study/BOREASFire/

Hoegh-Guldberg, O., 1999: Climate change, coral bleaching and the future of the world's coral reefs. *Marine and Freshwater Research*, **50(8)**, 839-66.

Hoegh-Guldberg, O., 2005: Marine ecosystems and climate change. In: *Climate Change and Biodiversity* [Lovejoy, T.E. and L. Hannah (eds.)]. Yale University Press, New Haven, CT, pp. 256-271.

Hoegh-Guldberg, O., R.J. Jones, S. Ward, W.K. Loh, 2002: Is coral bleaching really adaptive? *Nature*, **415(6872)**, 601-2.

Hoffman, A.A. and P.A. Parsons, 1997: *Extreme Environmental Change and Evolution.* Cambridge University Press, Cambridge, UK, and New York, 259 pp.

Inouye, D.W., 2000: The ecological and evolutionary significance of frost in the context of climate change. Ecology Letters, 3(5), 457-463.

IPCC (Intergovernmental Panel on Climate Change) 2007a: *Climate Change 2007: The Physical Science Basis.* Contribution of Working Group I to the Fourth Assessment Report (AR4) of the Intergovernmental Panel on Climate Change [Solomon, S., D. Qin, M. Manning, Z. Chen, M. Marquis, K.B. Averyt, M.Tignor, and H.L. Miller (eds.)]. Cambridge University Press, Cambridge, UK, and New York, 987 pp. Available at http://www.ipcc.ch

IPCC (Intergovernmental Panel on Climate Change) 2007b: Summary for Policy Makers. In: *Climate Change 2007: Impacts, Adaptation and Vulnerability.* Contribution of Working Group II to the Fourth Assessment Report (AR4) of the Intergovernmental Panel on Climate Change [Parry, M.L., O.F. Canziani, J.P. Palutikof, P.J. van der Linden and C.E. Hanson, (eds.)]. Cambridge University Press, Cambridge, UK, and New York, pp. 7-22. Available at http://www.ipcc.ch

ISRTC, 2007: Interagency Strategic Research Plan for Tropical Cyclone Research: The Way Ahead. FCM-P36-2007, Office of the Federal Coordinator for Meteorological Services and Supporting Research, Washington, DC. Available at <http://www.ofcm.gov/p36-isrtc/pdf/entire_p36_2007.pdf>

Janzen, F.J., 1994: Climate change and temperature-dependent sex determination in reptiles. *Proceedings of the National Academy of Sciences*, **91(16)**, 7487-7490.

Johnson, T., J. Dozier, and J. Michaelsen, 1999: Climate change and Sierra Nevada snowpack. In: *Interactions Between the Cryosphere, Climate and Greenhouse Gases.* [Tranter, M., R. Armstrong, E. Brun, G. Jones, M. Sharp, and M. Williams (eds.).] IAHS publication 256, International Association of Hydrological Sciences, Wallingford (UK), pp 63-70.

Kalkstein, L.S., J.S. Greene, D.M. Mills, A.D. Perrin, J.P. Samenow, and J.-C. Cohen, 2008: Analog European heat waves for U.S. cities to analyze impacts on heat-related mortality. Bulletin of the American Meteorological Society, 89(1), 75-85.

Karl, T.R., R.W. Knight, D.R. Easterling, and R.G. Quayle, 1996: Indices of climate change for the United States. *Bulletin of the American Meteorological Society*, **77(2)**, 279-292.

Kates, R.W., C.E. Colten, S. Laska, and S.P. Leatherman, 2006: Reconstruction of New Orleans after Hurricane Katrina: A research perspective. *Proceedings of the National Academy of Sciences*, **103(40)**, 14653-14660.

Kerry, M., G. Kelk, D. Etkin, I. Burton, and S. Kalhok, 1999: Glazed over: Canada copes with the ice storm of 1998. *Environment*, **41(1)**, 6-11; 28-33.

Kingdon, J.W., 1995: *Agendas, Alternatives and Public Policies.* HarperCollins Publishers, New York, 2nd ed., 254 pp.

Kunkel, K.E., R.A. Pielke, Jr., and S.A. Changnon, 1999: Temporal fluctuations in weather and climate extremes that cause economic and human health impacts: a review. *Bulletin of the American Meteorological Society*, **80(6)**, 1077-1098.

Lanzante, J.R., T.C. Peterson, F.J. Wentz, and K.Y. Vinnikov, 2006: What do observations indicate about the change of temperatures in the atmosphere and at the surface since the advent of measuring temperatures vertically? In: *Temperature Trends in the Lower Atmosphere: Steps for Understanding and Reconciling Differences.* [T.R. Karl, S.J. Hassol, C.D. Miller, and W.L. Murray (eds.)]. U.S. Climate Change Science Program, Washington, DC, pp. 47-70.

Lecomte, E., A.W. Pang, and J.W. Russell, 1998: *Ice Storm '98.* Institute for Catastrophic Loss Reduction, Ottawa, and Institute for Business and Home Safety, Boston, 39 pp.

Levitan, M., 2003: Climatic factors and increased frequencies of 'southern' chromosome forms in natural populations of *Drosophila robusta. Evolutionary Ecology Research*, **5(4)**, 597-604.

Logan, J.A., J. Regniere, and J.A. Powell, 2003: Assessing the impacts of global warming on forest pest dynamics. *Frontiers in Ecology and the Environment*, **1(3)**, 130-137.

Lott, N. and T. Ross, 2006: Tracking billion-dollar U.S. weather disasters. *Bulletin of the American Meteorological Society*, **87(5)**, 557-559. Available at ftp://ftp.ncdc.noaa.gov/pub/data/papers/2006nl557free.pdf

Luber, G.E., C.A. Sanchez, and L.M. Conklin, 2006: Heat-related deaths — United States, 1999–2003. *Mortality and Morbidity Weekly Report*, **55(29)**, 796-798. Available at http://www.cdc.gov/mmwr/preview/mmwrhtml/mm5529a2.htm

Magnuson, J.J., D.M. Robertson, B.J. Benson, R.H. Wynne, D.M. Livingstone, T. Arai, R.A. Assel, R.G. Barry, V. Card, E. Kuusisto, N.G. Granin, T.D. Prowse, K.M Stewart, and V.S. Vuglinski, 2000: Historical trends in lake and river ice cover in the Northern Hemisphere. *Science*, **289(5485)**, 1743- 1746.

Manzello, D.P., M. Brandt, T.B. Smith, D. Lirman, J.C. Hendee, and R.S. Nemeth, 2007: Hurricanes benefit bleached corals. *Proceedings of the National Academy of Sciences*, **104(29)**, 12035-12039.

Mattson, W.J. and R.A. Haack, 1987: The role of drought in outbreaks of plant-eating insects. *Bioscience*, **37(2)**, 110-118.

McMichael, A., D. Campbell-Lendrum, S. Kovats, *et al.* 2004: Global climate change. In: *Comparative Quantification of Health Risks: Global and Regional Burden of Disease due to Selected Major Risk Factors.* [Ezzati, M.J. et al. (eds.)]. World Health Organization, Geneva, pp. 1543-1649.

Meyer, P., M. Bisping, and M. Weber, 1997: *Tropical Cyclones.* Swiss Reinsurance Company, Zurich, 31 pp.

Mileti, D., 1999: *Disasters by Design.* Joseph Henry Press, Washington, DC, 351 pp.

Miller, N.L. and N.J. Schlegel, 2006: Climate change projected fire weather sensitivity: California Santa Ana wind occurrence. *Geophysical Research Letters,* **33**, doi:10.1029/2006GL025808.

Mills, E., 2005a: Insurance in a climate of change. *Science,* **309(5737)**, 1040-1044.

Mills, E., 2005b: Response to Pielke. *Science,* **310(5754)**, 1616.

Mote, P., A. Hamlet, M. Clark, and D. Lettenmaier, 2005: Declining mountain snowpack in western North America. *Bulletin of the American Meteorological Society,* **86(1)**, 39-49.

Munich Re, 2004: Topics Geo: Annual Review: Natural Catastrophes. Munich Re Group, Munich, Germany, 56 pp. Available at <http://www.munichre.com/en>

National Drought Mitigation Center, 2006: *Understanding and Defining Drought.* [Wedsite] Available at http://www.drought.unl.edu/whatis/concept.htm

National Research Council, 1999: *The Impacts of Natural Disasters: A Framework for Loss Estimation.* National Academy Press, Washington, DC, 80 pp.

Natural Resources Canada, 2007: *Mountain Pine Beetle Program.* [Website] Available at http://mpb.cfs.nrcan.gc.ca/index_e.html.

Nordhaus, W.D., 2006: *The Economics of Hurricanes in the United States.* National Bureau of Economic Research (NBER) Working Paper, Cambridge, MA, 46 pp. Available at http://www.nber.org/papers/w12813

O'Connor, D.R., 2002: *Part I Report of the Walkerton In□uiry: The Events of May 2000 and Related Issues.* Ontario Ministry of the Attorney General, Toronto, 500+ pp.

Oechel, W.C., S.J. Hastings, G. Vourlitis, and M. Jenkins, 1993: Recent change of arctic tundra ecosystems from net carbon dioxide sink to a source. *Nature,* **361(6412)**, 520-523.

Oechel, W.C., G.L. Vourlitis, S.J. Hastings, R.C. Zulueta, L. Hinzman, and D. Kane, 2000: Acclimation of ecosystem CO_2 exchange in the Alaska Arctic in response to decadal warming. *Nature,* **406(6799)**, 978-981.

Oedekoven, C.S., D.G. Ainley, and L.B. Spear, 2001: Variable responses of seabirds to changes in marine climate: California current 1985-1994. *Marine Ecology Progress Series,* **212**, 265-281.

OFCM, 2005: Proceedings of the Forum on Urban Meteorology: Meeting the Weather Needs of the Urban Community, September 21-23, 2004, Rockville Maryland. Office of the Federal Coordinator for Meteorological Services and Supporting Research, Washington DC. Available at <http://www.ofcm.noaa.gov/urbanmet/proceedings%202004/pdf/proceedings_of_the_forum_urban_met.pdf>

Parmenter, R.R., E.P. Yadav, C.A. Parmenter, P. Ettestad, and K.L. Gage, 1999: Incidence of plague associated with increased winter-spring precipitation in New Mexico. *American □ournal of Tropical Medicine and Hygiene,* **61(5)**, 814-821.

Parmesan, C., 1996: Climate and species range. *Nature,* **382(6592)**, 765-766.

Parmesan, C., 2005: Case study: detection at multiple levels: *Euphydryas editha* and climate change. In: *Climate Change and Biodiversity* [Lovejoy, T.E. and L. Hannah (eds.)]. Yale University Press, New Haven, CT, pp. 56-60.

Parmesan, C., 2006: Ecological and evolutionary responses to recent climate change. *Annual Reviews of Ecology, Evolution and Systematics,* **37**, 637-669.

Parmesan C. and H. Galbraith, 2004: *Observed Impacts of Global Climate Change in North America.* Pew Center for Global Climate Change, Arlington, VA, 56 pp.

Parmesan, C. and P. Martens, 2008: Climate change, wildlife and human health. Chapter 14 in: SCOPE Assessment: *"Biodiversity, Global Change and Human Health"*, [Sala, O.E., C. Parmesan, and L.A. Meyerson (eds.)]. Island Press, (in press).

Parmesan C. and G. Yohe, 2003: A globally coherent fingerprint of climate change impacts across natural systems. *Nature,* **421(6918)**, 37-42.

Parmesan, C., T.L. Root, and M.R. Willig, 2000: Impacts of extreme weather and climate on terrestrial biota. *Bulletin of the American Meteorological Society,* **81(3)**, 443-50.

Parmesan, C., S. Gaines, L. Gonzalez, D.M. Kaufman, J. Kingsolver, A.T. Peterson, and R. Sagarin, 2005: Empirical perspectives on species' borders: from traditional biogeography to global change. *Oikos,* **108(1)**, 58-75.

Patz, J.A., M.A. McGeehin, S.M. Bernard, K.L. Ebi, P.R. Epstein, A. Grambsch, D.J. Gubler, P. Reiter, E. Romieu, J.B. Rose, J.M. Samet, J. Trtanj, and T. F. Cecich, 2001: Potential consequences of climate variability and change for human health in the United States. In: *The U.S. National Assessment on Potential Conse□uences of Climate Variability and Climate Change.* United States Global Change Research Program, Cambridge University Press, Cambridge, UK, and New York, pp. 437-458.

Patz, J.A., A.K. Githeko, J.P. McCarty, S. Hussein, U. Confalonieri, and N. de Wet, 2003: Climate change and infectious diseases. In: *Climate Change and Human Health: Risks and Responses* [McMichael, A.J., D.H Campbell-Lendrum, C.F. Corvalán, K.L. Ebi, A.K. Githeko, J.D. Scheraga, and A. Woodward (eds.)]. World Health Organization, Geneva, Switzerland, pp. 103-132.

Patz, J.A., D. Campbell-Lendrum, T. Holloway, and J.A. Foley, 2005: Impact of regional climate change on human health. *Nature,* **438(7066)**, 310-317.

Peara, A. and E. Mills, 1999: Climate for change: an actuarial perspective on global warming and its potential impact on insurers. *Contingencies,* **11(1)**, 16-23.

Pickett, S.T.A. and P.S. White (eds.□1985: *The Ecology of Natural Disturbance and Patch Dynamics.* Academic Press, San Diego, CA, 472 pp.

Pielke, R.A., Jr., 2005: Are there trends in hurricane destruction? *Nature,* **438(7071)**, E11. doi:10.1038/nature04426

Pielke, R., 2007: Mistreatment of the economic impacts of extreme events in the Stern Review Report on the Economics of Climate Change. Global Environmental Change: Human and Policy Dimensions, 17(3-4), 302-310.

Pielke, R.A., Jr., and C.W. Landsea, 1998: Normalized hurricane damages in the United States: 1925-1995. *Weather and Forecasting,* **13(3)**, 621-631.

Pielke, R.A., Jr., J. Rubiera, C. Landsea, M. Fernández, and R.A. Klein, 2003: Hurricane vulnerability in Latin America and the

Caribbean: normalized damage and loss potentials. *Natural Hazards Review*, **4(3)**, 101-114.

Pielke, R.A., Jr., J. Gratz, C.W. Landsea, D. Collins, M. Saunders, and R. Musulin, 2008: Normalized hurricane damages in the United States: 1900-2005. Natural Hazards Review, 9(1), 29-42.

Precht, H., J. Christophersen, H. Hensel, and W. Larcher, 1973: *Temperature and Life.* Springer-Verlag, New York, 779 pp.

Pulwarty, R., K. Broad, and T. Finan, 2003: ENSO forecasts and decision making in Brazil and Peru. In: *Mapping Vulnerability: Disasters, Development and People* [Bankoff, G., G. Frerkes, and T. Hilhorst (eds.)]. Earthscan, London, pp. 83-98.

Pulwarty, R., K. Jacobs, and R. Dole, 2005: The hardest working river: drought and critical water problems on the Colorado. In: *Drought and Water Crises: Science, Technology and Management* [D. Wilhite (ed.)]. Taylor and Francis Press, Boca Raton, FL, pp. 249-285.

Pulwarty, R., U. Trotz, and L. Nurse, 2008: Risk and criticality-Caribbean Islands in a changing climate. In: □ey Vulnerable Regions and Climate Change [W. Hare and A. Battaglini (eds.)] (in press).

Riebsame, W.E., S.A. Changnon, Jr., and T.R. Karl, 1991: *Drought and Natural Resource Management in the United States.* Westview Press, Boulder, 174 pp.

Rodríguez-Trelles, F. and M.A. Rodríguez, 1998: Rapid microevolution and loss of chromosomal diversity in *Drosophila* in response to climate warming. *Evolutionary Ecolology,* **12(7)**, 829-838.

Root, T.L., J.T. Price, K.R. Hall, S.H. Schneider, C. Rosenzweig, and J.A. Pounds, 2003: Fingerprints of global warming on wild animals and plants. *Nature,* **421(6918)**, 57-60.

Rose, J.B., S. Daeschner, D.R. Easterling, F.C. Curriero, S. Lele, and J.A. Patz, 2000: Climate and waterborne outbreaks in the US: a preliminary descriptive analysis. □ournal of the American Water Works Association, **92(9)**, 77-87.

Rose, J.B., P.R. Epstein, E.K. Lipp, B.H. Sherman, S.M. Bernanrd, and J.A. Patz, 2001: Climate variability and change in the United States: potential impacts on water- and foodborne diseases caused by microbiologic agents. *Environmental Health Perpectives,* **109(Suppl 2)**, 211-221.

Rowan, R., 2004: Thermal adaptation in reef coral symbionts. *Nature,* **430(7001)**, 742.

Schär, C., P.L. Vidale, D. Lüthi, C. Frei, C. Häberli, M.A. Liniger, and C. Appenzeller, 2004: The role of increasing temperature variability for European summer heat waves. *Nature,* **427(6972)**, 332-336. doi:10.1038/nature02300

Singer, M.C. and P.R. Ehrlich, 1979: Population dynamics of the checkerspot butterfly *Euphydryas editha. Fortschritte der* □o-*ologie,* **25**, 53-60.

Sokolov, L.V., 2006: The influence of global warming on timing of migration and breeding of passerine bird in the twentieth century. □oologichesky □hurnal, **85**, 317-341.

Smit, B., I. Burton, R.J.T. Klein, and J. Wandel, 2000: An anatomy of adaptation to climate change and variability. *Climatic Change,* **45(1)**, 223-251.

Spear, L.B. and D.G. Ainley, 1999: Migration routes of sooty shearwaters in the Pacific ocean. *Condor,* **101(2)**, 205-218.

Stephenson, N.L., and D.J. Parsons, 1993: A research program for predicting the effects of climate change on the Sierra Nevada. In: *Proceeding of the Fourth Conference on Research in California's National Parks* [Veirs, S.D., Jr., T.J. Stohlgren, and C. Schonewald-Cox (eds.)]. National Park Service Transactions and Proceedings Series NPS/NRUX/NRTP-93/9, Cooperative Park Studies, Davis, CA, pp. 93-109.

Stern, N., 2006: *The Economics of Climate Change: The Stern Review,* Cambridge University Press, 712 pp.

Stern, N.H. and C. Taylor, 2007: Climate change: risks, ethics and the Stern Review. Science, 317(5835), 203-204.

Stott, P.A., D.A. Stone, and M.R. Allen, 2004: Human contribution to the European heat wave of 2003. *Nature,* **432(7017)**, 610-614.

Swetnam, T.W., 1993: Fire history and climate change in giant sequoia groves. *Science,* **262(5135)**, 885-889.

Taulman, J.F. and L.W. Robbins, 1996: Recent range expansion and distributional limits of the nine-banded armadillo (*Dasypus novemcinctus*) in the United States. □ournal of Biogeography, **23(5)**, 635-648.

Thomas, C.D., E.J. Bodsworth, R.J. Wilson, A.D. Simmons, Z.G. Davies, M. Musche, and L. Conradt, 2001: Ecological and evolutionary processes at expanding range margins. *Nature,* **411(6837)**, 577-581.

Tol, R.S.J. and G.W. Yohe. 2006. A review of the "Stern Review." World Economics, 7(4), 233-250.

Trenberth, K. E., A. Dai, R.M. Rasmussen, and D.B. Parsons, 2003: The changing character of precipitation. *Bulletin of the American Meteorological Society*, **84(9)**, 1205-1217.

van Vliet, A. and R. Leemans, 2006: Rapid species' responses to changes in climate require stringent climate protection targets. In *Avoiding Dangerous Climate Change* [Schellnhuber, H.J. (ed.)]. Cambridge University Press, Cambridge, UK, and New York, pp. 135-141.

Walker, L.R. (ed.), 1999: *Ecosystems of Disturbed Ground.* Elsevier, Amsterdam, New York, 868 pp.

Walther, G.-R., E. Post, P. Convery, A. Menzel, C. Parmesan, T.J.C. Beebee, J.-M. Fromentin, O. Hoegh-Guldberg, and F. Bairlein, 2002: Ecological responses to recent climate change. *Nature,* **416(6879)**, 389-95.

Warrick, R.A., 1980: Drought in the Great Plains: A case study of research on climate and society in the USA. In: *Climatic Constraints and Human Activities* [Ausubel, J. and A.K. Biswas (eds.)]. Pergamon, New York, pp. 93-123.

Weiser, W. (ed.), 1973: *Effects of Temperature on Ectothermic Organisms.* Springer-Verlag, New York and Berlin, 298 pp.

Werner, R.A., E.H. Holsten, S.M. Matsuoka, and R.E. Burnside, 2006: Spruce beetles and forest ecosystems in south-central Alaska: a review of 30 years of research. *Forest Ecology and Management,* **227(3)**, 195-206.

Westerling, A.L., H.G. Hidalgo, D.R. Cayan, and T.W. Swetnam, 2006: Warming and earlier spring increase in Western U.S. forest wildfire activity. *Science,* **313(5789)**, 940-943.

White, G.F., R.W. Kates, and I. Burton, 2001: Knowing better and losing even more: the use of knowledge in hazards management. *Global Environmental Change Part B: Environmental Hazards,* **3(3-4)**, 81-92.

WHO (World Health Organization), 2002: *World Health Report 2002: Reducing Risks, Promoting Healthy Life.* World Health Organization, Geneva, Switzerland, 248 pp.

WHO (World Health Organization), 2003: *Climate Change and Human Health: Risks and Responses.* [McMichael, A.J., D.H Campbell-Lendrum, C.F. Corvalán, K.L. Ebi, A.K. Githeko, J.D. Scheraga, and A. Woodward (eds.)]. World Health Organization, Geneva, Switzerland, 322 pp.

WHO (World Health Organization), 2004: *Using Climate to Predict Infectious Disease Outbreaks: A Review.* World Health Organization, Geneva, Switzerland, 55 pp.

Wilhite, D.A. (ed.), 2005: *Drought and Water Crises: Science, Technology and Management Issues.* Taylor and Francis Press, Boca Raton, FL, 432 pp.

Wilhite, D. and R. Pulwarty, 2005: Drought, crises and water management. In: *Drought and Water Crises: Science, Technology and Management Issues* [Wilhite, D.A. (ed.)]. Taylor and Francis Press, Boca Raton, FL, pp. 289-298.

Wilkinson, C.R. (ed.), 2000: *Global Coral Reef Monitoring Network: Status of Coral Reefs of the World in 2000.* Australian Institute of Marine Science, Townsville, Queensland, 363 pp.

Wilkinson, P., D.H. Campbell-Lendrum, and C.L. Bartlett, 2003: Monitoring the health effects of climate change. In: *Climate Change and Human Health: Risks and Responses* [McMichael, A.J., D.H Campbell-Lendrum, C.F. Corvalán, K.L. Ebi, A.K. Githeko, J.D. Scheraga, and A. Woodward (eds.)]. World Health Organization, Geneva, Switzerland, pp. 204-219.

WIST, 2002: Weather Information for Surface Transportation: National Needs Assessment. FCM-R18-2002, Office of the Federal Coordinator for Meteorological Services and Supporting Research, Washington, DC. Available at <http://www.ofcm.noaa.gov/wist_report/pdf/entire_wist.pdf>

CHAPTER 2 REFERENCES

Acuña-Soto, R., D.W. Stahle, M.K. Cleaveland, and M.D. Therrel, 2002: Megadrought and megadeath in 16th century Mexico. *Historical Review,* **8 (4),** 360-362.

Aguilar, E., T.C. Peterson, P. Ramírez Obando, R. Frutos, J.A. Retana, M. Solera, J. Soley, I. González García, R.M. Araujo, A. Rosa Santos, V.E. Valle, M. Brunet, L. Aguilar, L. Álvarez, M. Bautista, C. Castañón, L. Herrera, E. Ruano, J.J. Sinay, E. Sánchez, G.I. Hernández Oviedo, F. Obed, J.E. Salgado, J.L. Vázquez, M. Baca, M. Gutiérrez, C. Centella, J. Espinosa, D. Martínez, B. Olmedo, C.E. Ojeda Espinoza, R. Núñez, M. Haylock, H. Benavides, and R. Mayorga, 2005: Changes in precipitation and temperature extremes in Central America and northern South America, 1961-2003, *Journal of Geophysical Research,* **110,** D23107, doi:10.1029/2005JD006119.

Alexander, L.V., X. Zhang, T.C. Peterson, J. Caesar, B. Gleason, A.M.G. Klein Tank, M. Haylock, D. Collins, B. Trewin, F. Rahimzadeh, A. Tagipour, K. Rupa Kumar, J. Revadekar, G. Griffiths, L. Vincent, D.B. Stephenson, J. Burn, E. Aguilar, M. Brunet, M. Taylor, M. New, P. Zhai, M. Rusticucci, and J.L. Vazquez-Aguirre, 2006: Global observed changes in daily climate extremes of temperature and precipitation, *Journal of Geophysical Research,* **111,** D05109, doi:10.1029/2005JD006290.

Allan, J.C. and P.D. Komar, 2000: Are ocean wave heights increasing in the eastern North Pacific? *EOS, Transactions of the American Geophysical Union,* **47,** 561-567.

Allan, J.C. and P.D. Komar, 2006: Climate controls on US West Coast erosion processes. *Journal of Coastal Research,* **22(3),** 511-529.

Alley, W.M., 1984. The Palmer Drought Severity Index: limitations and assumptions. *Journal of Climate and Applied Meteorology,* **23(7),** 1100-1109.

An, S.I., J.S. Kug, A. Timmermann, I.S. Kang, and O. Timm, 2007: The influence of ENSO on the generation of decadal variability in the North Pacific. *Journal of Climate,* **20(4),** 667-680.

Andreadis, K.M. and D.P. Lettenmaier, 2006: Trends in 20th century drought over the continental United States. *Geophysical Research Letters,* **33,** L10403, doi:10.1029/2006GL025711.

Andreadis, K.M., E.A. Clark, A.W. Wood, A.F. Hamlet, and D.P. Lettenmaier, 2005: Twentieth-Century drought in the conterminous United States. *Journal of Hydrometeorology,* **6(6),** 985-1001.

Angel, J.R. and S.A. Isard, 1998: The frequency and intensity of Great Lake cyclones. *Journal of Climate,* **11(1),** 61-71.

Arguez, A. (ed.), 2007: State of the climate in 2006. *Bulletin of the American Meteorological Society,* **88(6),** s1-s135.

Assel, R.A., 2003: *Great Lakes ice cover, first ice, last ice, and ice duration.* NOAA Technical Memorandum GLERL-125, NOAA, Great Lakes Environmental Research Laboratory, Ann Arbor, MI, 49 pp.

Assel, R.A., 2005a: *Great Lakes ice cover climatology update: winters 200☐, 200☐, and 200☐.* NOAA Technical Memorandum GLERL-135, NOAA, Great Lakes Environmental Research Laboratory, Ann Arbor, MI, 21 pp.

Assel, R.A., 2005b: Classification of annual Great Lakes ice cycles: winters of 1973-2002. *Journal of Climate,* **18(22),** 4895-4905.

Assel, R.A., K. Cronk, and D.C Norton, 2003: Recent trends in Laurentian Great Lakes ice cover. *Climatic Change,* **57(1-2),** 185-204.

Bacon, S. and D.J.T. Carter, 1991: Wave climate changes in the North Atlantic and North Sea. *International Journal of Climatology,* **11(5),** 545-558.

Bell, D.B. and M. Chelliah, 2006: Leading tropical modes associated with interannual and multidecadal fluctuations in North Atlantic hurricane activity. *Journal of Climate,* **17(3),** 590-612.

Bernacchi, C.J., B.A. Kimball, D.R. Quarles, S.P. Long, and D R. Ort, 2007: Decreases in stomatal conductance of soybean under open-air elevation of [CO_2] are closely coupled with decreases in ecosystem evapotranspiration. *Plant Physiology,* **143,** 134-144.

Bonsal, B.R., X. Zhang, L. Vincent, and W. Hogg, 2001: Characteristics of daily and extreme temperatures over Canada. *Journal of Climate,* **14(9),** 1959-1976.

Bromirski, P.D., 2001, Vibrations from the "Perfect Storm". *Geochemistry Geophysics Geosystems,* **2(7),** doi:10.1029/2000GC000119.

Bromirski, P.D., R.E. Flick, and D.R. Cayan, 2003: Decadal storminess variability along the California coast: 1858-2000. *Journal of Climate,* **16(6),** 982-993.

Bromirski, P.D., D.R. Cayan, and R.E. Flick, 2005: Wave spectral energy variability in the northeast Pacific. *Journal of Geophysical Research*, **110**, C03005, doi:10.1029/2004JC002398.

Brooks, H.E., 2004: On the relationship of tornado path length and width to intensity. *Weather and Forecasting*, **19(2)**, 310-319.

Brooks, H.E., 2007: Development and use of climatologies of convective weather. In: *Atmospheric Convection: Research and Operational Forecasting Aspects* [Giaiotti, D.B., R. Steinacker, and F. Stel (eds.)]. Springer-Verlag, Wien, 222 pp.

Brooks, H.E. and C.A. Doswell, III, 2001: Some aspects of the international climatology of tornadoes by damage classification. *Atmospheric Research*, **56(1-4)**, 191-201.

Brooks, H.E. and N. Dotzek, 2008: The spatial distribution of severe convective storms and an analysis of their secular changes. In: *Climate Extremes and Society*. [H. F. Diaz and R. Murnane, (eds.)]. Cambridge University Press, Cambridge, UK, and New York, pp. 35-53.

Brooks, H.E., C.A. Doswell, III, and M.P. Kay, 2003a: Climatological estimates of local daily tornado probability. *Weather and Forecasting*, **18(4)**, 626-640.

Brooks, H.E., J.W. Lee, and J.P. Craven, 2003b: The spatial distribution of severe thunderstorm and tornado environments from global reanalysis data. *Atmospheric Research*, **67-68**, 73-94.

Burnett, A.W., M.E. Kirby, H.T. Mullins, and W.P. Patterson, 2003: Increasing Great Lake-effect snowfall during the twentieth century: a regional response to global warming? *Journal of Climate*, **16(21)**, 3535-3541.

Cavazos, T. and D. Rivas, 2004: Variability of extreme precipitation events in Tijuana, Mexico. *Climate Research*, **25(3)**, 229-243.

Cavazos, T., A.C. Comrie, and D.M. Liverman, 2002: Intraseasonal variability associated with wet monsoons in southeast Arizona. *Journal of Climate*, **15(17)**, 2477-2490.

Cayan, D.R., S.A. Kammerdiener, M.D. Dettinger, J.M. Caprio, and D.H. Peterson, 2001: Changes in the onset of spring in the Western United States. *Bulletin of the American Meteorological Society*, **82(3)**, 399-415.

Cayan, D.R., P.D. Bromirski, K. Hayhoe, M. Tyree, M. Dettinger, and R.E. Flick, 2008: Climate change projections of sea level extremes along the California coast. *Climatic Change*, (in press).

Chan, J.C.L., 2000: Tropical cyclone activity over the western North Pacific associated with El Niño and La Niña events. *Journal of Climate*, **13(16)**, 2960-2972.

Chan, J.C.L., 2006: Comment on "Changes in tropical cyclone number, duration, and intensity in a warming environment". *Science*, **311(5768)**, 1713.

Chan, J.C.L. and J.-E. Shi, 1996: Long-term trends and interannual variability in tropical cyclone activity over the western North Pacific. *Geophysical Research Letters*, **23(20)**, 2765-2767.

Chang, E.K.M. and Y. Fu, 2002: Inter-decadal variations in Northern Hemisphere winter storm track intensity. *Journal of Climate*, **15(6)**, 642-658.

Chang, E.K.M. and Y. Guo, 2007: Is the number of North Atlantic tropical cyclones significantly underestimated prior to the availability of satellite observations? *Geophysical Research Letters*, **34**, L14801, doi:10.1029/2007GL030169

Changnon, D., S.A. Changnon, and S.S. Changnon, 2001: A method for estimating crop losses from hail in uninsured periods and regions. *Journal of Applied Meteorology*, **40(1)**, 84-91.

Changnon, S.A., 1982: Trends in tornado frequency: fact or fallacy? In: *Preprints*, 12th Conference on Severe Local Storms, 11-15 January 1982, San Antonio, TX. American Meteorological Society, Boston, pp. 42-44.

Changnon, S.A. and D. Changnon, 2000: Long-term fluctuations in hail incidences in the United States. *Journal of Climate*, **13(3)**, 658-664.

Changnon, S. and T. Karl, 2003: Temporal and spatial variations in freezing rain in the contiguous U.S. *Journal of Applied Meteorology*, **42(9)**, 1302-1315.

Changnon, S.A., D. Changnon, and T.R. Karl, 2006: Temporal and spatial characteristics of snowstorms in the contiguous United States. *Journal of Applied Meteorology and Climatology*, **45(8)**, 1141-1155.

Chapin III, F.S., B.H. Walker, R.J. Hobbs, D.U. Hooper, J.H. Lawton, O.E. Sala, and D. Tilman, 1997: Biotic control over the functioning of ecosystems. *Science*, **277(5325)**, 500-504.

Chenoweth, M., 2003: *The 18th century climate of Jamaica derived from the journals of Thomas Thistlewood, 1750-1786.* American Philosophical Society, Philadelphia, PA, 153 pp.

Chu, J.-H., C.R. Sampson, A.S. Levine, and E. Fukada, 2002: *The Joint Typhoon Warning Center Tropical Cyclone Best Tracks, 1945-2000.* Naval Research Laboratory Reference Number NRL/MR/7540-02-16.

Clark, M.P., M.C. Serreze, and D.A. Robinson, 1999: Atmosphere controls on Eurasian snow cover extent. *International Journal of Climatology*, **19(1)**, 27-40.

Cleaveland, M.L., D.W. Stahle, M.D. Therrell, J. Villanueva-Diaz, and B.T. Burnes, 2004: Tree-ring reconstructed winter precipitation and tropical teleconnections in Durango, Mexico. *Climatic Change*, **59(3)**, 369-388.

Coles, S.G., 2001: *An Introduction to Statistical Modeling of Extreme Values.* Springer Verlag, New York, 208 pp.

Concannon, P.R., H.E. Brooks, and C.A. Doswell, III, 2000: Climatological risk of strong and violent tornadoes in the United States. In: *Preprints*, 2nd Symposium on Environmental Applications, 9-14 June 2000, Long Beach, CA. American Meteorological Society, Boston, pp. 212-219.

Cook, E.R., D.M. Meko, D.W. Stahle, and M.K. Cleaveland, 1999: Drought reconstructions for the continental United States. *Journal of Climate*, **12(4)**, 1145-1162.

Cook, E.R., R.D. D'Arrigo, and M.E. Mann, 2002: A well-verified, multiproxy reconstruction of the winter North Atlantic Oscillation index since A.D. 1400. *Journal of Climate*, **15(13)**, 1754-1764.

Cook, E.R., C.A. Woodhouse, C.M. Eakin, D.M. Meko, and D.W. Stahlo, 2004: Long-term aridity changes in the western United States. *Science*, **306(5698)**, 1015-1018.

Cooley, D., P. Naveau, and D. Nychka, 2007: Bayesian spatial modeling of extreme precipitation return levels. *Journal of the American Statistical Association*, **102(479)**, 824-840.

Cooter, E. and S. LeDuc, 1995: Recent frost date trends in the northeastern United States. *International Journal of Climatology*, **15(1)**, 65-75.

Dai, A.G., K.E. Trenberth, and T.T. Qian, 2004: A global dataset of Palmer Drought Severity Index for 1870-2002: relationship with soil moisture and effects of surface warming. *Journal of Hydrometeorology*, **5(6)**, 1117-1130.

Davis, R.E., R. Dolan, and G. Demme, 1993: Synoptic climatology of Atlantic coast north-easters. *International Journal of Climatology*, **13(2)**, 171-189.

Davison, A.C. and R.L. Smith, 1990: Models for exceedances over high thresholds (with discussion). *Journal of the Royal Statistical Society Series B*, **52(3)**, 393-442.

DeGaetano, A.T. and R.J. Allen, 2002: Trends in the twentieth century temperature extremes across the United States. *Journal of Climate*, **15(22)**, 3188-3205.

Delworth, T.L. and M.E. Mann, 2000: Observed and simulated multidecadal variability in the Northern Hemisphere. *Climate Dynamics*, **16(9)**, 661-676.

Deser, C., A.S. Phillips, and J.W. Hurrell, 2004: Pacific interdecadal climate variability: linkages between the tropics and the north Pacific during boreal winter since 1900. *Journal of Climate*, **17(16)**, 3109-3124.

Dolan, R., H. Lins, and B. Hayden, 1988: Mid-Atlantic coastal storms. *Journal of Coastal Research*, **4(3)**, 417-433.

Donnelly, J.P., 2005: Evidence of past intense tropical cyclones from backbarrier salt pond sediments: a case study from Isla de Culebrita, Puerto Rico, U.S.A. *Journal of Coastal Research*, **42(6)**, 201-210.

Donnelly, J.P. and T. Webb, III, 2004: Backbarrier sedimentary records of intense hurricane landfalls in the northeastern United States. In: *Hurricanes and Typhoons: Past, Present, and Future* [Murnane, R.J. and K-b. Liu (eds.)]. Columbia University Press, New York, pp. 58-95.

Donnelly, J.P. and J.D. Woodruff, 2007: Intense hurricane activity over the past 5,000 years controlled by El Niño and the West African monsoon. *Nature*, **447(7143)**, 465-468.

Donnelly, J.P., S.S. Bryant, J. Butler, J. Dowling, L. Fan, N. Hausmann, P. Newby, B. Shuman, J. Stern, K. Westover, and T. Webb, III, 2001a: A 700 yr. sedimentary record of intense hurricane landfalls in southern New England. *Geological Society of America Bulletin*, **113(6)**, 714-727.

Donnelly, J.P., S. Roll, M. Wengren, J. Butler, R. Lederer and T. Webb, III, 2001b: Sedimentary evidence of intense hurricane strikes from New Jersey. *Geology*, **29(7)**, 615-618.

Donnelly, J.P., J. Butler, S. Roll, M. Wengren, and T. Webb, III, 2004: A backbarrier overwash record of intense storms from Brigantine, New Jersey. *Marine Geology*, **210(1-4)**, 107-121.

Doswell, C.A., III, H.E. Brooks, and M.P. Kay, 2005: Climatological estimates of daily local nontornadic severe thunderstorm probability for the United States. *Weather and Forecasting*, **20(4)**, 577-595.

Doswell, C.A., III, R. Edwards, R.L. Thompson, and K.C. Crosbie, 2006: A simple and flexible method for ranking severe weather events. *Weather and Forecasting*, **21(6)**, 939-951.

Douglas, M.W., R.A. Maddox, K. Howard, and S. Reyes, 1993: The Mexican monsoon. *Journal of Climate*, **6(8)**, 1665-1677.

Easterling, D.R., 2002: Recent changes in frost days and the frost-free season in the United States. *Bulletin of the American Meteorological Society*, **83(19)**, 1327-1332.

Easterling, D.R., B. Horton, P.D. Jones, T.C. Peterson, T.R. Karl, D.E. Parker, M.J. Salinger, V. Razuvayev, N. Plummer, P. Jamison, and C.K. Folland, 1997: Maximum and minimum temperature trends for the globe. *Science*, **277(5324)**, 364-367.

Easterling, D.R., J.L. Evans, P.Ya. Groisman, T.R. Karl, K.E. Kunkel, and P. Ambenje, 2000: Observed variability and trends in extreme climate events: a brief review. *Bulletin of the American Meteorological Society*, **81(3)**, 417-425.

Easterling, D.R., T. Wallis, J. Lawrimore, and R. Heim, 2007: The effects of temperature and precipitation trends on U.S. drought. *Geophyical. Research Letters*, **34**, L20709, doi:10.1029/2007GL031541

Edwards, DC and T.B. McKee, 1997: *Characteristics of 20th Century Drought in the United States at Multiple Time Scales*. Department of Atmospheric Science, Colorado State University, Fort Collins, CO, 155 pp.

Eichler, T. and W. Higgins, 2006: Climatology and ENSO-related variability of North American extra-tropical cyclone activity. *Journal of Climate*, **19(10)**, 2076-2093.

Elsner, J.B., A.A. Tsonis, and T.H. Jagger, 2006: High-frequency variability in hurricane power dissipation and its relationship to global temperature. *Bulletin of the American Meteorological Society*, **87(6)**, 763-768.

Emanuel, K.A., 2005a: Increasing destructiveness of tropical cyclones over the past 30 years. *Nature*, **436(7051)**, 686-688.

Emanuel, K.A., 2005b: Emanuel replies. *Nature*, **438(7071)**, E13. doi:10.1038/nature04427

Emanuel, K.A., 2007: Environmental factors affecting tropical cyclone power dissipation. *Journal of Climate*, **20(22)**, 5497-5509.

Enfield, D.B. and L. Cid-Serrano, 2006: Projecting the risk of future climate shifts. *International Journal of Climatology*, **26(7)**, 885-895.

Englehart, P.J. and A.V. Douglas, 2001: The role of eastern North Pacific tropical storms in the rainfall climatology of western Mexico. *International Journal of Climatology*, **21(11)**, 1357-1370.

Englehart, P.J. and A.V. Douglas, 2002: México's summer rainfall patterns: an analysis of regional modes and changes in their teleconnectivity. *Atmósfera*, **15(3)**, 147-164.

Englehart, P.J. and A.V. Douglas, 2003: Urbanization and seasonal temperature trends: observational evidence from a data sparse part of North America. *International Journal of Climatology*, **23(10)**, 1253-1263.

Englehart, P.J. and A.V. Douglas, 2004: Characterizing Regional-Scale variations in monthly and seasonal surface air temperature over Mexico. *International Journal of Climatology*, **24(15)**, 1897-1909.

Englehart, P.J. and A.V. Douglas, 2005: Changing behavior in the diurnal range of surface air temperatures over Mexico. *Geophysical Research Letters*, **32**, L01701, doi:10.1029/2004GL021139.

Englehart, P.J. and A.V. Douglas, 2006: Defining intraseasonal variability within the North American monsoon. *Journal of Climate*, **19(17)**, 4243-4253.

Englehart, P.J., M.D. Lewis, and A.V. Douglas, 2008: Defining the frequency of near shore tropical cyclone activity in the eastern North Pacific from historical surface observations 1921-

2005. *Geophysical Research Letters*, **35**, L03706, doi:10.1029/2007GL032546.

Federal Emergency Management Agency, 1995: *National Mitigation Strategy: Partnerships for Building Safer Communities.* Federal Emergency Management Agency, Washington DC, 40 pp.

Fernández-Partagás, J. and H. F. Diaz, 1996: Atlantic hurricanes in the second half of the nineteenth century. *Bulletin of the American Meteorological Society*, **77(12)**, 2899-2906.

Ferretti, D.F., E. Pendall, J. A. Morgan, J. A.Nelson, D. LeCain, and A. R. Mosier, 2003: Partitioning evapotranspiration fluxes from a Colorado grassland using stable isotopes: seasonal variations and ecosystem implications of elevated atmospheric CO_2. *Plant and Soil*, **254(2)**, 291–303.

Feuerstein, B., N. Dotzek and J. Grieser, 2005: Assessing a tornado climatology from global tornado intensity distributions, *Journal of Climate*, **18(4)**, 585-596.

Folland, C.K., T.N. Palmer, and D.E. Parker, 1986: Sahel rainfall and worldwide sea temperatures, 1901-85. *Nature*, **320(6063)**, 602-607.

Frappier, A., D. Sahagian, S.J. Carpeter, L.A. Gonzalez, and B. Frappier, 2007: A stalagmite stable isotope record of recent tropical cyclone events. *Geology*, **35(2)**, 111-114. doi:10.1130/G23145A

Gandin L.S. and R.L. Kagan, 1976: *Statistical Methods of Interpretation of Meteorological Data.* (in Russian). Gidrometeoizdat, 359 pp.

García Herrera, R., F. Rubio, D. Wheeler, E. Hernández, M. R. Prieto, and L. Gimero, 2004: The use of Spanish and British documentary sources in the investigation of Atlantic hurricane incidence in historical times. In: *Hurricanes and Typhoons: Past, Present, and Future* [Murnane, R. J. and K-b. Liu (eds.)]. Columbia University Press, New York, pp. 149-176.

García Herrera, R., L. Gimeno, P. Ribera, and E. Hernández, 2005: New records of Atlantic hurricanes from Spanish documentary sources. *Journal of Geophysical Research*, **110**, D03109, doi:10.1029/2004JD005272.

Garriott, E.B., 1903: *Storms of the Great Lakes.* Bulletin K, US Department of Agriculture, Weather Bureau, Washington DC, 486 pp.

Geng, Q. and M. Sugi, 2001: Variability of the North Atlantic cyclone activity in winter analyzed from NCEP–NCAR reanalysis data. *Journal of Climate*, **14(18)**, 3863-3873.

Gershunov, A. and T.P. Barnett, 1998: Inter-decadal modulation of ENSO teleconnections. *Bulletin of the American Meteorological Society*, **79(12)**, 2715-2725.

Gershunov, A. and D.R. Cayan, 2003: Heavy daily precipitation frequency over the contiguous United States: sources of climate variability and seasonal predictability. *Journal of Climate*, **16(16)**, 2752-2765.

Gershunov, A. and H. Douville, 2008: Extensive summer hot and cold extremes under current and possible future climatic conditions: Europe and North America. In *Climate Extremes and Society* [Diaz, H.F. and R.J. Murnane, (eds.)]. Cambridge University Press, Cambridge, UK, and New York, pp. 74-98.

Goldenberg, S.B., C.W. Landsea, A.M. Mesta-Nuñez, and W. M. Gray, 2001: The recent increase in Atlantic hurricane activity: causes and implications. *Science*, **293(5529)**, 474-479.

Graham, N.E., 1994: Decadal-scale climate variability in the tropical and North Pacific during the 1970s and 1980s: observations and model results. *Climate Dynamics*, **10(3)**, 135-162.

Graham, N.E. and H.F. Diaz, 2001: Evidence for intensification of North Pacific winter cyclones since 1948, *Bulletin of the American Meteorological Society*, **82(7)**, 1869-1893.

Gray, S.T., L.J. Graumlich, J.L. Betancourt, and G.T. Pederson, 2004: A tree-ring based reconstruction of the Atlantic Multidecadal Oscillation since 1567 AD. *Geophysical Research Letters*, **31**, L12205, doi:10.1029/2004GL019932.

Grazulis, T.P., 1993: *Significant Tornadoes, 1680-1991.* Environmental Films, St. Johnsbury, VT, 1326 pp.

Groisman, P.Ya. and R.W. Knight, 2007: Prolonged dry episodes over North America: new tendencies emerged during the last 40 years. *Advances in Earth Science*, **22(11)**, 1191-1207.

Groisman, P.Ya. and R.W. Knight, 2008: Prolonged dry episodes over the conterminous United States: New tendencies emerging during the last 40 years. *Journal of Climate*, **21**, No. 9, 1850-1862.

Groisman, P.Ya., T.R. Karl, D.R. Easterling, R.W. Knight, P.B. Jamason, K.J. Hennessy, R. Suppiah, Ch.M. Page, J. Wibig, K. Fortuniak, V.N. Razuvaev, A. Douglas, E. Førland, and P.-M. Zhai, 1999: Changes in the probability of heavy precipitation: important indicators of climatic change. *Climatic Change*, **42(2)**, 243-283.

Groisman, P.Ya. R.W. Knight , and T.R. Karl, 2001: Heavy precipitation and high streamflow in the contiguous United States: trends in the 20th century. *Bulletin of the American Meteorological Society*, **82(2)**, 219-246.

Groisman, P.Ya., R.W. Knight, T.R. Karl, D.R. Easterling, B. Sun, and J.H. Lawrimore, 2004: Contemporary changes of the hydrological cycle over the contiguous United States: trends derived from *in situ* observations, *Journal of Hydrometeorology*, **5(1)**, 64-85.

Groisman, P.Ya, R.W. Knight, D.R. Easterling, T.R. Karl, G.C. Hegerl, and V.N. Razuvaev, 2005: Trends in intense precipitation in the climate record. *Journal of Climate*, **18(9)**, 1326-1350.

Groisman, P.Ya., B.G. Sherstyukov, V.N. Razuvaev, R.W. Knight, J.G. Enloe, N.S. Stroumentova, P.H. Whitfield, E. Førland, I. Hannsen-Bauer, H. Tuomenvirta, H. Aleksandersson, A.V. Mescherskaya, and T.R. Karl, 2007: Potential forest fire danger over Northern Eurasia: changes during the 20th century. *Global and Planetary Change*, **56(3-4)**, 371-386.

Gulev, S.K. and V. Grigorieva, 2004: Last century changes in ocean wind wave height from global visual wave data. *Geophysical Research Letters*, **31**, L24302, doi:10.1029/2004GL021040.

Gumbel, E.J., 1958: *Statistics of Extremes.* Columbia University Press, New York, 375 pp.

Guttman, N.B., 1998: Comparing the Palmer Drought Index and the Standardized Precipitation Index. *Journal of the American Water Resources Association*, **34(1)**, 113-121. doi:10.1111/j.1752-1688.1998.tb05964

Hallack-Alegria, M. and D.W. Watkins, Jr., 2007: Annual and warm season drought intensity-duration-frequency analysis for Sonora, Mexico. *Journal of Climate*, **20(9)**, 1897-1909.

Harman, J.R., R. Rosen, and W. Corcoran, 1980: Winter cyclones and circulation patterns on the western Great Lakes. *Physical Geography*, **1(1)**, 28-41.

Harnik, N. and E.K.M. Chang, 2003: Storm track variations as seen in radiosonde observations and reanalysis data. *Journal of Climate*, **16(3)**, 480-495.

Harper, B.A. and J. Callaghan, 2006: On the importance of reviewing historical tropical cyclone intensities. In: *27th Conference on Hurricanes and Tropical Meteorology*, 24-28 April, 2006, Monterey, CA. American Meteorological Society, Boston, Paper 2C.1. Extended abstract available at http://ams.confex.com/ams/pdfpapers/107768.pdf

Hartmann, D.L. and E.D. Maloney, 2001: The Madden-Julian oscillation, barotropic dynamics, and North Pacific tropical cyclone formation. Part II: stochastic barotropic modeling. *Journal of the Atmospheric Sciences*, **58(17)**, 2559-2570.

Heim, Jr., R.R., 2002: A review of twentieth-century drought indices used in the United States. *Bulletin of the American Meteorological Society*, **83(8)**, 1149-1165.

Herweijer, C., R. Seager. and E.R. Cook, 2006: North American droughts of the mid-to-late nineteenth century: a history, simulation and implication for mediaeval drought. *The Holocene*, **16(2)**, 159-171.

Herweijer, C., R. Seager, E.R. Cook, and J. Emile-Geay, 2007: North American droughts of the last millennium from a gridded network of tree-ring data. *Journal of Climate*, **20(7)**, 1353-1376.

Higgins, R.W., Y. Chen, and A.V. Douglas, 1999: Interannual variability of the North American warm season precipitation regime. *Journal of Climate*, **12(3)**, 653-680.

Hirsch, M.E., A.T. DeGaetano, and S.J. Colucci, 2001: An east coast winter storm climatology. *Journal of Climate*, **14(5)**, 882-899.

Hoegh-Guldberg, O., 2005: Low coral cover in a high-CO_2 world. *Journal of Geophysical Research*, **110**, C09S06, doi:10.1029/2004JC002528.

Holland, G.J., 1981: On the quality of the Australian tropical cyclone data base. *Australian Meteorological Magazine*, **29(4)**, 169-181.

Holland, G.J., 2007: Misuse of landfall as a proxy for Atlantic tropical cyclone activity. *EOS, Transactions of the American Geophysical Union*, **88(36)**, 349-350.

Holland, G.J. and P.J. Webster, 2007: Heightened tropical cyclone activity in the North Atlantic: natural variability or climate trend? *Philosophical Transactions of the Royal Society A*, **365(1860)**, 2695-2716. doi:10.1098/rsta.2007.2083

Holton, J.R., 1979: *An Introduction to Dynamic Meteorology*. Academic Press, New York, 2nd ed., 391 pp.

Hoyos, C.D., P.A. Agudelo, P.J. Webster and J.A. Curry, 2006: Deconvolution of the factors contributing to the increase in global hurricane intensity. *Science*, **312(5770)**, 94-97.

Hsu, S.A., M.F. Martin, and B.W. Blanchard, 2000: An evaluation of the USACE's deepwater wave prediction techniques under hurricane conditions during Georges in 1998. *Journal of Coastal Research*, **16(3)**, 823-829.

Huschke, R.E. (ed.), 1959: *Glossary of Meteorology*. American Meteorological Society, Boston, MA, pp. 106 and 419.

IPCC (Intergovernmental Panel on Climate Change), 2001: *Climate Change 2001: The Scientific Basis*. Contribution of Working Group I to the Third Assessment Report of the Intergovernmental Panel on Climate Change [Houghton, J.T., Y. Ding, D.J. Griggs, M. Noguer, P.J. van der Linden, X. Dai, K. Maskell, and C.A. Johnson (eds.)]. Cambridge University Press, Cambridge, United Kingdom and New York, NY, USA, 881 pp.

Jarvinen, B.R., C.J. Neumann, and M.A.S. Davis, 1984: *A tropical cyclone data tape for the North Atlantic Basin, 1886-1983: Contents, limitations and uses*. Technical Memorandum NWS NHC-22, NOAA, Washington DC, 21 pp.

Jones, G.V. and R.E. Davis, 1995: Climatology of Nor'easters and the 30 kPa jet. *Journal of Coastal Research*, **11(3)**, 1210-1220.

Jones, P.D. and A. Moberg, 2003: Hemispheric and large-scale surface air temperature variations: an extensive revision and an update to 2001. *Journal of Climate* **16(2)**, 206-223.

Jones, P.D., T.J. Osborn, and K.R. Briffa, 2003: Pressure-based measures of the North Atlantic Oscillation (NAO): A comparison and an assessment of changes in the strength of the NAO and in its influence on surface climate parameters. In: *The North Atlantic Oscillation: Climatic Significance and Environmental Impact* [Hurrell, J.W., Y. Kushnir, G. Ottersen, and M. Visbeck (eds.)]. American Geophysical Union, Washington, DC, pp. 51-62.

Junger, S., 1997: *The Perfect Storm: A True Story of Men Against the Sea*. Norton, New York.

Kalnay, E., M. Kanamitsu, R. Kistler, W. Collins, D. Deaven, L. Gandin, M. Iredell, S. Saha, G. White, J. Woollen, Y. Zhu, A. Leetmaa, B. Reynolds, M. Chelliah, W. Ebisuzaki, W. Higgins, J. Janowiak, K. Mo, C. Ropelewski, J. Wang, R. Jenne, and D. Joseph, 1996: The NCEP/NCAR 40-Year reanalysis project. *Bulletin of the American Meteorological Society*, **77(3)**, 437-471.

Kamahori, H., N. Yamazaki, N. Mannoji, and K. Takahashi, 2006: Variability in intense tropical cyclone days in the western North Pacific. *SOLA*, **2**, 104-107. doi:10.2151/sola.2006-027

Karl, T.R. and R.W. Knight. 1985. *Atlas of Monthly Palmer Hydrological Drought Indices 1931-1983 for the Contiguous United States*. Historical Climatology Series 3-7, National Climatic Data Center, Asheville, NC, 319 pp.

Karl, T.R. and R.W. Knight, 1998: Secular trends of precipitation amount, frequency, and intensity in the United States. *Bulletin of the American Meteorological Society*, **79(2)**, 231-241

Keetch, J.J. and G.M. Byram, 1968: *A Drought Index for Forest Fire Control*. U.S.D.A. Forest Service Research Paper SE-38, U.S. Department of Agriculture, Forest Service, Southeastern Forest Experiment Station, Asheville, NC, 35 pp. Available from: http://www.srs.fs.fed.us/pubs/

Key, J.R. and A.C.K. Chan, 1999: Multidecadal global and regional trends in 1000 mb and 500 mb cyclone frequencies. *Geophysical Research Letters*, **26(14)**, 2035-2056.

Kharin, V.V. and F.W. Zwiers, 2000: Changes in the extremes in an ensemble of transient climate simulation with a coupled atmosphere-ocean GCM. *Journal of Climate*, **13(2)**, 3760-3788.

Kharin, V.V., F.W. Zwiers, X. Zhang, and G.C. Hegerl, 2007: Changes in temperature and precipitation extremes in the IPCC ensemble of global coupled model simulations. *Journal of Climate*, **20(8)**, 1419-1444.

Kim, T-W., J.B. Valdes, and J. Aparicio, 2002: Frequency and spatial characteristics of droughts in the Conchos river basin, Mexico. *International Water Resources*, **27(3)**, 420-430.

Kimberlain, T.B. and J.B. Elsner, 1998: The 1995 and 1996 North Atlantic hurricane seasons: A return of the tropical-only hurricane. *Journal of Climate*, **11**, 2062-2069.

Klotzbach, P. J., 2006: Trends in global tropical cyclone activity over the past twenty years (1986-2005). *Geophysical Research Letters*, **33**, L10805, doi:10.1029/2006GL025881.

Knaff, J.A and C.R. Sampson, 2006: Reanalysis of West Pacific tropical cyclone intensity 1966-1987. In: *27th Conference on Hurricanes and Tropical Meteorology*, 24-28 April, 2006, Monterey, CA. American Meteorological Society, Boston, Paper #5B.5. Extended abstract available at: http://ams.confex.com/ams/pdfpapers/108298.pdf

Knight, J.R., R.J. Allan, C.K. Folland, M. Vellinga, and M.E. Mann, 2005: A signature of persistent natural thermohaline circulation cycles in observed climate. *Geophysical Research Letters*, **32**, L20708, doi:10.1029/2005GL024233.

Kocin, P.J., P.N. Schnumacher, R.F. Morales, Jr., and L.W. Uccellini, 1995: Overview of the 12-14 March 1993 superstorm. *Bulletin of the American Meteorological Society*, **76(2)**, 165-182.

Komar, P.D. and J.C. Allan, 2007a: Higher waves along U.S. east coast linked to hurricanes. *EOS, Transactions, American Geophysical Union*, **88**, 301.

Komar, P.D. and J.C. Allan, 2007b: A note on the depiction and analysis of wave-height histograms. *Shore □ Beach*, **75(4)**, 1-5.

Komar, P.D. and J.C. Allan, 2008: Increasing hurricane-generated wave heights along the U.S. East coast and their climate controls. Journal of Coastal Research, 24(2), 479-488.

Komar, P.D., J.C. Allan, and P. Ruggiero, 2008: Wave and near-shore-process climates: trends and variations due to Earth's changing climate. In: *Handbook of Coastal and Ocean Engineering*, [Y.C. Kim, (ed.)]. World Scientific Publishing Co., (in press).

Kossin, J.P. and D.J. Vimont, 2007: A more general framework for understanding Atlantic hurricane variability and trends. *Bulletin of the American Meteorological Society*, **88(11)**, 1767-1781.

Kossin, J.P., K.R. Knapp, D.J. Vimont, R.J. Murnane, and B.A. Harper, 2007a: A globally consistent reanalysis of hurricane variability and trends. *Geophysical Research Letters*, 34, L04815, doi:10.1029/2006GL028836.

Kossin, J.P., J.A. Knaff, H.I. Berger, D.C. Herndon, T.A. Cram, C.S. Velden, R.J. Murnane, and J.D. Hawkins, 2007b: Estimating hurricane wind structure in the absence of aircraft reconnaissance. *Weather and Forecasting*, **22(1)**, 89-101

Kunkel, K.E., 2003: North American trends in extreme precipitation. *Natural Hazards*, **29**, 291-305.

Kunkel, K.E., S.A. Changnon, and J.R. Angel, 1994: Climatic aspects of the 1993 Upper Mississippi River basin flood. *Bulletin of the American Meteorological Society*, **75(5)**, 811-822.

Kunkel, K.E., R.A. Pielke, Jr., and S.A. Changnon, 1999a: Temporal fluctuations in weather and climate extremes that cause economic and human health impacts: a review. *Bulletin of the American Meteorological Society*, **80(6)**, 1077-1098.

Kunkel, K.E., K. Andsager and D.R. Easterling, 1999b: Long-term trends in extreme precipitation events over the conterminous United States and Canada. *Journal of Climate*, **12(8)**, 2515-2527.

Kunkel, K.E., N.E. Westcott, and D.A.R. Kristovich, 2002: Assessment of potential effects of climate change on heavy lake-effect snowstorms near Lake Erie. *Journal of Great Lakes Research*, **28(4)**, 521-536.

Kunkel, K.E., D.R. Easterling, K. Redmond and K. Hubbard, 2003: Temporal variations of extreme precipitation events in the United States: 1895-2000. *Geophysical Research Letters*, **30(17)**, 1900, doi:10.1029/2003GL018052.

Kunkel, K.E., D.R. Easterling, K. Redmond, and K. Hubbard, 2004: Temporal variations in frost-free season in the United States: 1895-2000, *Geophysical Research* Letters, **31**, L03201, doi:10.1029/2003GL018624.

Kunkel, K.E., T.R. Karl, and D.R. Easterling, 2007a: A Monte Carlo assessment of uncertainties in heavy precipitation frequency variations. *Journal of Hydrometeorolgy*, **8(5)**, 1152-1160.

Kunkel, K.E., M. Palecki, L. Ensor, D. Robinson, K.Hubbard, D. Easterling, and K. Redmond, 2007b: Trends in 20th Century U.S. snowfall using a quality-controlled data set. In: *Proceedings, 75th Annual Meeting, Western Snow Conference*, Ominpress, 57-61.

Landsea, C.W,. R.A. Pielke, A. Mestas-Nunez, et al., 1999: Atlantic basin hurricanes: Indices of climatic changes. *Climatic Change*, **42(1)**, 89-129.

Landsea, C.W., 2005: Hurricanes and global warming. *Nature*, **438(7071)**, E11-E13. doi:10.1038/nature04477

Landsea, C., 2007: Counting Atlantic tropical cylones back in time. *EOS, Transactions of the American Geophysical Union*, **88(18)**, 197-203.

Landsea, C.W., C. Anderson, N. Charles, G. Clark, J. Dunion, J. Fernández-Partagás, P. Hungerford, C. Neumann, and M. Zimmer, 2004: The Atlantic hurricane database reanalysis project: documentation for the 1851-1910 alterations and additions to the HURDAT database. In *Hurricanes and Typhoons: Past, Present and Future* [Murnane, R. J. and K.-b. Liu (eds.)]. Columbia University Press, New York, pp. 177-221.

Landsea, C.W., B.A. Harper, K. Hoarau and J.A. Knaff, 2006: Can we detect trends in extreme tropical cyclones? *Science*. **313(5786)**, 452-454.

Latif, M., E. Roeckner, M. Botzet, M. Esch, H. Haak, S. Hagemann, J. Jungclaus, S. Legutke, S. Marsland, U. Mikolajewicz, and J. Mitchell, 2004: Reconstructing, monitoring, and predicting multidecadal-scale changes in the North Atlantic thermohaline circulation with sea surface temperature. *Journal of Climate*, **17(7)**, 1605-1614.

Lourensz, R.S., 1981: *Tropical Cyclones in the Australian Region □uly □909 to □une □98□ Met Summary.* Bureau of Meterology, PO Box 1289K, Melbourne, Vic 3001, Australia, 94 pp.

Lemke, P., J. Ren, R.B. Alley, I. Allison, J. Carrasco, G. Flato, Y. Fujii, G. Kaser, P. Mote, R.H. Thomas, and T. Zhang, 2007: Observations: changes in snow, ice and frozen ground. In: *Climate Change 2007: The Physical Basis*. Contribution of Working Group I to the Fourth Assessment Report of the Intergovernmental Panel on Climate Change [Solomon, S., D. Qin, M. Manning, Z. Chen, M. Marquis, K.B. Averyt, M. Tignor, and H.L. Miller (eds.)]. Cambridge University Press, Cambridge, UK, and New York, pp. 337-383.

Lewis, P.J., 1987: *Severe Storms over the Great Lakes: A Catalogue Summary for the Period ⬚9⬚7-⬚98⬚* Canadian Climate Center Report No. 87-13, Atmospheric Environment Service, Downsview, ON, Canada, 342 pp.

Lins, H.F. and J.R. Slack, 1999: Streamflow trends in the United States. *Geophysical Research Letters*, **26(2)**, 227-230.

Lins, H.F. and J.R. Slack, 2005: Seasonal and regional characteristics of U.S. streamflow trends in the United States from 1940 to 1999. *Physical Geography*, **26(6)**, 489-501.

Liu, K-b., 2004: Paleotempestology: principles, methods, and examples from Gulf Coast lake sediments. In: *Hurricanes and Typhoons: Past, Present, and Future* [Murnane, R. J. and K-b. Liu (eds.)]. Columbia University Press, New York, pp. 13-57.

Liu, K-b. and M.L. Fearn, 1993: Lake-sediment record of late holocene hurricane activities from coastal Alabama. *Geology*, **21(9)**, 793-796.

Liu, K-b. and M.L. Fearn, 2000: Reconstruction of prehistoric landfall frequencies of catastrophic hurricanes in northwestern Florida from lake sediment records. *⬚uaternary Research*, **54(2)**, 238-245.

Liu, K-b., C. Shen, and K.-s. Louie, 2001: A 1,000-year history of typhoon landfalls in Guangdong, southern China, reconstructed from Chinese historical documentary records. *Annals of the Association of American Geographers*, **91(3)**, 453-464.

Louie, K.-s. and K.-b. Liu, 2003: Earliest historical records of typhoons in China. *⬚ournal of Historical Geography*, **29(3)**, 299-316.

Louie, K.-s. and K.-b. Liu, 2004: Ancient records of typhoons in Chinese historical documents. In: *Hurricanes and Typhoons: Past, Present, and Future* [Murnane, R.J. and K-b. Liu (eds.)]. Columbia University Press, New York, pp. 222-248.

Ludlum, D.M., 1963: *Early American hurricanes, ⬚92-⬚870*. American Meteorological Society, Boston MA, 198 pp.

Madden, R. and P. Julian, 1971: Detection of a 40-50 day oscillation in the zonal wind in the tropical Pacific. *⬚ournal of the Atmospheric Sciences*, **28(5)**, 702-708.

Madden, R. and P. Julian, 1972: Description of global-scale circulation cells in the tropics with a 40-50 day period. *⬚ournal of the Atmospheric Sciences*, **29(6)**, 1109-1123.

Magnuson, J. J., D.M. Robertson, B.J. Benson, R.H. Wynne, D.M. Livingston, T. Arai, R.A. Assel, R. G. Barry, V. Card, E. Kuusisto, N.G. Granin, T.D. Prowse, K.M. Stewart, and V.S. Vuglinski, 2000: Historical trends in lake and river ice cover in the Northern Hemisphere. *Science*, **289(5485)**, 1743-1746.

Maloney, E.D. and D.L. Hartmann, 2000a: Modulation of eastern North Pacific hurricanes by the Madden-Julian oscillation. *⬚ournal of Climate*, **13(9)**, 1451-1460.

Maloney, E.D. and D.L. Hartmann, 2000b: Modulation of hurricane activity in the Gulf of Mexico by the Madden-Julian oscillation. *Science*, **287(5460)**, 2002-2004.

Maloney, E.D. and D.L. Hartmann, 2001: The Madden-Julian oscillation, barotropic dynamics, and North Pacific tropical cyclone formation. Part I: Observations. *⬚ournal of the Atmospheric Sciences*, **58(17)**, 2545-2558.

Mann, M.E. and K. Emanuel, 2006: Atlantic hurricane trends linked to climate change. *EOS, Transactions of the American Geophysical Union*, **87(24)**, 233, 238, 241.

Mann, M.E. and J. Park, 1994: Global-scale modes of surface temperature variability on interannual to century timescales. *⬚ournal of Geophysical Research*, **99(D12)**, 25819-25834.

Mann, M.E., T.A. Sabbatelli and U. Neu, 2007: Evidence for a modest undercount bias in early historical Atlantic tropical cyclone counts. *Geophysical Research Letters*, **34**, L22707, doi:10.1029/2007GL031781.

Mann, M.E., K.A. Emanuel, G.J. Holland and P.J. Webster, 2007: Atlantic tropical cyclones revisited. EOS, Transactions of the American Geophysical Union, 88(36), 349.

Manning, D.M. and R.E. Hart, 2007: Evolution of North Atlantic ERA40 tropical cyclone representation. *Geophysical Research Letters*, **34**, L05705, doi:10.1029/2006GL028266.

Maue, R.N. and R.E. Hart, 2007: Comment on "Low frequency variability in globally integrated tropical cyclone power dissipation" by Ryan Sriver and Matthew Huber. *Geophysical Research Letters*, **34**, L05705, doi:10.1029/2006GL028266.

McCabe, G.J., M.P. Clark, and M.C. Serreze, 2001: Trends in Northern Hemisphere surface cyclone frequency and intensity. *⬚ournal of Climate*, **14(12)**, 2763-2768.

McCarthy, D.W., J.T. Schaefer, and R. Edwards, 2006: What are we doing with (or to) the F-xcale? *Preprints, 23rd Conference on Severe Local Storms*, 5-10 November 2006, St. Louis, MO. American Meteorological Society, Boston, Paper 5.6. Extended abstract available online at http://ams.confex.com/ams/pdf-papers/115260.pdf

McKee, T.B., N.J. Doesken and J. Kleist, 1993: The relationship of drought frequency and duration to time scales. In: *Preprints of the Eight Conference on Applied Climatology*, 17-22 January 1993, Anaheim, CA. American Meteorological Society, Boston, pp. 179-184.

Michaels, P.J., P.C. Knappenberger, O.W. Frauenfeld, and R.E. Davis, 2004: Trends in precipitation on the wettest days of the year across the contiguous USA. *International ⬚ournal of Climatology* **24(15)**, 1873-1882.

Millás, J.C. and L. Pardue, 1968: *Hurricanes of the Caribbean and Adjacent Regions, ⬚⬚92-⬚800*. Academy of the Arts and Sciences of the Americas, Miami, FL, 328 pp.

Miller, A.J., Cayan, D., T. Barnett, N. Graham, and J. Oberhuber, 1994: The 1976-77 climate shift of the Pacific Ocean. *Oceanography*, **(7)1**, 21-26.

Miller, D. L., C.I. Mora, H.D. Grissino-Mayer, M.E. Uhle, and Z. Sharp, 2006: Tree-ring isotope records of tropical cyclone activity. *Proceedings of the National Academy of Sciences*, **103(39)**, 14294-14297.

Mo, K.C. and R.E. Livezey, 1986: Tropical-extratropical geopotential height teleconnections during the northern hemisphere winter. *Monthly Weather Review*, **114(12)**, 2488-2515.

Mock, C.J., 2004: Tropical cyclone reconstructions from documentary records: examples for South Carolina, United States. In: *Hurricanes and Typhoons: Past, Present, and Future* [Murnane, R.J. and K-b. Liu (eds.)]. Columbia University Press, New York, pp. 121-148.

Moon, I.J., I. Ginis, T. Hara, H.L. Tolman, C.W. Wright, and E.J. Walsh, 2003:, Numerical simulation of sea surface directional wave spectra under hurricane wind forcing. *Journal of Physical Oceanography*, **33(8)**, 1680-1706.

Morgan, J.A., D.R. LeCain, J.J. Read, H.W. Hunt, and W.G. Knight, 1998: Photosynthetic pathway and ontogeny affect water relations and the impact of CO_2 on *Bouteloua gracilis* (C_4) and *Pascopyrum smithii* (C_3). *Oecologia*, **114(4)**, 483-493.

Mueller, K.J., M. DeMaria, J. Knaff, J.P. Kossin, T.H. Vonder Haar, 2006: Objective estimation of tropical cyclone wind structure from infrared satellite data. *Weather and Forecasting*, **21(6)**, 990-1005.

Nakamura, H., 1992: Midwinter suppression of baroclinic wave activity in the Pacific. *Journal of the Atmospheric Sciences*, **49(17)**, 1629-1642.

Neelin, J.D., M. Münnich, H. Su, J.E. Meyerson, and C.E. Holloway, 2006: Tropical drying trends in global warming models and observations. *Proceedings of the National Academy of Sciences*, **103(16)**, 6110-6115.

Nelson, J.A., J.A. Morgan, D.R. LeCain, A.R. Mosier, D.G. Milchunas, B.A. Parton, 2004: Elevated CO_2 increases soil moisture and enhances plant water relations in a long-term field study in semi-arid shortgrass steppe of Colorado. *Plant and Soil*, **259(1/2)**, 169–179.

Neumann, C.J., B.R. Jarvinen, C.J. McAdie, and J.D. Elms, 1993: *Tropical Cyclones of the North Atlantic Ocean, 1871–1992.* Prepared by the National Climatic Data Center, Asheville, NC, in cooperation with the National Hurricane Center, Coral Gables, FL. NOAA, National Climatic Data Center, Asheville NC, 4th rev., 193 pp.

Nicholas, R. and D. S. Battisti, 2008: Drought recurrence and seasonal rainfall prediction in the Río Yaqui basin, Mexico. *Journal of Applied Meteorology and Climatology*, (in press).

Noel, J. and D. Changnon, 1998: A pilot study examining U.S. winter cyclone frequency patterns associated with three ENSO parameters. *Journal of Climate*, **11(8)**, 2152-2159.

Nyberg, J., B.A. Malmgren, A. Winter, M.R. Jury, K. Halimeda Kilbourne, and T.M. Quinn, 2007: Low Atlantic hurricane activity in the 1970s and 1980s compared to the past 270 years. *Nature*, **447(7145)**, 698-702.

Paciorek, C.J., J.S. Risbey, V. Ventura, and R.D. Rosen, 2002: Multiple indices of Northern Hemisphere cyclone activity, Winters 1949-99. *Journal of Climate*, **15(13)**, 1573-1590.

Palmer, W.C., 1965: *Meteorological Drought*. Research Paper No. 45, U.S. Department of Commerce, Weather Bureau, Washington DC, 58 pp.

Palmer, W.C., 1968: Keeping track of crop moisture conditions, nationwide: the new crop moisture index. *Weatherwise*, **21(4)**, 156-161.

Pavia, E.G. and A. Badan, 1998: ENSO modulates rainfall in the Mediterranean Californias. *Geophysical Research Letters*, **25(20)**, 3855-3858.

Peterson, T.C., 2003: Assessment of urban versus rural in situ surface temperatures in the contiguous United States: No difference found. *Journal of Climate*, **16(18)**, 2941-2959.

Peterson, T.C., M.A. Taylor, R. Demeritte, D.L. Duncombe, S. Burton, F. Thompson, A. Porter, M. Mercedes, E. Villegas, R. Semexant Fils, A. Klein Tank, A. Martis, R. Warner, A. Joyette, W. Mills, L. Alexander, B. Gleason, 2002: Recent changes in the climate extemes in the Caribbean region. *Journal of Geophysical Research*, **107(D21)**, 4601, doi:10.1029/2002JD002251.

Peterson, T.C., X. Zhang, M. Brunet-India, and J.L. Vázquez-Aguirre, 2008: Changes in North American extremes derived from daily weather data. Journal Geophysical Research, 113, D07113, doi:10.1029/2007JD009453.

Pickands, J., 1975: Statistical inference using extreme order statistics. *Annals of Statistics*, **3(1)**, 119-131.

Pielke, R.A., Jr., 2005: Are there trends in hurricane destruction? *Nature*, **438(7071)**, E11.

Robeson, S.M., 2004: Trends in time-varying percentiles of daily minimum and maximum temperature over North America. *Geophysical Research Letters*, **31**, L04203, doi:10.1029/2003GL019019.

Robinson, D.A., K.F. Dewey, and R.R. Heim, Jr., 1993: Global snow cover monitoring: an update. *Bulletin of the American Meteorological Society*, **74(9)**, 1689-1696.

Ropelewski, C., 1999: The great El Niño of 1997-1998: impacts on precipitation and temperature. *Consequences*, **5(2)**, 17-25.

Sanders, F. and J.R. Gyakum, 1980: Synoptic-dynamic climatology of the "bomb". *Monthly Weather Review*, **108(10)**, 1589-1606.

Santer, B.D., T.M.L. Wigley, P.J. Gleckler, C. Bonfils, M.F. Wehner, K. AchutaRao, T.P. Barnett, J.S. Boyle, W. Brueggemann, M. Fiorino, N. Gillett, J.E. Hansen, P.D. Jones, S.A. Klein, G.A. Meehl, S.C.B. Raper, R.W. Reynolds, K.E. Taylor, and W.M. Washington, 2006: Forced and unforced ocean temperature changes in Atlantic and Pacific tropical cyclogenesis regions. *Proceedings of the National Academy of Sciences*, **103(38)**, 13905-13910.

Schlesinger, M.E. and N. Ramankutty, 1994: An oscillation in the global climate system of period 65-70 years. *Nature*, **367(6465)**, 723-726.

SEMARNAT (Secretaría de Medio Ambiente y Recursos Naturales), 2000: *Programa nacional contra incendios forestales. Resultados 1999-2000*. SEMARNAT, Mexico, 263 pp.

Semenov, V.A. and L. Bengtsson, 2002: Secular trends in daily precipitation characteristics: greenhouse gas simulation with a coupled AOGCM. *Climate Dynamics*, **19(2)**, 123-140.

Serreze, M.C., F. Arse, R.G. Barry, and J.C. Rogers, 1997: Iclandic low cyclone activity. Climatological features, linkages with the NAO, and relationships with recent changes in the Northern Hemisphere circulation. *Journal of Climate*, **10(3)**, 453-464.

Seymour, R.J., R.R. Strange, D.R. Cayan, and R.A. Nathan, 1984: Influence of El Niños on California's wave climate. In: *Proceedings of the 19th International Conference on Coastal Engineering*, 3-7 September 1984, Houston, TX. American Society of Civil Engineers, New York, pp. 577-592.

Shabbar, A. and W. Skinner, 2004: Summer drought patterns in Canada and the relationship to global sea surface temperatures. *Journal of Climate*, **17(14)**, 2866-2880.

Shein, K. A. (ed.), 2006: State of the climate in 2005. *Bulletin of the American Meteorological Society*, **87(6)**, S1-S102.

Simmonds, I. and K. Keay, 2002: Surface fluxes of momentum and mechanical energy over the North Pacific and North Atlantic Oceans. *Meteorology and Atmospheric Physics*, **80(1-4)**, 1-18.

Sims, A.P., D.D.S. Niyogi, and S. Raman, 2002: Adopting drought indices for estimating soil moisture: A North Carolina case study. *Geophysical Research Letters*, **29(8)**, 1183.

Smith, R.L., 2003: Statistics of extremes, with applications in environment, insurance and finance. In: *Extreme Values in Finance, Telecommunications and the Environment* [Finkenstadt, B. and H. Rootzén (eds.)]. Chapman and Hall/CRC Press, Boca Raton, FL, pp. 1-78.

Smith, R.L., C. Tebaldi, D. Nychka, and L.O. Mearns, 2008: Bayesian modeling of uncertainty in ensembles of climate models. *Journal of the American Statistical Association,* (accepted).

Soja, A.J., N.M. Tchebakova, N.H.F. French, M.D. Flannigan, H.H. Shugart, B.J. Stocks, A.I. Sukhinin, E.I. Parfenova, F.S. Chapin, III, and P.W. Stackhouse, Jr., 2007: Climate-induced boreal forest change: predictions versus current observations. *Global and Planetary Change*, **56(3-4)**, 274-296.

Sriver, R. and M. Huber, 2006: Low frequency variability in globally integrated tropical cyclone power dissipation. *Geophysical Research Letters*, **33**, L11705, doi:10.1029/2006GL026167.

Stahl, K., R.D. Moore, and I.G. McKendry, 2006: Climatology of winter cold spells in relation to mountain pine beetle mortality in British Columbia, Canada. *Climate Research*, **32(1)**, 13-23.

Stahle, D. W., E.R. Cook, M.K. Cleaveland, M.D. Therrell, D.M. Meko, H.D. Grissino-Mayer, E. Watson, and B.H. Luckman, 2000: Tree-ring data document 16th century megadrought over North America. *EOS, Transactions of the American Geophysical Union*, **81(12)**, 121.

Stone, D.A., A.J. Weaver, and F.W. Zwiers, 2000: Trends in Canadian precipitation intensity. *Atmosphere-Ocean*, **38(2)**, 321-347.

Sun, B.M. and P.Ya. Groisman, 2004: Variations in low cloud cover over the United States during the second half of the twentieth century. *Journal of Climate*, **17(9)**, 1883-1888.

Therrell, M.D., D.W. Stahle, M.K. Cleaveland, and J. Villanueva-Diaz, 2002: Warm season tree growth and precipitation in Mexico. *Journal of Geophysical Research*, **107(D14)**, 4205, doi:10.1029/2001JD000851.

Trenberth, K.E., 1990: Recent observed interdecadal climate changes in the Northern Hemisphere. *Bulletin of the American Meteorological Society,* **71(7)**, 988-993.

Trenberth, K.E. and J.W. Hurrell, 1994: Decadal atmospheric-ocean variations in the Pacific. *Climate Dynamics*, **9(6)**, 303-319.

Trenberth, K.E. and D.J. Shea, 2006: Atlantic hurricanes and natural variability in 2005. *Geophysical Research Letters*, **33**, L12704, doi:10.1029/2006GL026894.

Trenberth, K.E., J.M. Caron, D.P. Stepaniak, and S. Worley, 2002: The evolution of El Niño-Southern Oscillation and global atmospheric surface temperatures. *Journal of Geophysical Research*, **107(D8)**, 4065, doi:10.1029/2000JD000298.

Uppala, S.M., P.W. Kållberg, A.J. Simmons, U. Andrae, V. da Costa Bechtold, M. Fiorino, J.K. Gibson, J. Haseler, A. Hernandez, G.A., Kelly, X. Li, K. Onogi, S. Saarinen, N. Sokka, R.P. Allan, E. Andersson, K. Arpe, M.A. Balmaseda, A.C.M. Beljaars, L. van de Berg, J. Bidlot, N. Bormann, S. Caires, F. Chevallier, A. Dethof, M. Dragosavac, M. Fisher, M. Fuentes, S. Hagemann, E. Hólm, B.J. Hoskins, L. Isaksen, P.A.E.M. Janssen, R. Jenne, A.P. McNally, J.-F. Mahfouf, J.-J. Morcrette, N.A. Rayner, R.W. Saunders, P. Simon, A. Sterl, K.E. Trenberth, A. Untch, D. Vasiljevic, P. Viterbo, and J. Woollen, 2005: The ERA-40

re-analysis. *Quarterly Journal of the Royal Meteorological Society,* **131(612)**, 2961-3012. doi:10.1256/qj.04.176

Vecchi, G.A. and T.R. Knutson, 2008: On estimates of historical North Atlantic tropical cyclone activity. Journal of Climate, Early online release doi:10.1175/2008JCLI2178.1.

Verbout, S.M., H.E. Brooks, L.M. Leslie, and D.M. Schultz, 2006: Evolution of the US tornado database: 1954-2003. *Weather and Forecasting*, **21(1)**, 86-93.

Vimont, D.J. and J.P. Kossin, 2007: The Atlantic meridional mode and hurricane activity. *Geophysical Research Letters*, **34**, L07709, doi:10.1029/2007GL029683.

Vincent, L.A. and E. Mekis, 2006: Changes in daily and extreme temperature and precipitation indices for Canada over the 20th century. *Atmosphere-Ocean*, **44(2)**, 177-193.

Vörösmarty, C., D. Lettenmaier, C. Leveque, M. Meybeck, C. Pahl-Wostl, J. Alcamo, W. Cosgrove, H. Grassl, H. Hoff, P. Kabat, F. Lansigan, R. Lawford ,R. Naiman, as members of The Framing Committee of the Global Water System Project, 2004: Humans transforming the global water system. *EOS, Transactions of the American Geophysical Union*, **85(48)**, 509, 513-514.

Wang, D.W., D.A. Mitchell, W.J. Teague, E. Jarosz, and M.S. Hulbet, 2005: Extreme waves under Hurricane Ivan. *Science,* **309(5736)**, 896.

Wang, W.L. and V.R. Swail, 2001: Changes of extreme wave heights in Northern Hemisphere oceans and related atmospheric circulation regimes. *Journal of Climate,* **14(10)**, 2201-2204.

Wang, X.L., V.R. Swail, and F.W. Zwiers, 2006a: Climatology and changes of extratropical storm tracks and cyclone activity: comparison of ERA-40 with NCEP/NCAR reanalysis for 1958-2001. *Journal of Climate,* **19(13)**, 3145-3166.

Wang, X.L., H. Wan, and V.R. Swail, 2006b: Observed changes in cyclone activity in Canada and their relationships to major circulation regimes. *Journal of Climate,* **19(6)**, 895-916.

Webster, P.J., G.J. Holland, J.A. Curry, and H.-R. Chang, 2005: Changes in tropical cyclone number, duration, and intensity in a warming environment. *Science,* **309(5742)**, 1844-1846.

Wehner, M.F., 2004: Predicted 21st century changes in seasonal extreme precipitation events in the Parallel Climate Model. *Journal of Climate,* **17(21)**, 4281-4290.

Wehner, M., 2005: Changes in daily precipitation and surface air temperature extremes in the IPCC AR4 models. *US CLIVAR Variations,* **3(3)**, 5-9.

Westerling, A.L., H.G. Hidalgo, and D.R. Cayan, 2006. Warming and earlier spring increases in western U.S. forest wildfire activity. *Science* **313(5789)**, 940-943.

Woodhouse, C.A. and J.T. Overpeck, 1998: **2000 Years of drought variability in the central United States.** *Bulletin of the American Meteorological Society,* **79(12)**, 2693-2714.

Wu, L., B. Wang, and S. Geng, 2005: Growing typhoon influence on east Asia. *Geophysical Research Letters*, **32**, L18703, doi:10.1029/2005GL022937.

Wu, C.-C., K.-H. Chou, Y. Wang, Y.-H. Kuo, 2006: Tropical cyclone initialization and prediction based on four-dimensional variational data assimilation. *Journal of the Atmospheric Sciences*, **63(9)**, 2383-2395.

Xie, L., L.J. Pietrafesa, J.M. Morrison, and T. Karl, 2005: Climatological and interannual variability of North Atlantic hurricane tracks. *Journal of Climate*, **18(24)**, 5370-5381.

Zhang, R., T.L. Delworth, and I.M. Held, 2007: Can the Atlantic Ocean drive the observed multidecadal variability in Northern Hemisphere mean temperature? *Geophysical Research Lett*ers, **34**, L02709, doi:10.1029/2006GL028683.

Zhang, X., W.D. Hogg, and E. Mekis, 2001: Spatial and temporal characteristics of heavy precipitation events over Canada. *Journal of Climate*, **14(9)**, 1923-1936.

Zhang, X., J.E. Walsh, J. Zhang, U.S. Bhatt and M. Ikeda, 2004: Climatology and inter-annual variability of Arctic cyclone activity, 1948-2002. *Journal of Climate*, **17(12)**, 2300-2317.

Zwiers, F.W. and V.V. Kharin, 1998: Changes in the extremes of the climate simulated by CCC GCM2 under CO_2 doubling. *Journal of Climate*, **11(9)**, 2200-2222.

CHAPTER 3 REFERENCES

Alexander, L.V., X. Zhang, T.C. Peterson, J. Caesar, B. Gleason, A.M.G. Klein Tank, M. HaylockD. Collins, B. Trewin, F. Rahimzadeh, A. Tagipour, K. Rupa Kumar, J. Revadekar, G. Griffiths, L. Vincent, D.B. Stephenson, J. Burn, E. Aguilar, M. Brunet, M. Taylor, M. New, P. Zhai, M. Rusticucci, and J.L. Vazquez-Aguirre, 2006: Global observed changes in daily climate extremes of temperature and precipitation. *Journal of Geophysical Research*, **111**, D05109, doi:10.1029/2005JD006290.

Allan, J.C. and P.D. Komar, 2000: Are ocean wave heights increasing in the eastern North Pacific? *EOS, Transaction of the American Geophysical Union*, **81**, 561-567.

Allan, J.C. and P.D. Komar, 2006: Climate controls on US West Coast erosion processes. *Journal of Coastal Research*, **22(3)**, 511-529.

Allen, M., 2003: Liability for climate change: will it ever be possible to sue anyone for damaging the climate? *Nature*, **421(6926)**, 891-892.

Allen, M.R. and W.J. Ingram, 2002: Constraints on future changes in climate and the hydrological cycle. *Nature*, **419(6903)**, 224-232.

Andreas, E.L. and K.A. Emanuel, 2001: Effects of sea spray on tropical cyclone intensity. *Journal of the Atmospheric Sciences*, **58(24)**, 3741-3751.

Arora, V.K. and G.J. Boer, 2001: Effects of simulated climate change on the hydrology of major river basins. *Journal of Geophysical Research*, **106(D4)**, 3335-3348.

Arzayus, L.F. and W.J. Skirving, 2004: Correlations between ENSO and coral reef bleaching. In: *Tenth International Coral Reef Symposium*, 28 June - 2 July 2004, Okinawa, Japan. Japanese Coral Reef Society, Tokyo.

Arzel, O., T. Fichefet, and H. Goosse, 2006: Sea ice evolution over the 20th and 21st centuries as simulated by current AOGCMs. *Ocean Modelling*, **12(3-4)**, 401-415.

Baik, J.-J. and J.-S. Paek, 1998: A climatology of sea surface temperature and the maximum intensity of western North Pacific tropical cyclones. *Journal of the Meteorological Society of Japan*, **76(1)**, 129-137.

Barnett, D.N., S.J. Brown, J.M. Murphy, D.M.H. Sexton, and M.J. Webb, 2006: Quantifying uncertainty in changes in extreme event frequency in response to doubled CO_2 using a large

ensemble of GCM simulations. *Climate Dynamics*, **26(5)**, 489-511.

Bell, G.D. and M. Chelliah, 2006: Leading tropical modes associated with interannual and multidecadal fluctuations in North Atlantic hurricane activity. *Journal of Climate*, **19(4)**, 590-612.

Bell, J.L. and L.C. Sloan, 2006: CO_2 Sensitivity of extreme climate events in the western United States. *Earth Interactions*, **10(15)**, 1-17. doi:10.1175/EI181.1

Bell, J.L., L.C. Sloan, and M.A. Snyder, 2004: Regional changes in extreme climatic events: a future climate scenario. *Journal of Climate*, **17(1)**, 81-87.

Bender, M.A. and I. Ginis, 2000: Real-case simulations of hurricane-ocean interaction using a high-resolution coupled model: effects on hurricane intensity. *Monthly Weather Review*, **128(4)**, 917-946.

Bengtsson, L., M. Botzet, and M. Esch, 1996: Will greenhouse gas-induced warming over the 50 years lead to a higher frequency and greater intensity of hurricanes? *Tellus A*, **48(1)**, 57-73.

Bengtsson, L., K. Hodges, and E. Roeckner, 2006: Storm tracks and climate change. *Journal of Climate*, **19(15)**, 3518-3543.

Bengtsson, L., K.I. Hodges, M. Esch, N. Keenlyside, L. Kornblueh, J.-J. Luo, and T. Yamagata, 2007: How may tropical cyclones change in a warmer climate. *Tellus A*, **59(4)**, 539-561.

Beniston, M., 2004: The 2003 heat wave in Europe: a shape of things to come? An analysis based on Swiss climatological data and model simulations. *Geophysical Research Letters*, **31**, L02202, doi:10.1029/2003GL018857.

Bister, M. and K.A. Emanuel, 1997: The genesis of hurricane Guillermo: TEXMEX analyses and a modeling study. *Monthly Weather Review*, **125(10)**, 2662-2682.

Brabson, B.B., D.H. Lister, P.D. Jones, and J.P. Palutikof, 2005: Soil moisture and predicted spells of extreme temperatures in Britain. *Journal of Geophysical Research*, **110**, D05104, doi:10.1029/2004JD005156.

Breshears, D.D., N.S. Cobb, P.M. Rich, K.P. Price, C.D. Allen, R.G. Balice, W.H. Romme, J.H. Kastens, M.L. Floyd, J. Belnap, J.J. Anderson, O.B. Myers, and C.W. Meyer 2005: Regional vegetation die-off in response to global-change-type drought. *Proceedings of the National Academy of Sciences*, **102(42)**, 15144-15148.

Broccoli, A.J. and S. Manabe, 1990: Can existing climate models be used to study anthropogenic changes in tropical cyclone climate? *Geophysical Research Letters*, **17(11)**, 1917-1920.

Brooks, H.E. and N. Dotzek, 2008: The spatial distribution of severe convective storms and an analysis of their secular changes. In: *Climate Extremes and Society*. [Diaz, H. F. and R. Murnane (eds.)]. Cambridge University Press, Cambridge, UK, and New York, pp. 35-53.

Brooks, H.E., J.W. Lee, and J.P. Craven, 2003: The spatial distribution of severe thunderstorm and tornado environments from global reanalysis data. *Atmospheric Research*, **67-68**, 73-94.

Brunner, R.D., A.H. Lynch, J.C. Pardikes, E.N. Cassano, L.R. Lestak, and J.M. Vogel., 2004: An Arctic disaster and its policy implications. *Arctic*, **57(4)**, 336-346.

Burke, E.J., S.J. Brown, and N. Christidis, 2006: Modelling the recent evolution of global drought and projections for the 21st century with the Hadley Centre climate model. *Journal of Hydrometeorology*, **7(5)**, 1113-1125.

Burkholder, B.A. and D.J. Karoly, 2007: An assessment of US climate variability using the Climate Extremes Index. In: *Nineteenth Conference on Climate Variability and Change*, 15-18 January, 2007, San Antonio, TX. American Meteorological Society, Boston, Paper 2B.9. Extended abstract available at http://ams.confex.com/ams/pdfpapers/117942.pdf

Caesar, J., L. Alexander, and R. Vose, 2006: Large-scale changes in observed daily maximum and minimum temperatures: creation and analysis of a new gridded data set. *Journal of Geophysical Research*, **111**, D05101, doi:10.1029/2005JD006280.

Caires, S. and A. Sterl, 2005: 100-year return value estimates for wind speed and significant wave height from the ERA-40 data. *Journal of Climate*, **18(7)**, 1032-1048.

Caires, S., V.R. Swail, and X.L. Wang, 2006: Projection and analysis of extreme wave climate. *Journal of Climate*, **19(21)**, 5581-5605.

Camargo, S., A.G. Barnston, and S.E. Zebiak, 2005: A statistical assessment of tropical cyclone activity in atmospheric general circulation models. *Tellus A*, **57(4)**, 589-604. doi:10.1111/j.1600-0870.2005.00117

Camargo, S., K.A. Emanuel, and A.H. Sobel, 2006: ENSO and genesis potential index in reanalysis and AGCMs. In: *27th Conference on Hurricanes and Tropical Meteorology*, 24-28 April, 2006, Monterey, CA. American Meteorological Society, Boston, Paper #15C.2. Extended abstract available at http://ams.confex.com/ams/pdfpapers/108038.pdf

Camargo, S.J., A.H. Sobel, A.G. Barnston, and K. A. Emanuel, 2007: Tropical cyclone genesis potential index in climate models. *Tellus A*, **59(4)**, 428-442.

Cassano, E.N., A.H. Lynch, J.J. Cassano, and M.R. Koslow, 2006: Classification of synoptic patterns in the western Arctic associated with extreme events at Barrow, Alaska, USA. *Climate Research*, **30(2)**, 83-97.

Cayan, D.R., P.D. Bromirski, K. Hayhoe, M. Tyree, M.D. Dettinger, and R.E. Flick, 2008: Climate change projections of sea level extremes along the California coast. *Climatic Change* (in press).

Chan, J.C.L., 1985: Tropical cyclone activity in the northwest Pacific in relation to the El Niño/Southern Oscillation phenomenon. *Monthly Weather Review*, **113(4)**: 599-606.

Chauvin, F., J.-F. Royer and M. Déqué, 2006: Response of hurricane-type vortices to global warming as simulated by ARPEGE-Climat at high resolution. *Climate Dynamics*, **27(4)**, 377-399.

Christensen, J.H. and O.B. Christensen, 2003: Severe summertime flooding in Europe. *Nature*, **421(6925)**, 805-806.

Christensen, J.H., B. Hewitson, A. Busuioc, A. Chen, X. Gao, I. Held, R. Jones, R.K. Kolli, W.-T. Kwon, R. Laprise, V. Magaña Rueda, L. Mearns, C.G. Menéndez, J. Räisänen, A. Rinke, A. Sarr, and P. Whetton, 2007: Regional climate projections. In: *Climate Change 2007: The Physical Basis*. Contribution of Working Group I to the Fourth Assessment Report of the Intergovernmental Panel on Climate Change [Solomon, S., D. Qin, M. Manning, Z. Chen, M. Marquis, K.B. Averyt, M. Tignor, and H.L. Miller (eds.)]. Cambridge University Press, Cambridge, UK, and New York, pp. 847-940.

Christensen, O.B. and J.H. Christensen, 2004: Intensification of extreme European summer precipitation in a warmer climate. *Global Planetary Change*, **44(1-4)**, 107-117.

Christidis, N., P.A. Stott, S. Brown, G. Hegerl, and J. Caesar, 2005: Detection of changes in temperature extremes during the second half of the 20th century. *Geophysical Research Letters*, **32**, L20716, doi:10.1029/2005GL023885.

Clark, R., S. Brown, and J. Murphy, 2006: Modelling northern hemisphere summer heat extreme changes and their uncertainties using a physics ensemble of climate sensitivity experiments. *Journal of Climate*, **19(17)**, 4418-4435.

Cook, E.R., C.A. Woodhouse, C.M. Eakin, D.M. Meko, and D.W. Stahle, 2004: Long-term aridity changes in the western United States. *Science*, **306(5698)**, 1015-1018.

Corbosiero, K.L. and J. Molinari, 2003: The relationship between storm motion, vertical wind shear, and convective asymmetries in tropical cyclones. *Journal of the Atmospheric Sciences*, **60(2)**, 366-376.

Cubasch, U., G.A. Meehl, G.J. Boer, R.J. Stouffer, M. Dix, A. Noda, C.A. Senior, S. Raper, and K.S. Yap, 2001: Projections of future climate change. In: *Climate Change 2001: The Scientific Basis*. Contribution of Working Group I to the Third Assessment Report of the Intergovernmental Panel on Climate Change [Houghton, J.T.,Y. Ding, D.J. Griggs, M. Noguer, P.J. van der Linden, X. Dai, K. Maskell, and C.A. Johnson (eds.)]. Cambridge University Press, Cambridge, pp. 525-582.

Dai, A., K.E. Trenberth, and T. Qian, 2004: A global data set of Palmer Drought Severity Index for 1870-2002: relationship with soil moisture and effects of surface warming. *Journal of Hydrometeorology*, **5(6)**, 1117-1130.

Davis, C.A and L.F. Bosart, 2001: Numerical simulations of the genesis of hurricane Diana (1984). Part I: control simulation. *Monthly Weather Review*, **129(8)**,1859-1881.

Davis, C.A. and L.F. Bosart, 2006: The formation of hurricane Humberto (2001): the importance of extra-tropical precursors. *Quarterly Journal of the Royal Meteorological Society*, **132(619)**, 2055-2085.

Del Genio, A.D., M.-S. Yao, and J. Jonas, 2007: Will moist convection be stronger in a warner climate? *Geophysical Research Letters*, **34**, L16703, doi:10.1029/2007GL030525.

Delworth, T.L., J.D. Mahlman, and T.R. Knutson, 1999: Changes in heat index associated with CO_2-induced global warming. *Climatic Change*, **43(2)**, 369-386.

DeMaria, M., 1996: The effect of vertical shear on tropical cyclone intensity change *Journal of the Atmospheric Sciences*, **53(14)**, 2076-2088.

DeMaria, M. and J. Kaplan, 1994: Sea-surface temperature and the maximum intensity of Atlantic tropical cyclones. *Journal of Climate*, **7(9)**, 1324-1334.

Déry, S.J. and E.F. Wood, 2006: Analysis of snow in the 20th and 21st century Geophysical Fluid Dynamics Laboratory coupled climate model simulations. *Journal of Geophysical Research*, **111**, D19113. doi:10.1029/2005JD006920.

Deser, C., A.S. Phillips, and J.W. Hurrell, 2004: Pacific decadal interdecadal climate variability: linkages between the tropics and the North Pacific during boreal winter since 1900. *Journal of Climate*, **17(16)**, 3109-3124.

Donner, S.D., T.R. Knutson, and M. Oppenheimer, 2007: Model-based assessment of the role of human-induced climate change in the 2005 Caribbean coral bleaching event. *Proceedings of the National Academy of Sciences*, **104(13)**, 5483-5488.

Dunion, J.P and C.S. Velden, 2004: The impact of the Saharan air layer on Atlantic tropical cyclone activity. *Bulletin of the American Meteorological Society*, **85(3)**, 353-365.

Dunn, G.E., 1940: Cyclogenesis in the tropical Atlantic. *Bulletin of the American Meteorological Society*, **21(6)**, 215-229.

Easterling, D.R., G.A. Meehl, C. Parmesan, S.A. Changnon, T.R. Karl, and L.O. Mearns, 2000: Climate extremes: observations, modeling and impacts. *Science*, **289(5487)**, 2068-2074.

Eichler, T. and W. Higgins, 2006: Climatology and ENSO-related variability of North American extratropical cyclone activity. *Journal of Climate*, **19(10)**, 2076-2093.

Emanuel, K.A., 1987: The dependence of hurricane intensity on climate. *Nature*, **326(6112)**, 483-485.

Emanuel K.A., 1995: Sensitivity of tropical cyclones to surface exchange coefficients and a revised steady-state model incorporating eye dynamics. *Journal of the Atmospheric Sciences*, **52(22)**, 3969-3976.

Emanuel, K., 2000: A statistical analysis of tropical cyclone intensity. *Monthly Weather Review*, **128(4)**, 1139-1152.

Emanuel, K.A., 2005: Increasing destructiveness of tropical cyclones over the past 30 years. *Nature*, **436(7051)**, 686-688.

Emanuel, K., 2006: Environmental influences on tropical cyclone variability and trends. In *27th Conference on Hurricanes and Tropical Meteorology*, 24-28 April, 2006, Monterey, CA. American Meteorological Society, Boston, Paper #4C.2. Extended abstract available at: http://ams.confex.com/ams/pdfpapers/107575.pdf

Emanuel, K.A., 2007: Environmental factors affecting tropical cyclone power dissipation. *Journal of Climate*, **20(22)**, 5497-5509.

Emanuel, K. and D.S. Nolan, 2004: Tropical cyclone activity and the global climate system. In: *26h Conference on Hurricanes and Tropical Meteorology*, 2-7 May 2004, Miami, FL. American Meteorological Society, Boston, Paper, #10A.2, pp. 240-241. Extended abstract available at http://ams.confex.com/ams/pdfpapers/75463.pdf

Emanuel, K., S.Ravela, E.Vivant, and C.Risi. 2006: A statistical deterministic approach to hurricane risk assessment. *Bulletin of the American Meteorological Society,*: **87(6)**, 299-314.

Emanuel, K., R. Sundararajan, and J. Williams, 2008: Hurricanes and global warming: results from downscaling IPCC AR4 simulations. Bulletin of the American Meteorological Society, 89(3), 347-367.

Emori, S. and S.J. Brown, 2005: Dynamic and thermodynamic changes in mean and extreme precipitation under changed climate. *Geophysical Research Letters*, **32**, L17706, doi:10.1029/2005GL023272.

Evan, A.T., J. Dunion, J.A. Foley, A.K. Heidinger, and C.S. Velden, 2006: New evidence for a relationship between Atlantic tropical cyclone activity and African dust outbeaks. *Geophysical Research Letters*, **33**, L19813, doi:10.1029/2006GL026408.

Ferreira, R.N. and W.H. Schubert. 1997: Barotropic aspects of ITCZ breakdown. *Journal of the Atmospheric Sciences*, **54(2)**, 261-285.

Fiorino, M. and R.L. Elsberry, 1989: Contributions to tropical cyclone motion by small, medium and large scales in the initial vortex. *Monthly Weather Review*, **117(4)**, 721-727.

Fischer-Bruns, I., H. Von Storch, J.F. Gonzalez-Rouco, and E. Zorita, 2005: Modelling the variability of midlatitude storm activity on decadal to century time scales. *Climate Dynamics*, **25(5)**, 461-476.

Frank, N.L. and G. Clark, 1980: Atlantic tropical systems of 1979. *Monthly Weather Review*, **108(7)**, 966-972.

Frank, W.M. and E.A. Ritchie, 1999: Effects of environmental flow upon tropical cyclone structure. *Monthly Weather Review*, **127(9)**, 2044-2061.

Frei, A. and G. Gong, 2005: Decadal to century scale trends in North American snow extent in coupled atmosphere-ocean general circulation models. *Geophysical Research Letters*, **32**, L18502, doi:10.1029/2005GL023394.

Frei, C., C. Schär, D. Lüthi, and H.C. Davies, 1998: Heavy precipitation processes in a warmer climate. *Geophysical Research Letters*, **25(9)**, 1431-1434.

Frich, P., L.V. Alexander, P. Della-Marta, B. Gleason, M. Haylock, A.M.G.K. Tank, and T. Peterson, 2002: Observed coherent changes in climatic extremes during the second half of the twentieth century. *Climate Research*, **19(3)**, 193-212.

Fyfe, J.C., 2003: Extratropical southern hemisphere cyclones: Harbingers of climate change? *Journal of Climate*, **16(17)**, 2802-2805.

Gao, S.B. and H.G. Stefan, 2004: Potential climate change effects on ice covers of five freshwater lakes. *Journal of Hydrologic Engineering*, **9(3)**, 226-234.

Gedney, N., P.M. Cox, R.A. Betts, O. Boucher, C. Huntingford, and P.A. Stott, 2006a: Detection of a direct carbon dioxide effect in continental river runoff records. *Nature*, **439(7078)**, 835-838.

Gedney, N., P.M. Cox, R.A. Betts, O. Boucher, C. Huntingford, and P.A. Stott, 2006b: Continental runoff - a quality-controlled global runoff data set - Reply. *Nature*, **444(7120)**, E14-E15. doi:10.1038/nature05481

Gershunov, A. and D.R. Cayan, 2003: Heavy daily precipitation frequency over the contiguous United States: sources of climatic variability and seasonal predictability. *Journal of Climate*, **16(16)**, 2752-2765.

Gillett, N.P., 2005: Northern Hemisphere circulation. *Nature*, **437(7058)**, 496.

Gillett, N.P., G.C. Hegerl, M.R. Allen, and P.A. Stott, 2000: Implications of changes in the Northern Hemispheric circulation for the detection of anthropogenic climate change. *Geophysical Research Letters*, **27(7)**, 993-996.

Gillett, N.P., M.R. Allen, R.E. McDonald, C.A. Senior, D.T. Shindell, and G.A. Schmidt, 2002: How linear is the Arctic Oscillation response to greenhouse gases? *Journal of Geophysical Research*, **107(D3)**, 4022, doi:10.1029/2001JD000589.

Gillett, N.P., H.F. Graf, and T.J. Osborn, 2003: Climate change and the North Atlantic Oscillation. In: *The North Atlantic Oscillation: Climatic Significance and Environmental Impact* [Hurrell, J.W., Y Kushnir, G. Ottersen and M. Visbeck (eds.)]. American Geophysical Union, Washington, DC, pp. 193-209.

Gillett, N.P., A.J. Weaver, F.W. Zwiers, and M.F. Wehner, 2004: Detection of volcanic influence on global precipitation. *Geophysical Research Letters*, **31(12)**, L12217, doi:10.1029/2004GL020044.

Gillett, N. P., R.J. Allan, and T.J. Ansell, 2005: Detection of external influence on sea level pressure with a multi-model ensemble. *Geophysical Research Letters*, **32(19)**, L19714, doi:10.1029/2005GL023640.

Goldenberg, S.B., C.W. Landsea, A.M. Mesta-Nuñez, and W. M. Gray, 2001: The recent increase in Atlantic hurricane activity: causes and implications. *Science*, **293(5529)**, 474-479.

Graham, N.E. and H.F. Diaz, 2001: Evidence for intensification of North Pacific winter cyclones since 1948. *Bulletin of the American Meteorological Society*, **82(9)**, 1869-1893.

Gray, W.M., 1968: Global view of the origin of tropical disturbances and storms. *Monthly Weather Review*, **96(10)**, 669-700.

Gray, W.M., 1979: Hurricanes: their formation, structure, and likely role in the tropical circulation. In *Meteorology Over the Tropical Oceans* [Shaw, D. B. (ed.)]. Royal Meteorological Society, Bracknell, UK, pp. 155-218.

Gray, W.M., 1984: Atlantic seasonal hurricane frequency. Part II: forecasting its variability. *Monthly Weather Review*, **112(9)**, 1669-1683.

Gray, W.M., 1990: Strong association between West African rainfall and U.S. landfall of intense hurricanes. *Science*, **249(4974)**, 1251-1256.

Groisman, P.Ya., T.R. Karl, D.R. Easterling, R.W. Knight, P.F. Jamason, K.J. Hennessy, R. Suppiah, C.M. Page, J. Wibig, K. Fortuniak, V.N. Razuvaev, A. Douglas, E. Førland, and P.-M. Zhai, 1999: Changes in the probability of heavy precipitation: important indicators of climatic change. *Climatic Change*, **42(1)**, 243-283.

Groisman, P.Ya., R.W. Knight, T.R. Karl, D.R. Easterling, B. Sun, and J.H. Lawrimore, 2004: Contemporary changes of the hydrological cycle over the contiguous United States: trends derived from *in situ* observations. *Journal of Hydrometeorology*, **5(1)**, 64-85.

Groisman, P.Ya., R.W. Knight, D.R. Easterling, T.R. Karl, G.C. Hegerl, and V.N. Razuvaev, 2005: Trends in intense precipitation in the climate record. *Journal of Climate*, **18(9)**, 1326-1350.

Gulev, S.K. and V. Grigorieva, 2004: Last century changes in ocean wind wave height from global visual wave data. *Geophysical Research Letters*, **31,** L24302, doi:10.1029/2004FL021040.

Haarsma, R.J., J.F.B. Mitchell, and C.A. Senior, 1993: Tropical disturbances in a GCM. *Climate Dynamics*, **8(5)**, 247-257.

Hasegawa, A. and S. Emori, 2005: Tropical cyclones and associated precipitation over the Western North Pacific: T106 atmospheric GCM simulation for present-day and doubled CO_2 climates. *SOLA*, **1,** 145-148. doi:10.2151/sola.2005-038

Hayhoe, K, C.P. Wake, T.G. Huntington, L.F. Luo, M.D. Schwartz, J. Sheffield, E. Wood, B. Anderson, J. Bradbury, A. DeGaetano, T.J. Troy, and D. Wolfe, 2007: Past and future changes in climate and hydrological indicators in the US Northeast. *Climate Dynamics*, **28(4)**, 381-407.

Hegerl, G.C., F.W. Zwiers, V.V. Kharin, and P.A. Stott, 2004: Detectability of anthropogenic changes in temperature and precipitation extremes. *Journal of Climate*, **17(19)**, 3683-3700.

Hegerl, G.C., F.W. Zwiers, P. Braconnot, N.P. Gillett, Y. Luo, J. Marengo Orsini, N. Nicholls, J.E. Penner, and P.A, Stott, 2007: Understanding and attributing climate change. In: *Climate Change 2007: The Physical Basis*. Contribution of Working Group I to the Fourth Assessment Report of the Intergovernmental Panel on Climate Change [Solomon, S., D. Qin, M.

Manning, Z. Chen, M. Marquis, K.B. Averyt, M. Tignor, and H.L. Miller (eds.)]. Cambridge University Press, Cambridge, UK, and New York, pp. 663-745.

Held, I.M. and B.J. Soden, 2006: Robust responses of the hydrological cycle to global warming. *Journal of Climate*, **19(21)**, 5686-5699.

Henderson-Sellers, A., H. Zhang, G. Berz, K. Emanuel, W. Gray, C. Landsea, G. Holland, J. Lighthill, S-L. Shieh, P. Webster, and K. McGuffie, 1998: Tropical cyclones and global climate change: a post-IPCC assessment. *Bulletin of the American Meteorological Society*, **79(1)**, 19-38.

Hendricks, E.A., M.T. Montgomery, and C.A. Davis, 2004: The role of "vortical" hot towers in the formation of tropical cyclone Diana (1984). *Journal of the Atmospheric Sciences*, **61(11)**, 1209-1232.

Hodgkins, G.A., I.C. James, and T.G. Huntington, 2002: Historical changes in lake ice-out dates as indicators of climate change in New England, 1850-2000. *International Journal of Climatology*, **22(15)**, 1819-1827.

Hodgkins, G.A., R.W. Dudley, and T. G. Huntington, 2003: Changes in the timing of high river flows in New England over the 20th century. *Journal of Hydrology*, **278(1-4)**, 244-252.

Hoegh-Guldberg, O., 1999: Climate change, coral bleaching and the future of the world's coral reefs. *Marine and Freshwater Research*, **50(8)**, 839-66.

Hoegh-Guldberg, O., 2005: Marine ecosystems and climate change. In: *Climate Change and Biodiversity* [Lovejoy, T.E. and L. Hannah (eds.)]. Yale University Press, New Haven, CT, pp. 256-271.

Hoerling, M.P. and A. Kumar, 2003: The perfect ocean for drought. *Science*, 299(5607), 691-694.

Hoerling, M.P., J.W. Hurrell, T. Xu, G.T. Bates, and A.S. Phillips, 2005: Twentieth century North Atlantic climate change. Part II: Understanding the effect of Indian Ocean warming. *Climate Dynamics*, **23(3-4)**, 391-405.

Hoerling, M., J. Eischeid, X. Quan, and T.Y. Xu, 2007: Explaining the record US warmth of 2006. *Geophysical Research Letters*, **24**, L17704, doi:10.1029/2007GL030643.

Holland, G.J., 1984: Tropical cyclone motion. a comparison of theory and observation. *Journal of the Atmospheric Sciences*, **41(1)**, 68-75.

Holland, G.J., 1995: Scale interaction in the western Pacific monsoon. *Meteorological and Atmospheric Physics*, **56(1-2)**, 57-79.

Holland, G.J., 1997: The maximum potential intensity of tropical cyclones. *Journal of the Atmospheric Sciences*, **54(21)**, 2519-2541.

Holland, G.J., 2007: Misuse of landfall as a proxy for Atlantic tropical cyclone activity. *EOS, Transactions of the American Geophysical Union*, **88(36)**, 349-350.

Holland, G.J. and P.J. Webster, 2007: Heightened tropical cyclone activity in the North Atlantic: natural variability or climate trend? *Philosophical Transactions of the Royal Society Series A*, **365(1860)**, 2695-2716. doi:10.1098/rsta.2007.2083

Holland, M.M., C.M. Bitz, and B. Tremblay, 2006: Future abrupt reductions in the summer Arctic sea ice. *Geophysical Research Letters*, **33,** L23503, doi:10.1029/2006GL028024.

Houze, R.A., Jr., 1977: Structure and dynamics of a tropical squall-line system. *Monthly Weather Review*, **105(12)**, 1540-1567.

Huntington, T.G., G.A. Hodgkins, and R.W. Dudley, 2003: Historical trend in river ice thickness and coherence in hydroclimatological trends in Maine. *Climatic Change,* **61(1-2),** 217-236.

Hurrell, J.W., 1995: Decadal trends in the North-Atlantic oscillation - regional temperatures and precipitation. *Science,* **269(5224),** 676-679.

Hurrell, J.W., 1996: Influence of variations in extratropical wintertime teleconnections on Northern Hemisphere temperature. *Geophysical Research Letters,* **23(6),** 665-668.

Hurrell, J.W., M.P. Hoerling, A.S. Phillips, and T. Xu, 2005: Twentieth century North Atlantic climate change. Part I: Assessing determinism. *Climate Dynamics,* **23(3-4),** 371-389.

IADAG (International Ad Hoc Detection and Attribution Group), 2005: Detecting and attributing external influences on the climate system: a review of recent advances. *Journal of Climate,* **18(9),** 1291-1314.

IPCC (Intergovernmental Panel on Climate Change), 2001. *Climate Change 2001: The Scientific Basis.* Contribution of Working Group I to the Third Assessment Report of the Intergovernmental Panel on Climate Change [Houghton, J. T., Y. Ding, D.J. Griggs, M. Noguer, P.J. van der Linden, X. Dai, K. Maskell, and C.A. Johnson (eds.)]. Cambridge University Press, Cambridge, UK, and New York, 881 pp.

IPCC (Intergovernmental Panel on Climate Change), 2007: Summary for policymakers. In: *Climate Change 2007: The Physical Science Basis.* Contribution of Working Group I to the Fourth Assessment Report of the Intergovernmental Panel on Climate Change [Solomon, S., D. Qin, M. Manning, Z. Chen, M. Marquis, K.B. Averyt, M.Tignor, and H.L. Miller (eds.)]. Cambridge University Press, Cambridge, UK, and New York, pp. 1-18.

Jagger, T.H. and J.B. Elsner, 2006: Climatology models for extreme hurricane winds near the United States. *Journal of Climate,* **19(13),** 3220-3236.

Jones, S.C., P.A. Harr, J. Abraham, L.F. Bosart, P.J. Bowyer, J.L. Evans, D.E. Hanley, B.N. Hanstrum, R.E. Hart, F. Lalaurette, M.R. Sinclair, R.K. Smith, and C. Thorncroft, 2003: The extratropical transition of tropical cyclones: forecast challenges, current understanding, and future directions. *Weather and Forecasting,* **18(6),** 1052-1092.

Karl, T.R. and R.W. Knight, 1997: The 1995 Chicago heat wave: How likely is a recurrence? *Bulletin of the American Meteorological Society,* **78(6),** 1107-1119.

Karl, T.R. and R.W. Knight, 1998: Secular trends of precipitation amount, frequency, and intensity in the USA. *Bulletin of the American Meteorological Society,* **79(2),** 231-241.

Karl, T.R. and K.E. Trenberth, 2003: Modern global climate change. *Science,* **302(5651),** 1719-1723.

Karoly, D.J. and Q. Wu, 2005: Detection of regional surface temperature trends. *Journal of Climate,* **18(21),** 4337-4343.

Karoly, D.J., K.Braganza, P.A. Stott, J.M. Arblaster, G.A. Meehl, A.J. Broccoli, and K.W. Dixon, 2003: Detection of a human influence on North American climate. *Science,* **302(5648),** 1200-1203.

Katz, R.W., 1999: Extreme value theory for precipitation: sensitivity analysis for climate change. *Advances in Water Researches,* **23(2),** 133-139.

Katz, R.W. and B.G. Brown, 1992: Extreme events in a changing climate: variability is more important than averages. *Climatic Change,* **21(3),** 289-302.

Kenyon, J. and G.C. Hegerl, 2008: Influence of modes of climate variability on global temperature extremes. Journal of Climate, Early online release doi:10.1175/2008JCLI2125.1.

Kharin, V.V. and F.W. Zwiers, 2005: Estimating extremes in transient climate change simulations. *Journal of Climate,* **18(8),** 1156-1173.

Kharin, V., F.W. Zwiers, X. Zhang and G.C. Hegerl, 2007: Changes in temperature and precipitation extremes in the IPCC ensemble of global coupled model simulations. *Journal of Climate,* **20(8),** 1419-1444.

Kiktev, D., D. Sexton, L. Alexander, and C. Folland, 2003: Comparison of modelled and observed trends in indices of daily climate extremes. *Journal of Climate,* **16(22),** 3560-3571.

Knutson, T.R. and R.E. Tuleya, 1999: Increased hurricane intensities with CO_2-induced warming as simulated using the GFDL hurricane prediction system. *Climate Dynamics,* **15(7),** 503-519.

Knutson, T.R. and R.E. Tuleya, 2004: Impact of CO_2-induced warming on simulated hurricane intensity and precipitation: sensitivity to the choice of climate model and convective parameterization. *Journal of Climate,* **17(18),** 3477-3495.

Knutson, T.R. and R.E. Tuleya, 2008: Tropical cyclones and climate change: revisiting recent studies at GFDL. In *Climate Extremes and Society* [Diaz, H. and R. Murnane (eds.)]. Cambridge University Press, Cambridge, UK, and New York, pp. 120-144.

Knutson, T.R., R.E. Tuleya, and Y. Kurihara, 1998: Simulated increase of hurricane intensities in a CO2-warmed climate. *Science,* **279(5353),** 1018-1020.

Knutson, T.R., R.E. Tuleya, W. Shen, and I. Ginis, 2001: Impact of CO_2-induced warming on hurricane intensities as simulated in a hurricane model with ocean coupling. *Journal of Climate,* **14(11),** 2458-2468.

Knutson, T.R., T.L. Delworth, K.W. Dixon, I.M. Held, J. Lu, V. Ramaswamy, D. Schwarzkopf, G. Stenchikov, and R.J. Stouffer, 2006: Assessment of twentieth-century regional surface temperature trends using the GFDL CM2 coupled models. *Journal of Climate,* **19(9),** 1624-1651.

Knutson, T.R., J.J. Sirutis, S.T. Garner, I.M. Held, and R.E. Tuleya, 2007: Simulation of the recent multidecadal increase of Atlantic hurricane activity using an 18-km grid regional model. *Bulletin of the American Meteorological Society,* **88(10),** 1549-1565.

Kossin, J.P. and D.J. Vimont, 2007: A more general framework for understanding Atlantic hurricane variability and trends. *Bulletin of the American Meteorological Society,* **88(11),** 1767-1781.

Kumar, A., F. Yang, L. Goddard, and S. Schubert, 2004: Differing trends in the tropical surface temperatures and precipitation over land and oceans. *Journal of Climate,* **17(3),** 653-664.

Kunkel, K.E., N.E. Westcott, and D.A.R. Kristovich, 2002: Assessment of potential effects of climate change on heavy lake-effect snowstorms near Lake Erie. *Journal of Great Lakes Research,* **28(4),** 521-536.

Kunkel, K.E., X.-Z. Liang, J. Zhu, and Y. Lin, 2006: Can CGCMS simulate the twentieth century "warming hole" in the central United States? *Journal of Climate,* **19(7)**, 4137-4153.

Lambert, F.H., P.A. Stott, M.R. Allen, and M.A. Palmer, 2004: Detection and attribution of changes in 20th century land precipitation. *Geophysical Research Letters,* **31(10)**, L10203, doi:10.1029/2004GL019545

Lambert, S.J. and J.C. Fyfe, 2006: Changes in winter cyclone frequencies and strengths simulated in enhanced greenhouse warming experiments: results from the models participating in the IPCC diagnostic exercise. *Climate Dynamics,* **26(7-8)**, 713-728. doi:10.1007/s00382-006-0110-3

Lander, M.A., 1994a: Description of a monsoon gyre and its effects on the tropical cyclones in the western North Pacific during August 1991. *Weather and Forecasting,* **9(4)**, 640-654.

Lander, M.A., 1994b: An exploratory analysis of the relationship between tropical storm formation in the western North Pacific and ENSO. *Monthly Weather Review,* **122(4)**, 636-651.

Leung, L.R., Y. Qian, X.D. Bian, W.M. Washington, J.G. Han, and J.O. Roads, 2004: Mid-century ensemble regional climate change scenarios for the western United States. *Climatic Change,* **62(1-3)**, 75-113.

Liebmann, B., H.H. Hendon, and J.D. Glick, 1994: The relationship between tropical cyclones of the western Pacific and Indian Oceans and the Madden-Julian oscillation. *Journal of the Meteorological Society of Japan,* **72(3)**, 401-412.

Liu, G., A.E. Strong, W. Skirving, and L.F. Arzayus, 2006: Overview of NOAA Coral Reef Watch Program's near-real-time satellite global coral bleaching monitoring activities. In: *Proceedings of the 10th International Coral Reef Symposium,* 28 June - 2 July, 2004, Okinawa. Japanese Coral Reef Society, Tokyo, pp. 1783-1793.

Lozano, I. and V. Swail, 2002: The link between wave height variability in the North Atlantic and the storm track activity in the last four decades. *Atmosphere-Ocean,* **40(4)**, 377-388.

Lynch, A.H., J.A. Curry, R.D. Brunner, and J.A. Maslanik, 2004: Toward an integrated assessment of the impacts of extreme wind events on Barrow, Alaska. *Bulletin of the American Meteorological Society,* **85(2)**, 209-221.

Maloney, E.D. and D.L. Hartmann, 2000: Modulation of hurricane activity in the Gulf of Mexico by the Madden-Julian oscillation. *Science,* **287(5460)**, 2002-2004.

Manabe, S. and R.J. Stouffer, 1980: Sensitivity of a global climate model to an increase of CO_2 concentration in the atmosphere. *Journal of Geophysical Research,* **85(C10)**, 5529-5554.

Manabe, S., Wetherald R.T., Stouffer, R.J., 1981: Summer dryness due to an increase of atmospheric CO_2 concentration. *Climate Change,* **3(4)**, 347-386.

Mann, M. and K. Emanuel, 2006: Atlantic hurricane trends linked to climate change. *EOS, Transactions of the American Geophysical Union,* **87(24)**, 233, 238, 241.

Marchok, T., R. Rogers, and R. Tuleya, 2007: Validation schemes for tropical cyclone quantitative precipitation forecasts: evaluation of operational models for U.S. landfalling cases. *Weather and Forecasting,* **22(4)**, 726-746.

Mars, J.C., and D.W. Houseknecht, 2007: Quantitative remote sensing study indicates doubling of coastal erosion rate in past 50 yr along a segment of the Arctic coast of Alaska. *Geology,* **35(7)**, 583-586.

Marsh, P.T., H.E. Brooks, and D.J. Karoly, 2007: Assessment of the severe weather environment in North America simulated by a global climate model. Atmospheric Science Letters, 8(4), 100-106.

McCabe, G.J., M.P. Clark, and M.C. Serreze, 2001: Trends in Northern Hemisphere surface cyclone frequency and intensity. *Journal of Climate,* **14(12)**, 2763-2768.

McDonald, R.E., D.G. Bleaken, D.R. Cresswell, V.D. Pope, and C.A. Senior, 2005: Tropical storms: representation and diagnosis in climate models and the impacts of climate change. *Climate Dynamics,* **25(1)**, 19-36. doi:10.1007/s00382-004-0491-0

Mearns, L.O., R.W. Katz, and S.H. Schneider, 1984: Extreme high temperature events: changes in their probabilites with changes in mean temperature. *Journal of Climate and Applied Meteorology,* **23(12)**, 1601-1613.

Meehl, G.A. and C. Tebaldi, 2004: More intense, more frequent, and longer lasting heat waves in the 21st century. *Science,* **305(5686)**, 994-997.

Meehl, G.A., C. Tebaldi, and D. Nychka, 2004: Changes in frost days in simulations of twenty-first century climate. *Climate Dynamics,* **23(5)**, 495-511.

Meehl, G.A., J.M. Arblaster, and C. Tebaldi, 2005: Understanding future patterns of precipitation extremes in climate model simulations. *Geophysical Research Letters,* **32**, L18719, doi:10.1029/2005GL023680

Meehl, G.A., W.M. Washington, B.D. Santer, W.D. Collins, J.M. Arblaster, A. Hu, D.M. Lawrence, H. Teng, L.E. Buja, and W.G. Strand 2006: Climate change projections for the twenty-first century and climate change commitment in the CCSM3. *Journal of Climate,* **19(11)**, 2597-2616.

Meehl, G.A., T.F. Stocker, W.D. Collins, P. Friedlingstein, A.T. Gaye, J.M. Gregory, A. Kitoh, R. Knutti, J.M. Murphy, A. Noda, S.C.B. Raper, I.G. Watterson, A.J. Weaver, and Z.-C. Zhao, 2007a: Global climate projections. In: *Climate Change 2007: The Physical Basis.* Contribution of Working Group I to the Fourth Assessment Report of the Intergovernmental Panel on Climate Change [Solomon, S., D. Qin, M. Manning, Z. Chen, M. Marquis, K.B. Averyt, M. Tignor, and H.L. Miller (eds.)]. Cambridge University Press, Cambridge, UK, and New York, pp. 747-845.

Meehl, G.A., J.M. Arblaster, and C. Tebaldi, 2007b: Contributions of natural and anthropogenic forcing to changes in temperature extremes over the U.S. *Geophysical Research Letters* **34**, L19709, doi:10.1029/2007GL030948.

Meehl, G.A., C. Tebaldi, H. Teng, and T.C. Peterson, 2007c: Current and future U.S. weather extremes and El Niño. *Geophysical Research Letters,* **34**, L20704, doi:10.1029/2007GL031027.

Mendelssohn, R., S.J. Bograd, F.B. Schwing, and D.M. Palacios, 2005: Teaching old indices new tricks: A state-space analysis of El Niño related climate indices. *Geophysical Research Letters,* **32**, L07709.

Merryfield, W.J., 2006: Changes to ENSO under CO_2 doubling in a multimodel ensemble. *Journal of Climate,* **19(16)**, 4009-4027.

Miller, N.L. and N.J. Schlegel, 2006: Climate change projected fire weather sensitivity: California Santa Ana wind occurrence. *Geophysical Research Letters,* **33(15)**, L15711, doi:10.1029/2006GL025808.

Miller, N.L., K.E. Bashford, and E. Strem, 2003: Potential impacts of climate change on California hydrology. *Journal of the American Water Resources Association,* **39(4),** 771-784.

Milly, P.C.D., R.T. Wetherald, K.A. Dunne, and T.L. Delworth, 2002: Increasing risk of great floods in a changing climate. *Nature,* **415(6871),** 514-517.

Milly, P.C.D., K.A. Dunne, and A.V. Vecchia, 2005: Global pattern of trends in streamflow and water availability in a changing climate. *Nature,* **438(7066),** 347-350.

Molinari, J. and D. Vollaro, 2000: Planetary and synoptic-scale influence on eastern Pacific tropical cyclogenesis. *Monthly Weather Review,* **128(9),** 3296-3307.

Montgomery, M.T., M.E. Nicholls, T.A. Cram, and A.B. Saunders, 2006: A vortical hot tower route to tropical cyclogenesis. *Journal of the Atmospheric Sciences,* **63(1),** 355-386.

Morris, K., M. Jeffries, and C. Duguay, 2005: Model simulation of the effects of climate variability and change on lake ice in central Alaska, USA. *Annals of Glaciology,* **40(1),** 113-118.

Neal, A.B. and G.J. Holland, 1976: *The Australian Tropical Cyclone Forecasting Manual.* Bureau of Meteorology, Melbourne, Victoria, Australia, 274 pp.

Neelin, J.D., M. Münnich, H. Su, J.E. Meyerson, and C. Holloway, 2006: Tropical drying trends in global warming models and observations. *Proceedings of the National Academy of Sciences,* **103(16),** 6110-6115.

Neu, H.J.A., 1984: Interannual variations and longer-term changes in the sea state of the North Atlantic from 1970 to 1982. *Journal of Geophysical Research,* **89(C4):** 6397-6402.

Nolan, D.S., E.D. Rappin, and K.A. Emanuel, 2006: Could hurricanes form from random convection in a warmer world? In: *27th Conference on Hurricanes and Tropical Meteorology,* 24-28 April, 2006, Monterey, CA. American Meteorological Society, Boston, Paper 1C.8. Extended abstract available at: http://ams.confex.com/ams/pdfpapers/107936.pdf.

Oouchi, K., J.Yoshimura, H. Yoshimura, R. Mizuta, S. Kusunoki, and A. Noda, 2006: Tropical cyclone climatology in a global-warming climate as simulated in a 20km-mesh global atmospheric model: frequency and wind intensity analysis. *Journal of the Meteorological Society of Japan,* **84(2),** 259-276.

Osborn, T.J., 2004: Simulating the winter North Atlantic Oscillation: the roles of internal variability and greenhouse gas forcing. *Climate Dynamics,* **22(6-7),** 605-623.

Osborn, T.J. and M. Hulme, 1997: Development of a relationship between station and grid-box rainday frequencies for climate model evaluation. *Journal of Climate,* **10(8),** 1885-1908.

Osborn, T.J., K.R. Briffa, S.F.B. Tett, P.D. Jones, and R.M. Trigo, 1999: Evaluation of the North Atlantic Oscillation as simulated by a coupled climate model. *Climate Dynamics,* **15(9),** 685-702.

Ostermeier, G.M. and J.M. Wallace, 2003: Trends in the North Atlantic Oscillation-Northern Hemisphere annular mode during the twentieth century. *Journal of Climate,* **16(2),** 336-341.

Paciorek, C.J., J.S. Risbey, V. Ventura, and R.D. Rosen, 2002: Multiple indices of Northern Hemisphere cyclone activity, winters 1949-99. *Journal of Climate,* **15(13),** 1573-1590.

Pal, J.S., F. Giorgi, and X. Bi, 2004: Consistency of recent European summer precipitation trends and extremes with future regional climate projections. *Geophysical Research Letters,* **31,** L13202, doi:10.1029/2004GL019836.

Pall, P., M.R. Allen, and D.A. Stone, 2007: Testing the Clausius-Clapeyron constraint on changes in extreme precipitation under CO_2 warming . *Climate Dynamics,* **28(4),** 351-363.

Palmer, T.N. and J. Räisänen, 2002: Quantifying the risk of extreme seasonal precipitation events in a changing climate. *Nature,* **415(6871),** 514-517.

Pasch, R.J., L.A. Avila, and J-G. Jiing, 1998: Atlantic tropical systems of 1994 and 1995: a comparison of a quiet season to a near-record-breaking one. *Monthly Weather Review,* **126(5),** 1106-1123.

Peel, M.C. and T.A. McMahon, 2006: A quality-controlled global runoff data set. *Nature,* **444(7120),** E14. doi:10.1038/nature05480

Peterson, T.C., X. Zhang, M. Brunet-India, and J.L. Vázquez-Aguirre, 2008: Changes in North American extremes derived from daily weather data. Journal Geophysical Research, 113, D07113, doi:10.1029/2007JD009453.

Pinto, J.G., U. Ulbrich, G.C. Leckebusch, T. Spangehl, M. Reyers, and S. Zacharias, 2007: Changes in storm track and cyclone activity in three SRES ensemble experiments with the ECHAM5/MPI-OM1 GCM. *Climate Dynamics,* **29(2-3),** 195-210. doi:10.1007/s00382-007-0230-4

Price, J.F., 1981: Upper ocean response to a hurricane. *Journal of Physical Oceanography,* **11(2),** 153-175.

Qian, T., A. Dai, K.E. Trenberth, and K.W. Oleson, 2006: Simulation of global land surface conditions from 1948 to 2002: Part I: forcing data and evaluations. *Journal of Hydrometeorology,* **7(5),** 953-975.

Räisänen, J., 2005: Impact of increasing CO_2 on monthly-to-annual precipitation extremes: analysis of the CMIP2 experiments. *Climate Dynamics,* **24(2-3),** 309-323.

Randall, D.A., R.A. Wood, S. Bony, R. Colman, T. Fichefet, J. Fyfe, V. Kattsov, A. Pitman, J. Shukla, J. Srinivasan, R.J. Stouffer, A. Sumi, and K.E. Taylor, 2007: Climate models and their evaluation. In: *Climate Change 2007: The Physical Science Basis.* Contribution of Working Group I to the Fourth Assessment Report of the Intergovernmental Panel on Climate Change [Solomon, S., D. Qin, M. Manning, Z. Chen, M. Marquis, K.B. Averyt, M.Tignor and H.L. Miller (eds.)]. Cambridge University Press, Cambridge, UK, and New York, pp. 589-662.

Ritchie, E.A., 2003: Some aspects of midlevel vortex interaction in tropical cyclogenesis. In: *Cloud Systems, Hurricanes, and the Tropical Rainfall Measuring Mission (TRMM) A Tribute to Dr. Joanne Simpson* [Tao, W.-K. and R. Adler (eds.)]. Meteorological Monograph, 29(51), American Meteorological Society, Boston, MA, pp. 165-174.

Ritchie, E.A. and G.J. Holland, 1997: Scale interactions during the formation of typhoon Irving. *Monthly Weather Review,* **125(7),** 1377-1396.

Ritchie, E.A. and G.J. Holland, 1999: Large-scale patterns associated with tropical cyclogenesis in the western Pacific. *Monthly Weather Review,* **127(9),** 2027-2043.

Rosenzweig, C., G. Casassa, D.J. Karoly, A. Imeson, C. Liu, A. Menzel, S. Rawlins, T.L. Root, B. Seguin, and P. Tryjanowski, 2007: Assessment of observed changes and responses in natural and managed systems. In: *Climate Change 2007: Impacts, Adaptation and Vulnerability.* Contribution of Working Group

II to the Fourth Assessment Report of the Intergovernmental Panel on Climate Change [Parry, M.L., O.F. Canziani, J.P. Palutikof, P.J. van der Linden, and C.E. Hanson, (eds)]. Cambridge University Press, Cambridge, UK, and New York, pp. 79-131.

Rotunno, R. and K.A. Emanuel, 1987: An air-sea interaction theory for tropical cyclones. Part II: evolutionary study using a nonhydrostatic axisymmetric numerical model. *Journal of the Atmospheric Sciences*, **44(3)**, 542-561.

Rowell, D.P., and R.G. Jones, 2006: Causes and uncertainty of future summer drying over Europe. *Climate Dynamics*, **27(2-3)**, 281-299.

Royer. J.-F., F. Chauvin, B. Timbal, P. Araspin, and D. Grimal, 1998: A GCM study of the impact of greenhouse gas increase on the frequency of occurrence of tropical cyclones. *Climatic Change*, **38(3)**, 307-343.

Ryan, B.F., I.G. Watterson, and J.L. Evans, 1992: Tropical cyclones frequencies inferred from Gray's yearly genesis parameter: validation of GCM tropical climate. *Geophysical Research Letters*, **19(18)**, 1831-1834.

Santer, B.D., T.M.L. Wigley, P.J. Gleckler, C. Bonfils, M.F. Wehner, K. AchutaRao, T.P. Barnett, J.S. Boyle, W. Brüggemann, M. Fiorino, N. Gillett, J.E. Hansen, P.D. Jones, S.A. Klein, G.A. Meehl, S.C.B. Raper, R.W. Reynolds, K.E. Taylor, and W.M. Washington, 2006: Forced and unforced ocean temperature changes in Atlantic and Pacific tropical cyclogenesis regions. *Proceedings of the National Academy of Sciences*, **103(38)**, 13905-13910. doi:10.1073/pnas.0602861103

Santer, B.D., C. Mears, F.J. Wentz, K.E. Taylor, P.J. Gleckler, T.M.L. Wigley, T.P. Barnett, J.S. Boyle, W. Brüggemann, N.P. Gillett, S.A. Klein, G.A. Meehl, T. Nozawa, D.W. Pierce, P.A. Stott, W.M. Washington, and M.F. Wehner, 2007: Identification of human-induced changes in atmospheric moisture content. *Proceedings of the National Academy of Sciences*, **104(39)**, 15248-15253.

Scaife, A.A., J.R. Knight, G.K. Vallis, and C.K. Folland, 2005: A stratospheric influence on the winter NAO and North Atlantic surface climate. *Geophysical Research Letters*, **32**, L18715, doi:10.1029/2005GL023226.

Schade, L.A. and K.A. Emanuel, 1999: The ocean's effect on the intensity of tropical cyclones: results from a simple coupled atmosphere-ocean model. *Journal of the Atmospheric Sciences*, **56(4)**, 642-651.

Schär, C., P.L. Vidale, D. Lüthi, C. Frei, C. Häberli, M.A. Liniger, and C. Appenzeller, 2004: The role of increasing temperature variability in European summer heat waves. *Nature*, **427(6972)**, 332-336.

Schubert, W.H., P.E. Ciesielski, D.E. Stevens, and H-C. Kuo, 1991: Potential vorticity modeling of the ITCZ and the Hadley circulation. *Journal of the Atmospheric Sciences*, **48(12)**, 1493-1509.

Schubert, S.D., M.J. Suarez, P.J. Region, R.D. Koster, and J.T. Bacmeister, 2004: Causes of long term drought in the U. S. Great Plains. *Journal of Climate*, **17(3)**, 485-503.

Seager, R., Y. Kushnir, C. Herweijer, N. Naik, and J. Velez, 2005: Modeling of tropical forcing of persistent droughts and pluvials over western North America: 1856-2000. *Journal of Climate*, **18(19)**, 4065-4088.

Semenov, V.A. and L. Bengtsson, 2002: Secular trends in daily precipitation characteristics: greenhouse gas simulation with a coupled AOGCM. *Climate Dynamics*, **19(2)**, 123-140.

Shapiro, L.J., 1982: Hurricane climatic fluctuations. Part II: relation to large-scale circulation. *Monthly Weather Review*, **110(8)**, 1014-1023.

Shapiro, L.J. and S.B. Goldenberg, 1998: Atlantic sea surface temperatures and tropical cyclone formation. *Journal of Climate*, **11(4)**, 578-590.

Shiogama, H., M. Watanabe, M. Kimoto, and T. Nozawa, 2005: Anthropogenic and natural forcing impacts on ENSO-like decadal variability during the second half of the 20th century. *Geophysical Research Letters*, **32**, L21714, doi:10.1029/2005GL023871.

Simpson, J., E.A. Ritchie, G.J. Holland, J. Halverson, and S. Stewart, 1997: Mesoscale interactions in tropical cyclogenesis. *Monthly Weather Review*, **125(10)**, 2643-2661.

Stone, D.A. and A.J. Weaver, 2002: Daily maximum and minimum temperature trends in a clime model. *Geophysical Research Letters*, **29(9)**, 1356, doi:10.1029/2001GL014556.

Stott, P.A., 2003: Attribution of regional-scale temperature changes to anthropogenic and natural causes. *Geophysical Research Letters*, **30(14)**, 1724, doi:10.1029/2003GL017324.

Stott, P.A. and S.F.B. Tett, 1998: Scale-dependent detection of climate change. *Journal of Climate*, **11(12)**, 3282-3294.

Stott, P.A., D.A. Stone, and M.R. Allen, 2004: Human contribution to the European heatwave of 2003. *Nature*, **432(7017)**, 610-614.

Strong, A.E., F. Arzayus, W. Skirving, and S.F. Heron, 2006: Identifying coral bleaching remotely via coral reef watch-improved integration and implications for changing climate. In: *Corals and Climate Change: Science and Management* [Phinney, J.T., A. Strong, W. Skrving, J. Kleypas, and O. Hoegh-Guldberg, (eds.)]. Coastal and Estuarine Studies 61, American Geophysical Union, Washington DC, pp. 163-180.

Sugi, M., A. Noda, and N. Sato, 2002: Influence of global warming on tropical cyclone climatology: an experiment with the JMA global model. *Journal of the Meteorological Society of Japan*, **80(2)**, 249-272. doi:10.2151/jmsj.80.249

Tebaldi, C., K. Hayhoe, J.M. Arblaster, and G.A. Meehl, 2006: Going to the extremes. *Climatic Change*, **79(3-4)**, 185-211.

Thompson, D.W.J. and J.M. Wallace, 1998: The Arctic oscillation signature in the wintertime geopotential height and temperature fields. *Geophysical Research Letters*, **25(9)**, 1297-1300.

Thompson, D.W.J. and J.M. Wallace, 2001: Regional climate impacts of the Northern Hemisphere annular mode. *Science*, **293(5527)**, 85-89.

Thompson, D., J.M. Wallace, and G.C. Hegerl, 2000: Annular modes in the extratropical circulation: Part II, trends. *Journal of Climate*, **13(5)**, 1018-1036.

Thorncroft, C. and K. Hodges, 2001: African easterly wave variability and its relationship to Atlantic tropical cyclone activity. *Journal of Climate*, **14(6)**, 1166-1179.

Timmermann, A., J. Oberhuber, A. Bacher, M. Esch, M. Latif, And E. Roeckner, 1999: Increased El Niño frequency in a climate model forced by future greenhouse warming. *Nature*, **398(6729)**, 694-696.

Tonkin, H., G.J. Holland, N. Holbrook, and A. Henderson-Sellers, 2000: An evaluation of thermodynamic estimates of climatological maximum potential tropical cyclone intensity. *Monthly Weather Review*, **128(3)**, 746-762.

Trapp, R.J., N.S. Diffenbaugh, H.E. Brooks, M.E. Baldwin, E.D. Robinson, and J.S. Pal, 2007: Changes in severe thunderstorm frequency during the 21st century caused by anthropogenically enhanced global radiative forcing. Proceedings of the National Academies of Science, 140(50, 19719-19723.

Trenberth, K., 1999: Conceptual framework for changes of extremes of the hydrological cycle with climate change. *Climatic Change*, **42(1)**, 327-339.

Trenberth, K.E. and D.J. Shea, 2006: Atlantic hurricanes and natural variability in 2005. *Geophysical Research Letters*, **33**, L12704, doi:10.1029/2006GL026894.

Trenberth, K.E., J. Fasullo, and L. Smith, 2005: Trends and variability in column-integrated atmospheric water vapor. *Climate Dynamics*, **24(7-8)**, 741-758.

Trenberth, K.E., P.D. Jones, P. Ambenje, R. Bojariu, D. Easterling, A. Klein Tank, D. Parker, F. Rahimzadeh, J.A. Renwick, M. Rusticucci, B. Soden, and P. Zhai, 2007: Observations: surface and atmospheric climate change. In: *Climate Change 2007: The Physical Basis.* Contribution of Working Group I to the Fourth Assessment Report of the Intergovernmental Panel on Climate Change [Solomon, S., D. Qin, M. Manning, Z. Chen, M. Marquis, K.B. Averyt, M. Tignor, and H.L. Miller (eds.)]. Cambridge University Press, Cambridge, UK, and New York, pp. 235-335.

Tsutsui, J., 2002: Implications of anthropogenic climate change for tropical cyclone activity: a case study with the NCAR CCM2. *Journal of the Meteorological Society of Japan*, **80(1)**, 45-65. doi:10.2151/jmsj.80.45

Vavrus, S.J., J.E. Walsh, W.L. Chapman, and D. Portis, 2006: The behavior of extreme cold air outbreaks under greenhouse warming. *International Journal of Climatology*, **26(9)**, 1133-1147.

Vecchi, G.A. and B.J. Soden, 2007: Increased tropical Atlantic wind shear in model projections of global warming. *Geophysical Research Letters*, **34**, L08702, doi:10.1029/2006GL028905.

Vecchi, G.A. and T.R. Knutson, 2008: On estimates of historical North Atlantic tropical cyclone activity. Journal of Climate, Early online release doi:10.1175/2008JCLI2178.1.

Vitart, F. and J. L. Anderson, 2001: Sensitivity of Atlantic tropical storm frequency to ENSO and interdecadal variability of SSTs in an ensemble of AGCM integrations. *Journal of Climate*, **14(4)**, 533-545.

Vitart, F., J.L. Anderson, and W.F. Stern, 1997: Simulation of inter-annual variability of tropical storm frequency in an ensemble of GCM integrations. *Journal of Climate*, **10(4)**, 745-760.

Vitart, F., J.L. Anderson, J. Sirutis, and R.E. Tuleya, 2001: Sensitivity of tropical storms simulated by a general circulation model to changes in cumulus parametrization. *Quarterly Journal of the Royal Meteorological Society*, **127(571)**, 25-51.

Voss, R., W. May, and E. Roeckner, 2002: Enhanced resolution modeling study on anthropogenic climate change: changes in the extremes of the hydrological cycle. *International Journal of Climatology*, **22(7)**, 755-777.

Wang, B. and J.C.L. Chan, 2002: How strong ENSO events affect tropical storm activity over the western North Pacific. *Journal of Climate*, **15(13)**, 1643-1658.

Wang, G., 2005: Agricultural drought in a future climate: Results from 15 global climate models participating in the IPCC 4th assessment. *Climate Dynamics*, **25(7-8)**, 739-753.

Wang, X.L., F.W. Zwiers, and V.R. Swail, 2004: North Atlantic ocean wave climate change scenarios for the twenty-first century. *Journal of Climate*, **17(12)**, 2368-2383.

Wang X.L., V.R. Swail, and F.W. Zwiers, 2006: Climatology and changes of extra-tropical storm tracks and cyclone activity: Comparison of ERA-40 with NCEP/NCAR Reanalysis for 1958-2001. *Journal of Climate*, **19(13)**, 3145-3166.

Wang, Y., 2002: Vortex Rossby waves in a numerically simulated tropical cyclone. Part II: the role in tropical cyclone structure and intensity changes. *Journal of the Atmospheric Sciences*, **59(7)**, 1239-1262.

Wang, Y., J.D. Kepert, and G.J. Holland, 2001; The effect of sea spray evaporation on tropical cyclone boundary layer structure and intensity. *Monthly Weather Review*, **129(10)**, 2481-2500.

WASA Group (Waves and Storms in the North Atlantic), 1998: Changing waves and storms in the northeast Atlantic? *Bulletin of the American Meteorological Society*, **79(5)**, 741-760.

Watterson, I.G., 2005: Simulated changes due to global warming in the variability of precipitation, and their interpretation using a gamma-distributed stochastic model. *Advances in Water Resources*, **28(12)**, 1368-1381.

Watterson, I.G. and M.R. Dix, 2003: Simulated changes due to global warming in daily precipitation means and extremes and their interpretation using the gamma distribution. *Journal of Geophysical Research*, **108(D13)**, 4379, doi:10.1029/2002JD002928

Webster, P.J. and H-R. Chang, 1988: Equatorial energy accumulation and emanation regions: impacts of a zonally varying basic state. *Journal of the Atmospheric Sciences*, **45(5)**, 803-829.

Webster, P.J., G.J. Holland, J.A. Curry, and H.-R. Chang, 2005: Changes in tropical cyclone number, duration, and intensity in a warming environment. *Science*, **309(5742)**, 1844-1846.

Wehner, M., 2005: Changes in daily precipitation and surface air temperature extremes in the IPCC AR4 models. *US CLIVAR Variations*, **3(3)**, 5-9.

Weisheimer, A. and T.N. Palmer, 2005: Changing frequency of occurrence of extreme seasonal-mean temperatures under global warming. *Geophysical Research Letters*, **32**, L20721, doi:10.1029/2005GL023365.

Wentz, F. J., L. Ricciardulli, K. Hilburn, and C. Mears, 2007: How much more rain will global warming bring? *Science*, **317(5835)**, 233-235. doi: 10.1126/science.1140746

Wettstein, J.J. and L.O. Mearns, 2002: The influence of the North Atlantic-Arctic Oscillation on mean, variance, and extremes of temperature in the northeastern United States and Canada. *Journal of Climate*, **15(24)**, 3586-3600.

Whitney, L.D. and J.S. Hobgood, 1997: The relationship between sea surface tempertures and maximum intensities of tropical cyclones in the eastern North Pactific Ocean. *Journal of Climate*, **10(11)**, 2921-2930.

Wilby, R.L., and T.M.L. Wigley, 2002: Future changes in the distribution of daily precipitation totals across North America. *Geophysical Research Letters*, **29(7)**, 1135, doi:10.1029/2001GL013048.

Wilkinson, C.R. (ed.), 2000: *Global Coral Reef Monitoring Network: Status of Coral Reefs of the World in 2000*. Australian Institute of Marine Science, Townsville, Queensland, 363 pp.

Willett, K.M., N.P. Gillett, P.D. Jones, and P.W. Thorne, 2007: Attribution of observed surface humidity changes to human influence. *Nature*, **449(7163)**, 710-712. doi:10.1038/nature06207

Williams, G., K.L. Layman, H.G. Stefan, 2004: Dependence of lake ice covers on climatic, geographic and bathymetric variables. *Cold Regions Science and Technology*, **40(3)**, 145-164.

WMO (World Meteorological Organization), 2006: Atmospheric Research and Environment Programme. *Statement on Tropical Cyclones and Climate Change*, 13 pp., http://www.wmo.int/pages/prog/arep/tmrp/documents/iwtc_summary.pdf and *Summary Statement on Tropical Cyclones and Climate Change*, 1 p., http://www.wmo.int/pages/prog/arep/tmrp/documents/iwtc_statement.pdf

Wu, L. and B. Wang, 2004: Assessing impacts of global warming on tropical cyclone tracks. *Journal of Climate*, **17(8)**, 1686-1698.

Wu, P.L., R. Wood, and P. Stott, 2005: Human influence on increasing Arctic river discharges. *Geophysical Research Letters*, **32**, L02703, doi:10.1029/2005GL021570.

Yin, J.H., 2005: A consistent poleward shift of the storm tracks in simulations of 21st century climate. *Geophysical Research Letters*, **32**, L18701, doi:10.1029/2005GL023684.

Yonetani, T. and H.B. Gordon, 2001: Simulated changes in the frequency of extremes and regional features of seasonal/annual temperature and precipitation when atmospheric CO_2 is doubled. *Journal of Climate*, **14(8)**, 1765-1779.

Yoshimura, J., M. Sugi and A. Noda, 2006: Influence of greenhouse warming on tropical cyclone frequency. *Journal of the Meteorological Society of Japan*, **84(3)**, 405-428.

Yu, B. and F.W. Zwiers, 2007: The impact of combined ENSO and PDO on the PNA climate: a 1,000-year climate modeling study. *Climate Dynamics*, **29(7-8)**, 837-851. doi:10.1107/s00382-007-0267-4

Yu, B., A. Shabbar, and F.W. Zwiers, 2007: The enhanced PNA-like climate response to Pacific interannual and decadal variability. *Journal of Climate*, **20(13)**, 5285-5300. doi:10.1175/2007JCLI1480.1

Zelle, H., G.J. Van Oldenborgh, G. Burgers, and H. Dijkstra, 2005: El Niño and greenhouse warming: Results from ensemble simulations with the NCAR CCSM. *Journal of Climate*, **18(22)**, 4669-4683.

Zhang, R. and T.L. Delworth, 2006: Impact of Atlantic multidecadal oscillations on India/Sahel rainfall and Atlantic hurricanes. *Geophysical Research Letters*, **33**, L17712, doi:10.1029/2006GL026267.

Zhang, X.D. and J.E. Walsh, 2006: Toward a seasonally ice-covered Arctic Ocean: scenarios from the IPCC AR4 model simulations. *Journal of Climate*, **19(9)**, 1730-1747.

Zhang, X., F.W. Zwiers, and P.A. Stott, 2006: Multi-model multi-signal climate change detection at regional scale. *Journal of Climate*, **19(17)**, 4294-4307.

Zhang, X., F.W. Zwiers, G.C. Hegerl, N. Gillett, H. Lambert, and S. Solomon, 2007: Detection of human influence on twentieth-century precipitation trends. *Nature*, **448(7152)**, 461-465.

Zipser, E.J., 1977: Mesoscale and convective-scale downdrafts as distinct components of squall-line structure. *Monthly Weather Review*, **105(12)**, 1568-1589.

Zwiers, F.W. and X. Zhang, 2003: Toward regional scale climate change detection. *Journal of Climate*, **16(5)**, 793-797.

CHAPTER 4 REFERENCES

AGU (American Geophysical Union), 2006: *Hurricanes and the U.S. Gulf Coast: Science and Sustainable Rebuilding*. American Geophysical Union, Washington DC, 29 pp. Available at http://www.agu.org/report/hurricanes/

Cassano, J.J., P. Uotila, and A. Lynch, 2006: Changes in synoptic weather patterns in the polar regions in the 20th and 21st centuries, Part 1: Arctic. *International Journal of Climate*, **26(8)**, 1027-1049.

Cook, E., C.A. Woodhouse, C.M. Eakin, D.M. Meko, and D.W. Stahle, 2004: Long-term aridity changes in the western United States. *Science*, **306(5698)**, 1015-1018.

Elsner, J.B., K-B. Liu, and B. Kocher, 2000: Spatial variations in major U.S. hurricane activity: statistics and a physical mechanism. *Journal of Climate*, **13(13)**, 2293-2305.

HiFi, 2006: *A Program for Hurricane Intensity Forecast Improvements and Impacts Projections (HiFi) Science Strategy*. Available at http://www.nova.edu/ocean/hifi/hifi_science_strategy.pdf

Lynch, A., P. Uotila, and J.J. Cassano, 2006: Changes in synoptic weather patterns in the polar regions in the 20th and 21st centuries, Part 2: Antarctic. *International Journal of Climate*, **26(9)**, 1181-1199.

Mann, M.E., R.S. Bradley, and M.K. Hughes, 1999: Northern Hemisphere temperatures during the past millennium: inferences, uncertainties, and limitations. *Geophysical Research Letters*, **26(6)**, 759-762.

Meehl, G.A. and A. Hu, 2006: Megadroughts in the Indian monsoon region and southwest North America and a mechanism for associated multi-decadal Pacific sea surface temperature anomalies. *Journal of Climate*, **19(9)**, 1605-1623.

Meehl, G.A. and C. Tebaldi, 2004: More intense, more frequent, and longer lasting heat waves in the 21st century. *Science*, **305(5686)**, 994-997.

Meehl, G.A., C. Tebaldi, and D. Nychka, 2004: Changes in frost days in simulations of twenty-first century climate. *Climate Dynamics*, **23(5)**, 495-511.

NOAA SAB (Science Advisory Board), 2006: *Hurricane Intensity Research Working Group Final Report*. [62 pp.] Available at http://www.sab.noaa.gov/Reports/HIRWG_final73.pdf

NRC (National Research Council) of the National Academies, Board of Atmospheric Science and Climate, 2006: *Completing the Forecast: Characterizing and Communicating Uncertainty for Better Decisions Using Weather and Climate Forecasts*. Recommendation 3.4, National Academies Press, Washington DC, pp. 49.

NSB (National Science Board), 2006: *Hurricane Warning: The Critical Need for a National Hurricane Research Initiative*. NSB-06-115, National Science Foundation, Arlington, VA, 36 pp. Available at http://www.nsf.gov/nsb/committees/hurricane/initiative.pdf

Oouchi, K., J. Yoshimura, H. Yoshimura, R. Mizuta, S. Kusunoki, and A. Noda, 2006: Tropical cyclone climatology in a global-warming climate as simulated in a 20km-mesh global atmospheric model: frequency and wind intensity analyses. *Journal of the Meteorological Society of Japan*, **84(2)**, 259-276.

Pulwarty, R.S., D.A. Wilhite, D.M. Diodato, and D.I. Nelson, 2007: Drought in changing environments: creating a roadmap, vehicles and drivers. *Natural Hazard Observer*, **31(5)**, 10-12.

Randall, D., 2005: Counting the clouds. *Journal of Physics: Conference Series*, **16**, 339-342. doi:10.1088/1742-6596/16/1/046 Available at http://www.iop.org/EJ/toc/1742-6596/16/1

Stott, P.A., D.A. Stone, and M.R. Allen, 2004: Human contribution to the European heatwave of 2003. *Nature*, **432(7017)**, 610-614.

Thompson, D.W.J. and J.M. Wallace, 2001: Regional climate impacts of the Northern Hemisphere annular mode and associated climate trends. *Science*, **293(5527)**, 85-89.

Wehner, M., L. Oliker, and J. Shalf, 2008: Towards ultra-high resolution models of climate and weather. *International Journal of High Performance Computing Applications*, (in press).

Woodhouse, C. and J. Overpeck, 1998: 2000 years of drought variability in the central United States. *Bulletin of the American Meteorological Society*, **79(12)**, 2693-2714.

APPENDIX A REFERENCES

Brockwell, P.J. and R.A. Davis, 2002: *Introduction to Time Series and Forecasting*. Springer, New York, 2nd ed., 434 pp.

R Development Core Team, 2007: R: A language and environment for statistical computing. R Foundation for Statistical Computing, Vienna, Austria. Available at http://www.R-project.org

Vecchi, G.A. and T.R. Knutson, 2008: On estimates of historical North Atlantic tropical cyclone activity. Journal of Climate, Early online release doi:10.1175/2008JCLI2178.1.

Wigley, T.M.L., 2006: Appendix A: Statistical issues regarding trends. In: *Temperature Trends in the Lower Atmosphere: Steps for Understanding and Reconciling Differences* [Karl, T.R., S.J. Hassol, C.D. Miller, and W.L. Murray (eds.)]. Climate Change Science Program, Washington DC, pp. 129-139.

PHOTOGRAPHY CREDITS

Cover/Title Page/Table of Contents
Cover, Image 2, (Breaking waves), Ric Tomlinson

Executive Summary
Page 3. Image 1 (Red sun), USDA photo
Page 8, Collage includes (Breaking waves), Ric Tomlinson

Chapter 1
Page 9, Chapter heading, (Breaking waves), Ric Tomlinson,
Page 9, Image 1, (Smashed truck), Weather Stock photo, Copyright© 1993, Warren Faidley/Weather Stock, serial #000103.
Page 13, Text Box insert, (Pine beetle damage), Copyright©, Province of British Columbia. Reprinted with permission of the Province of British Columbia. www.ipp.gov.bc.ca
Page 21, Text Box insert, (Butterfly), Camille Parmesan, University of Texas at Austin.
Page 22, Text Box insert, (Coral), Andy Bruckner, NOAA, National Marine Fisheries Service.
Page 24, Text Box image, (Flooding), Grant Goodge, STG. Inc., Asheville, N.C.
Page 26, Image 1, (Aerial storm damage), Grant Goodge, STG. Inc., Asheville, N.C.
Page 29, Image 1, (Helicopter and wreckage), Weather Stock photo, Copyright© 1993, Warren Faidley/Weather Stock, serial #000103.

Chapter 2
Page 33, Image 1, (Palm trees, windy beach), Weather Stock photo, Copyright© 1993, Warren Faidley/Weather Stock, serial #000103.
Page 34, Image 1, (Upturned cars), Weather Stock photo, Copyright© 1993, Warren Faidley/Weather Stock, serial #000103
Page 40, Image 1, (Butte sunset), Weather Stock photo, Copyright© 1993, Warren Faidley/Weather Stock, serial #000103
Page 41, Image 1, (Forest fire), Weather Stock photo, Copyright© 1993, Warren Faidley/Weather Stock, serial #000103
Page 46, Image 1, (Rainy traffic), Jesse Enloe, NOAA, NCDC, Asheville, N.C.
Page 46, Image 2, (CRN station), Courtesy of ATDD Oak Ridge, Oak Ridge, Tennessee.
Page 51, Image 1, (Bridge and floodwaters), Grant Goodge, STG. Inc., Asheville, N.C.
Page 52, Image 1, (Clean up crew), Weather Stock photo, Copyright© 1993, Warren Faidley/Weather Stock, serial #000103
Page 53, Image 1, (Hurricane, top view), Weather Stock photo, Copyright© 1993, Warren Faidley/Weather Stock, serial #000103
Page 70, Image 1, (High waves and cliffs), Weather Stock photo, Copyright© 1993, Warren Faidley/Weather Stock, serial #000103
Page 72, Image 1, (Snowy path), Weather Stock photo, Copyright© 1993, Warren Faidley/Weather Stock, serial #000103
Page 75, Image 1, (Storm damage), Weather Stock photo, Copyright© 1993, Warren Faidley/Weather Stock, serial #000103
Page 75, Image 2, (Night lightning), Grant Goodge, STG. Inc., Asheville, N.C.

Page 79, Image 1, (Foggy red sunset), Grant Goodge, STG. Inc., Asheville, N.C.

Chapter 3

Page 82, Image 1, (CRN Installation), Grant Goodge, STG. Inc., Asheville, N.C.

Page 94, Image 1, (Marina storm damage), Weather Stock photo, Copyright© 1993, Warren Faidley/Weather Stock, serial #000103

Page 98, Image 1, (Distant storm), Grant Goodge, STG. Inc., Asheville, N.C.

Page 114, Image 1, (Rocky coastline), Weather Stock photo, Copyright© 1993, Warren Faidley/Weather Stock, serial #000103

Chapter 4

Page 117, Image 1, (CRN instruments), Grant Goodge, STG. Inc., Asheville, N.C.

Page 118, Image 1, (CRN anemometer), Courtesy of ATDD Oak Ridge, Oak Ridge, Tennessee.

Contact Information

Global Change Research Information Office
c/o Climate Change Science Program Office
1717 Pennsylvania Avenue, NW
Suite 250
Washington, DC 20006
202-223-6262 (voice)
202-223-3065 (fax)

The Climate Change Science Program incorporates the U.S. Global Change Research Program and the Climate Change Research Initiative.

To obtain a copy of this document, place an order at the Global Change Research Information Office (GCRIO) web site: http://www.gcrio.org/orders

Climate Change Science Program and the Subcommittee on Global Change Research

William J. Brennan, Chair
Department of Commerce
National Oceanic and Atmospheric Administration
Acting Director, Climate Change Science Program

Jack Kaye, Vice Chair
National Aeronautics and Space Administration

Allen Dearry
Department of Health and Human Services

Anna Palmisano
Department of Energy

Mary Glackin
National Oceanic and Atmospheric Administration

Patricia Gruber
Department of Defense

William Hohenstein
Department of Agriculture

Linda Lawson
Department of Transportation

Mark Myers
U.S. Geological Survey

Jarvis Moyers
National Science Foundation

Patrick Neale
Smithsonian Institution

Jacqueline Schafer
U.S. Agency for International Development

Joel Scheraga
Environmental Protection Agency

Harlan Watson
Department of State

EXECUTIVE OFFICE AND OTHER LIAISONS

Stephen Eule
Department of Energy
Director, Climate Change Technology Program

Katharine Gebbie
National Institute of Standards & Technology

Stuart Levenbach
Office of Management and Budget

Margaret McCalla
Office of the Federal Coordinator for Meteorology

Rob Rainey
Council on Environmental Quality

Gene Whitney
Office of Science and Technology Policy

www.ingramcontent.com/pod-product-compliance
Lightning Source LLC
Chambersburg PA
CBHW080638180526
45168CB00008B/3220